Biotic Resources

This reference book provides advanced knowledge on sustainable biogenic waste management. It covers innovative waste processing technologies to produce biofuels, energy products, and biochemicals. To create a circular bioeconomy, it is imperative to develop processes where the waste generated through one process acts as a feedstock for the other. This book discusses the latest developments in biochemical and thermochemical methods of conversion and covers the potential of different kinds of biomass in more decentralized biorefineries. It describes sustainable solutions for a greener supplement to fossil resources. The book is meant for microbiologists, chemists, and biotechnologists.

Biotic Resources

Circular Bioeconomy Perspective

Edited by
Thallada Bhaskar
Sunita Varjani
Bhavya B. Krishna
Ashok Pandey

CRC Press
Taylor & Francis Group
Boca Raton London New York

CRC Press is an imprint of the
Taylor & Francis Group, an **informa** business

Cover Image Credit: Shutterstock

First Edition published 2023
by CRC Press
6000 Broken Sound Parkway NW, Suite 300, Boca Raton, FL 33487-2742

and by CRC Press
4 Park Square, Milton Park, Abingdon, Oxon, OX14 4RN

CRC Press is an imprint of Taylor & Francis Group, LLC

ISBN: 978-1-032-37100-9 (hbk)
ISBN: 978-1-032-37186-3 (pbk)
ISBN: 978-1-003-33574-0 (ebk)

DOI: 10.1201/9781003335740

Typeset in Times New Roman
by Apex CoVantage, LLC

Contents

Preface

The world is increasingly moving towards the use of renewable sources of energy fueled by the issues of climate change and geopolitical situations across the globe. In the wake of the decarbonization goal to be reached around 2050–2070, different countries are adopting different methodologies to reach this target. Various energy sources are being exploited depending on their intensity and availability in different areas of the world. Solar, wind, geothermal, tidal, hydroelectric energies, etc., are being explored for renewable electricity production.

Renewable organic carbon is still required for the production of chemicals and materials in a sustainable manner. Biomass (aka biotic resources) are a good source of this renewable organic carbon. There are different sources of biomass feedstocks: edible sources of biomass such as sugarcane, beet, sorghum, edible oils, etc., and also nonedible sources. They could be lignocellulosic biomass or algal biomass. These biomasses can be used for the production of biomaterials, biochemicals, biofuels, and bioenergy. These feedstocks can be processed in a biorefinery in single or multiple streams. A combination of biotechnological and/or thermochemical methods of conversion is used for processing and further refining. The circular economy concept would ensure that there is complete and effective utilization of all the components of biotic resources in an economically viable and environmentally friendly manner.

This book encompasses a collection of chapters that discuss different unconventional biotic sources and the methods by which they can be used for the production of value-added products. The chapters aim to provide an overview of the different processes being developed for these feedstocks and the areas in which future research can be directed.

The editors thank the authors for providing an insight into these important areas of research and development. The editors also take this opportunity to thank CRC Press/Tayor & Francis Group for providing a platform to publish this book on a circular economy concept for biotic resources.

Editor's Biographies

Dr. Thallada Bhaskar is Senior Principal Scientist and Head, Material Resource Efficiency Division, at the CSIR-Indian Institute of Petroleum, Dehradun, India. He completed his PhD at Osmania University, Hyderabad, India. He carried out postdoctoral research at Okayama University, Okayama, Japan, after which he worked there as an assistant professor for 5 years. He has been conferred awards such as Scientist of the Year, Raman Research Fellowship, CAS Presidential Award for the Foreign Fellows, and Distinguished Researcher Award, to name a few. He has more than 200 publications in journals of international repute and 38 book chapters. He has 15 patents to his credit with an h-index of 59 (total citations: ~10800).

Dr. Sunita Varjani is a scientific officer at the Gujarat Pollution Control Board, Gandhinagar, Gujarat, India. Her major areas of research are industrial and environmental biotechnology, wastewater treatment and process engineering, bioprocess technology, and waste management. She has worked as a visiting scientist at EPFL, Lausanne, Switzerland. Dr. Varjani has authored more than 350 publications, including research and review papers, books, book chapters, and conference communications. She has been listed as a highly cited researcher (top 2% in the world), Elsevier Citation Report (2021). She was selected as a member of the National Academy of Sciences, India (NASI) in 2021.

Dr. Bhavya B. Krishna is a Principal scientist and head of the Sustainability Impact Assessment Area, Material Resource Efficiency Division at CSIR-Indian Institute of Petroleum, Dehradun. She obtained her PhD in 2015 for her work on pyrolysis of lignocellulosic biomass. Her research experience of around 10 years encompasses works utilizing several biomass feedstocks such as agricultural and forest residues, aquatic biomass, deoiled cakes, etc. She is an author of more than 40 papers in peer-reviewed international journals, 15 book chapters in books by publishers of international repute, CSIR Young Scientist, recipient of awards at international conferences, and has filed one Indian patent.

Professor Ashok Pandey is currently a distinguished scientist at the Centre for Innovation and Translational Research, CSIR-Indian Institute of Toxicology Research, Lucknow, India, and Distinguished Professor, Sustainability Cluster, School of Engineering, University of Petroleum and Energy Studies, Dehradun, India. His major research and technological development interests are industrial and environmental biotechnology and energy biosciences focusing on biomass to biofuels and chemicals, waste to wealth and energy, industrial enzymes, etc. He has ~1650 publications/communications, which include 16 patents, 110 books, and ~900 papers, book chapters, etc., with an h-index of 122 and >64,000 citations on Google Scholar.

Contributors

Shashi Arya
CSIR-National Environmental Engineering Research Institute (CSIR-NEERI)
Nehru Marg, Nagpur, Maharashtra
India and Academy of Scientific and Innovative Research (AcSIR)
Ghaziabad, Uttar Pradesh, India

Pritam Bardhan
Department of Molecular Biology and Biotechnology
Tezpur University
Tezpur, Assam, India

Bunushree Behera
Department of Biotechnology & Medical Engineering
National Institute of Technology Rourkela
Rourkela, Odisha, India

Digambar Chavan
CSIR-National Environmental Engineering Research Institute (CSIR-NEERI)
Nehru Marg, Nagpur, Maharashtra
and
India and Academy of Scientific and Innovative Research (AcSIR)
Ghaziabad, Uttar Pradesh, India

Diptarka Dasgupta
Material Resource Efficiency Division
CSIR-Indian Institute of Petroleum
Mohkampur, Haridwar Road, Dehradun, Uttarakhand
India and Academy of Scientific and Innovative Research (AcSIR)
Ghaziabad, Uttar Pradesh, India

Baishali Dey
Department of Biotechnology & Medical Engineering
National Institute of Technology Rourkela
Rourkela, Odisha, India

Huma Fatima
Enzyme and Microbial Biochemistry Laboratory
Department of Chemistry
Indian Institute of Technology Delhi
New Delhi, Delhi, India

Debashish Ghosh
Material Resource Efficiency Division
CSIR-Indian Institute of Petroleum Mohkampur
Dehradun, Uttarakhand, India
and
Academy of Scientific and Innovative Research (AcSIR)
Ghaziabad, Uttar Pradesh, India

Jasneet Grewal
Enzyme and Microbial Biochemistry Laboratory
Department of Chemistry
Indian Institute of Technology Delhi
New Delhi, Delhi, India

Kuldeep Gupta
Department of Molecular Biology and Biotechnology
Tezpur University
Tezpur, Assam, India

Sharareh Harirchi
Swedish Centre for Resource Recovery
University of Borås
Borås, Sweden
and
Department of Cell and Molecular Biology & Microbiology
Faculty of Biological Science and Technology
University of Isfahan
Isfahan, Iran

Haoquan Hu
State Key Laboratory of Fine Chemicals, Institute of Coal Chemical Engineering
School of Chemical Engineering
Dalian University of Technology
Dalian, China

Arijit Jana
Material Resource Efficiency Division
CSIR-Indian Institute of Petroleum Mohkampur
Dehradun, Uttarakhand, India

Sivaraman Jayaraman
Department of Biotechnology & Medical Engineering
National Institute of Technology Rourkela
Rourkela, Odisha, India

Lijun Jin
State Key Laboratory of Fine Chemicals
Institute of Coal Chemical Engineering, School of Chemical Engineering
Dalian University of Technology
Dalian, China

Rupam Kataki
Department of Energy, Tezpur University
Tezpur, Assam, India

Sunil Kumar Khare
Enzyme and Microbial Biochemistry Laboratory
Department of Chemistry
Indian Institute of Technology Delhi
New Delhi, Delhi, India

Sumit Kumar
Enzyme and Microbial Biochemistry Laboratory
Department of Chemistry
Indian Institute of Technology Delhi
New Delhi, Delhi, India
and
Amity Institute of Biotechnology
Amity University
Noida, Uttar Pradesh, India

Sunil Kumar
CSIR-National Environmental Engineering Research Institute (CSIR-NEERI)
Nagpur, Maharashtra, India
and
Academy of Scientific and Innovative Research (AcSIR)
Ghaziabad, Uttar Pradesh, India

Andrzej Listowski
Sydney Olympic Park Authority
Sydney, Australia

Manabendra Mandal
Department of Molecular Biology and Biotechnology
Tezpur University
Tezpur, Assam, India

Kaustubha Mohanty
School of Energy Science and Engineering
and
Department of Chemical Engineering
Indian Institute of Technology Guwahati
Guwahati, Assam, India

Chandan Mukherjee
Department of Pharmaceutical Biotechnology
Maharana Pratap College of Pharmacy
Kanpur, Uttar Pradesh, India

Long D. Nghiem
Centre for Technology in Water and Wastewater
University of Technology Sydney
Ultimo, New South Wales, Australia

Balasubramanian Paramasivan
Department of Biotechnology & Medical Engineering
National Institute of Technology Rourkela
Rourkela, Odisha, India

Karthika Parvathy K. R
Food and Microbiology Laboratory, Department of Life Science
National Institute of Technology Rourkela
Rourkela, Odisha, India

Madonna Roy
School of Energy Science and Engineering
Indian Institute of Technology Guwahati
Guwahati, Assam, India

Taner Sar
Swedish Centre for Resource Recovery
University of Borås
Borås, Sweden

Mohammad J. Taherzadeh
Swedish Centre for Resource Recovery
University of Borås
Borås, Sweden

Shilpa Vishwakarma
CSIR-National Environmental Engineering Research Institute (CSIR-NEERI)
Nagpur, Maharashtra, India

Guijin Wang
State Key Laboratory of Fine Chemicals
Institute of Coal Chemical Engineering, School of Chemical Engineering
Dalian University of Technology
Dalian, China

Fatemeh Yazdian
Department of Life Science Engineering
Faculty of New Sciences and Technologies
University of Tehran
Tehran, Iran

Hongyou Yuan
Key Laboratory of Renewable Energy, CAS, Guangdong Key Laboratory of New and Renewable Energy Research and Development
Guangzhou Institute of Energy Conversion
Chinese Academy of Sciences
Guangzhou, China

1 Potential Benefits of Waste Biorefinery into Value-Added Products

An Approach toward Sustainable Development Goals

Shashi Arya, Digambar Chavan,
Shilpa Vishwakarma, and Sunil Kumar

CONTENTS

1.1 INTRODUCTION

Waste management is one of the most challenging issues due to its unavoidable generation over the years. With due time and several factors, such as reducing waste burden, resource conservation, health and safety, and a clean environment, the concept of value out of waste is proliferating rapidly. Biorefineries are one of the most common sectors where a huge amount of waste is generated in the

DOI: 10.1201/9781003335740-1

form of residues and effluents. The biorefinery waste generated in bulk quantities is readily available as feedstock material due to its nutrient enrichment (Devesa-Rey et al., 2011; Lin et al., 2014). As stated by Wysokinska (2016), the nutrient-rich waste material can be recycled for energy production and the development of value-added materials. Nowadays, this approach is being explored drastically to meet the Sustainable Development Goals (SDGs) in an environmentally friendly way through a sustainable waste management value chain, as presented in Figure 1.1. The European Union (EU) emphasized and drafted legislation toward the promotion of energy recovery from waste residues to achieve sustainable bioeconomy growth. Also, the Renewable Energy Directive (2009/28/EC) established that the value-added products and energy generation from the valorization of the waste residue should be environmentally friendly.

With reference to the existing practices and research towards resourceful technologies, the key principles associated with waste management and building a sustainable society are crucial. This chapter is aimed at exploring the significant benefits of volarizing waste biorefineries into value-added products. The topic presented in this chapter demonstrates the insights with respect to the present development, future trends, and potential benefits in the management of a waste biorefinery. The content in this chapter also emphasizes and highlights the importance of this concept, but also tries to promote knowledge on the development of innovative sustainable bioenergy systems.

Key factors that contribute to sustainable waste management are stakeholders (local authorities, nongovernmental organizations [NGOs], informal sectors etc.,), elements (following a waste management hierarchy), and critical factors (technical support, finance, advanced technologies) that help to meet the targets of SDGs.

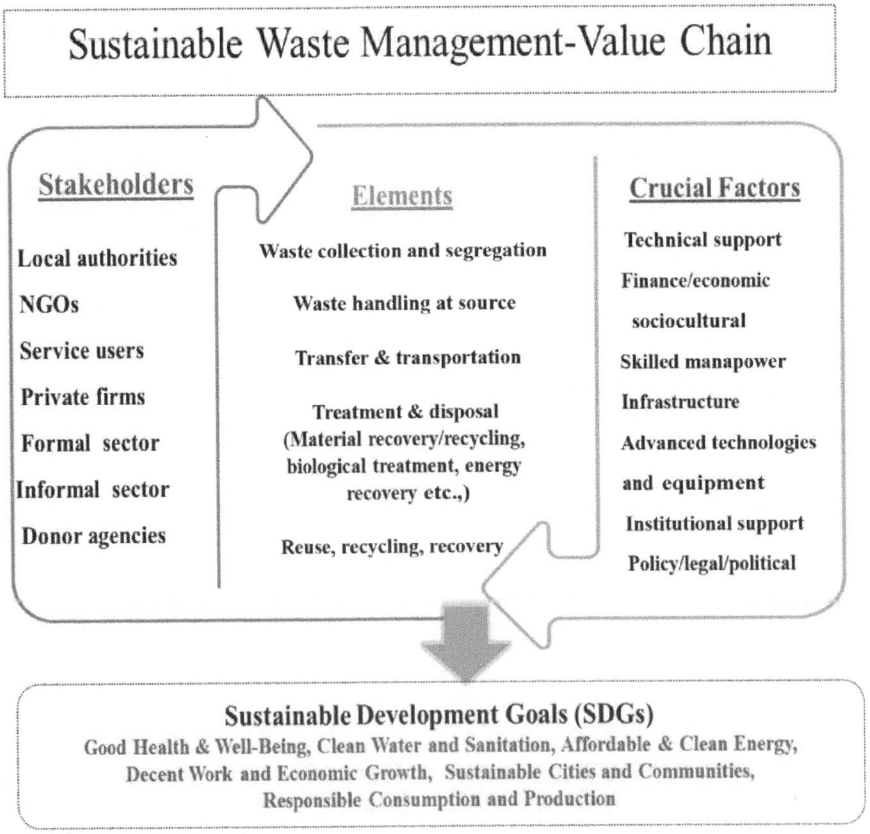

FIGURE 1.1 Sustainable waste management factors towards meeting SDGs.

1.1.1 Critical Factors Associated with Waste Management

The rapid growth of the global economy and increased population along with competitive market strategies have prompted development in all the sectors, which have created a burden for the supply of food, fuel, and other essential end-products (Ghisellini et al., 2016). Increased demand for market goods and modern living standards have accelerated the rate of fossil fuel consumption. Food and fuel consumption patterns are interlinked with each other and are key elements for understanding the progress of any economy. In the past few years food production has been shifted from traditional farming to industrial farming, which consumes a larger amount of energy in the form of power and fuel, which results in increased fossil fuel consumption, especially in developed countries (the United States and European Union) and developing countries like China and India. Along with industrial farming, agro-based industries like fertilizer production, making of agro-euipment and vehicles, and the storage and transportation sector established in the agro-economy also require a large amount of energy (70–75%), and ultimately it poses a problem of natural resource depletion (Obulisamy et al., 2017). The account of greenhouse gas (GHGs) emissions in the food sector from various processes like food production, processing, and transportation is up to 30%, which contributes to global climate change and its associated impacts. This altered pattern of food and essential goods consumption led to the highest per capita waste generation in most of the developed and developing countries. The increased rate of waste generation and limited infrastructure in developing countries poses a challenge to waste management. Urban local bodies (ULBs) in developing countries are dealing with waste and do not have enough funds to tackle the huge amount of waste and to incorporate advanced waste treatment processes like anaerobic digestion (AD), incineration, pyrolysis, gasification, etc., because the majority of the waste is being dumped at open dumpsites (Brunner and Rechberger, 2015). The approach of sustainable waste management is still lacking in most of the ULBs of developing countries due to limited allocated budgets and maintenance facilities (Nizami et al., 2016, Tozlu et al., 2016; Tan et al., 2015). Waste biorefineries, on the other hand, have tremendous potential for converting the existing waste materials into value-added products, fuels, heat, and power, which can solve the waste disposal problem in many developing countries (Nizami et al., 2017). More than 50% of the waste composition in most of the developing countries is organic (Kumar et al., 2017), and it is a significant source of bioenergy, recyclable materials, biomass, and chemicals. Use of this extensive waste material in biorefineries like AD, fermentation, pyrolysis, and gasification can convert waste into value-added products, which will play a vital role in the growth of the developing economies. In India and other neighboring countries, the idea of waste biorefineries is not so popular and relevant due to the current waste management challenges. Conversion of this organic waste biomass into value-added products through waste biorefineries might play a very important role to boost the economy in semi-urban and rural parts of developing countries, which has significant potential to create new markets, businesses, and employment. The aim of the present work is to explore the potential benefits of waste biorefineries in developing countries, which will help in the rate of installing waste biorefineries, which in turn serve as a solution to their current waste disposal issues and as facilities to produce fuels, power, heat, value-added materials, and chemicals to enable circular economies.

1.1.2 Potential Options for Biorefining of Different Types of Waste Biomass

This high volume of waste biomass can be converted into an enormous amount of energy and raw products. Biomass is available in different forms like wood, yard trimmings, raw organics from the food processing industry, agricultural waste, and animal waste, and municipal solid waste (MSW) can also be converted into value-added products. GHG emissions can be reduced through the combustion of these types of waste biomass, which has great potential to produce heat and electrical energy (https://www.hurstboiler.com/biomass_boiler_systems/biomass_fuel_types). In

India more than 150 million tons (Mt) of MSW are generated, of which more than 50% is biodegradable waste and around 90% of the generated MSW is dumped at open landfills, which cover around 60,000 ha of land, which is a potential source of GHGs emissions (http://www. sahyog-Europa-India. eu/inventories). Further, treatment of the domestic sewage at Indian sewage treatment plants (STPs) produces 10 million tons of excessive sludge generation, which contributes to more than 20 million tons of CO_2 emissions. The existing scenario indicates that the biomass from the waste sector has not yet tapped even a fraction of its potential (Venkata Mohan, 2016). Waste to energy (WtE) plants for the processing of the huge amounts of everyday MSW generated in Indian cities and dumped at MSW landfills have great potential to exploit the clean heat energy source and to minimize the overall cost of energy production. AD and composting of organic waste, which is abundantly available (more than 50%) in typical Indian MSWs, would play a significant role in the promotion of a sustainable circular economy in rural as well as urban India. Agricultural waste residues like raw crops, rice husk, straw, leaves, pulp, peel, stalks, etc., are available in huge quantities can be transferred into the form of energy, which can substantially displace nonrenewable-based fossil fuels (https://www.altenergymag.com/article/2009/08/biomass-wastes/530/).

Until now it is an unutilized source of waste biomass for its conversion into value-added products. If utilized properly, it would have a greater impact on the development of the rural economy. The status of the SDGs in rural economies in developing countries is moderately dependent on the potential for the conversion of waste biomass into renewable energy sources and value-added products. Various types of biomass waste like MSW, agricultural residues and food waste, and industrial raw organics are available in abundance, and the supply is steady in rural India, which could be a motivation for the new biorefinery startups. With the increased demand for food and energy, developing countries are looking towards the rural economies for their potential to uplift the circular economy and to achieve the targets of the SDGs. Apart from this, biorefineries can serve the purpose of mitigating GHG emissions and to combat the national and global climate change crises. Waste biorefineries for the conversion of waste biomass into value-added products or renewable energy sources can give a boost to the financial growth of farmers without compromising conventional farming. Conversion of different types of waste biomass through techno-economical biorefinery options could be an attractive option for the production of fuel, heat, and electrical energy with fewer environmental impacts and lead to the creation of employment, which may help for the overall growth of the circular economy, especially in rural areas.

1.2 ENVIRONMENTAL SOUND MANAGEMENT OF WASTE FOR ACHIEVING SDGS

The waste residues come from a wide spectrum through several activities such as municipal, industrial, commercial, agricultural, etc., that consist of sufficient organic carbon fractions coupled with intrinsic energy value. Considering the significant potential of waste management, intensive research and development are being explored in terms of sustainable and cost-effective approaches. The valorization of waste is one among them, wherein a cradle-to-cradle approach is being adopted, i.e., the processed waste residue discharged is utilized as feedstock material for another process and maximum treatment efficiency is achieved. During the last few years biorefinery waste has evinced tremendous progress in research innovations. The biorefineries are exploited significantly for optimizing and utilizing the organic biomass towards maximum production of value-added products (chemicals and energy). The processes adopted also withstand the norms towards environmental sustainability that also advocates the circular economy.

Figure 1.2 shows how environmentally sound waste management helps in achieving the target and SDGs, namely Goal 3 (Good Health and Well-Being), Goal 6 (Clean Water and Sanitation), Goal 7 (Affordable and Clean Energy), Goal 11 (Sustainable Cities and Communities), and Goal 12 (Sustainable Consumption and Production).

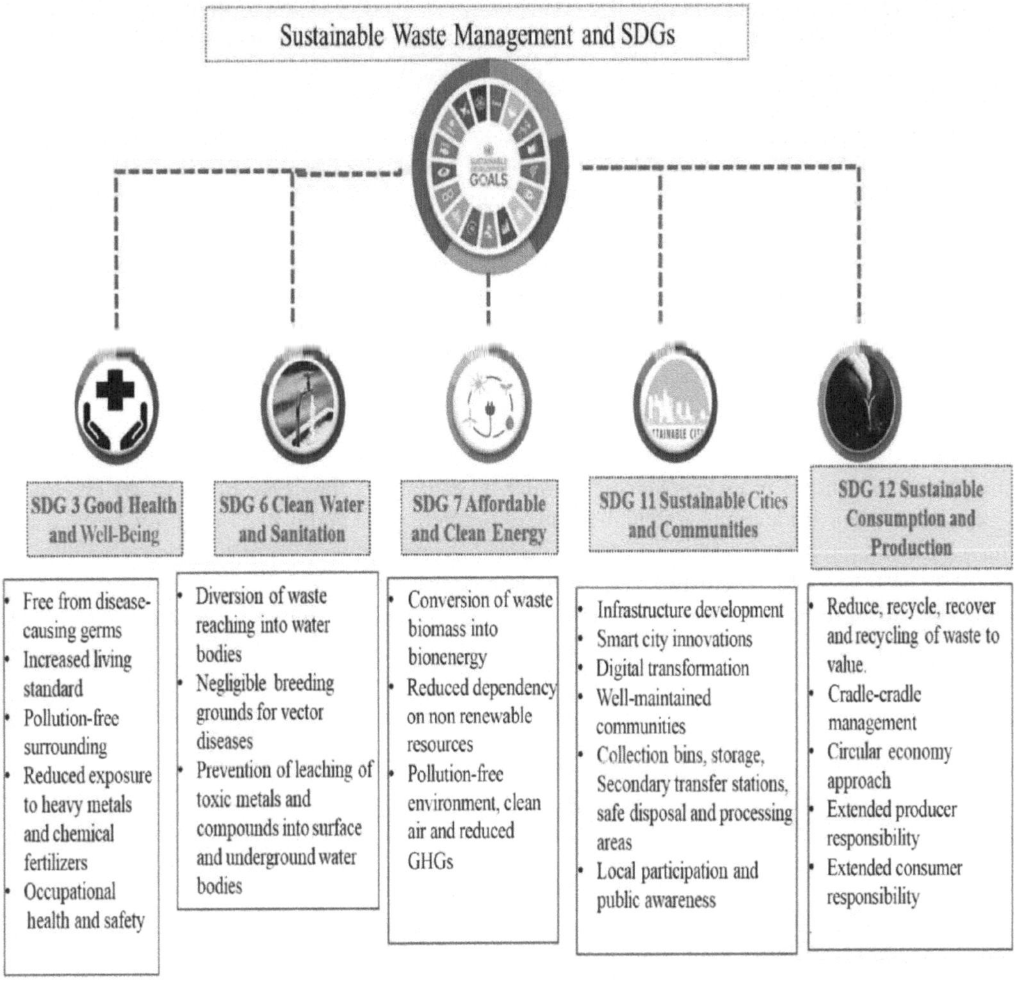

FIGURE 1.2 Environmentally sound waste management meeting SDGs.

1.2.1 Biorefining of Waste: Recycling, Reuse, and Recovery of Waste Biomass

Recycling of biorefinery waste is assumed to play a prominent role in maintaining the ecological footprint. Also, volarizing waste biorefineries through integrating different strategies and approaches leads to futuristic and sustainable solutions followed by the development of various value-added products.

The tremendous consumption of natural resources is affected critically by the existing demand and supply trends. The environmental concerns and good health and well-being are the most influential components that get affected due to overconsumption of resources and huge waste generation. Therefore, in such a situation, an alternative solution is crucial and immediate to mitigate the menace of the waste challenges and reduced natural excavation of resources. For example, the replacement of fossil fuel consumption with waste biomass could be a sustainable and viable option towards reducing waste menace and dependency on fossil fuels (Smith et al., 2011).

Renewable alternatives to fossil fuels are areas of intense research to boost energy security and reduce GHG pollution (Lin et al., 2013). Waste to energy conversion is globally recognized for its significant contribution capacity to overcome those two critical global problems (Lin et al., 2013). In addition to energy security and waste management, clean water and adequate nutrition are also pressing problems for India and globally (Bene et al., 2015). For example, India ranks among the third-largest CO_2 emitter in the world, with CO_2 emissions predicted to increase to 8% of the total global emissions by 2035, 63% of which is contributed by coal combustion (Ahn and Graczyk, 2012). In 2009, 289 million had no access to electricity and 836 million had no access to modern fuels for cooking and heating.

Similarly, the developed countries generate more than 4 Mt (avg.) of solid food waste per year, commonly deposited in landfills and contributing substantially to GHG emissions (11 Mt CO_2 equivalent) (Lou et al., 2013; Parfitt et al., 2010). In such a case, biofertilizer production from organic waste would offer an intensive energy- and cost-efficient solution to improving agricultural productivities of low-carbon soils, due to the high carbon and nitrogen contents of AD solid residues.

1.2.2 Conversion of Waste Biomass into Value-Added Products

The crucial components that are exploited from biomass waste are typically oils, polysaccharides, aromatics, nonfood sugars, etc. The processes are cost-effective, i.e., reduced expenditures over equipment, low energy requirements, concerning overall utilization of feedstock materials, improvised and efficient productivity, and high standard marketing are viably explored through volarizing the waste materials. The conversion of waste to value involves valorization of waste biomass followed by pretreating the waste and fractioning it into various derivatives and ultimately high-value products, as presented in Figure 1.3 and Table 1.1. It is an emerging and immediate measure to meet the SDGs, simultaneously leading to a clean and safe environment. A few of the biomaterials that

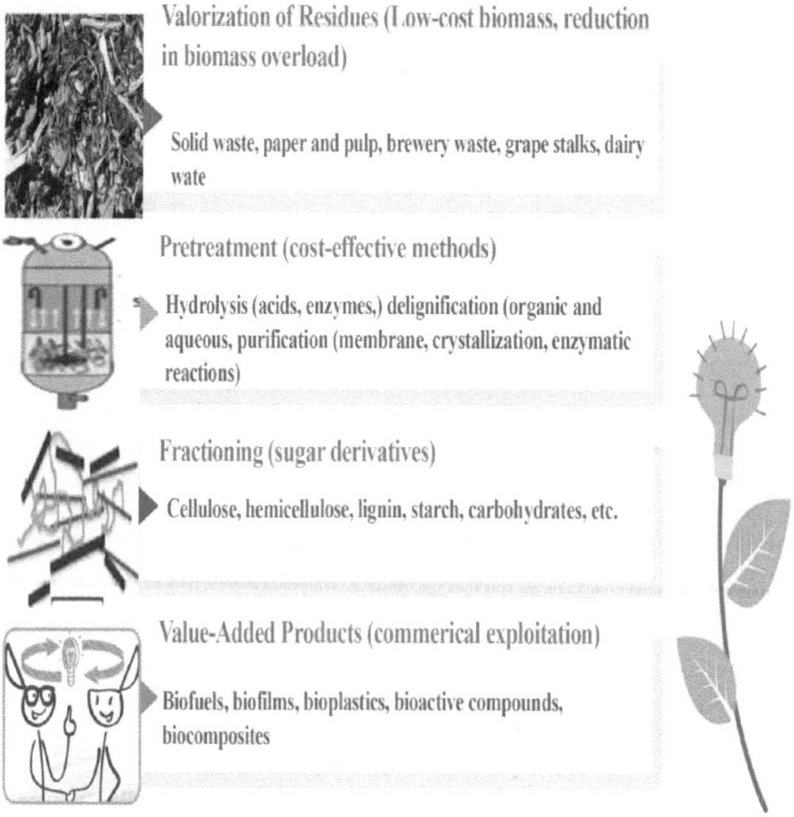

FIGURE 1.3 Schematic presentation of value out of waste.

TABLE 1.1
Valorization of Common Waste Residues Exploited Extensively for Value-Added Products

Waste Residues	Raw Material	Pretreatment	Fraction	Value-Added Product	References
Valorization of residues from pulp and paper mills	Eucalyptus chips, eucalyptus bark	Auto-hydrolysis treatment, acid hydrolysis, fermentation, supercritical CO_2 fluid extraction	Cellulose, hemicelluloses, extractives	Bioethanol, oligosaccharides, triterpenic acid	(De Melo et al., 2014; Gullón et al., 2008; Rodríguez-López et al., 2012; Romaní et al., 2010)
Valorization of residues from paper mills	Paper sludge	Enzymatic scarification, enzymatic hydrolysis, fermentation, fermentation for biogas production	Cellulose, organic matter	Lactic acid, ethanol, volatile fatty acids	(Bayr and Rintala, 2012; Gomes et al., 2016; Romaní et al., 2008)
Valorization of residues from pulp and paper mills	Pine sawdust, pine residues	NaOH delignification pretreatment, pulp treatment, TEMPO-mediated oxidation, mechanical treatment, compressed CO_2 extraction	Cellulose, lignin	Nanocellulose, phenolic and lipophilic compounds	(Conde et al., 2013)
Valorization of residues from pulp and paper mills	Waste fiber sludge	Enzymatic hydrolysis and fermentation	Glucose from cellulose hydrolysis	Bacterial cellulose	(Cavka et al., 2013)
Valorization of residues from wine-making industry	Vine pruning	Auto-hydrolysis treatment, acid hydrolysis, alkali extraction, alkaline pretreatment, enzymatic hydrolysis, fermentation	Hemicellulose, lignin, cellulose	Oligosaccharides, phenolic acids, glucose, bioadsorbents, xylitol and lactic acid	(Cotana et al., 2015; Dávila et al., 2016; Max et al., 2009; Portilla-Rivera et al., 2008; Vecino et al., 2014)
Valorization of residues from wine-making industry	Grape stalk and grape pomace and grape skin from grape pomace	Extraction with ethanol, methanol, hexane and water, acid hydrolysis, and fermentation	Cellulose, cellulose extractives containing soluble sugars, hemicellulose	Dietary fiber, phenolic-rich extract, biocomposites ethanol, lactic acid and intracellular biosurfactants	(García-Becerra et al., 2016; González-Centeno et al., 2010; Mendes et al., 2013; Rivera et al., 2007)
Valorization of residues from wine-making industry	Breweries spent grain	Sequential water and alkali extractions, steam explosion, enzymatic saccharification, CO_2 supercritical extraction, enzymatic hydrolysis, auto-hydrolysis treatment, fungal consolidated bioprocessing	Hemicellulose, cellulose	Arabinoxylan, bioethanol, glucose, lactic acid, phenolic compound and flavonoids	(Coelho et al., 2016; Djukić-Vuković et al., 2016; Michelin and Teixeira, 2016; Spinelli et al., 2016; Vieira et al., 2016)
Valorization of brewery wastes	Spent brewer's yeast, brewery yeast waste	Alkaline treatment, aqueous two-phase system	Yeast cell walls, intracellular	β-glucan, invertase recovery	(DeLeón-González et al., 2016; Romaní et al., 2018)
Valorization of dairy wastes	Cheese whey	Ultrafiltration and membrane filtration technology, chemical and enzymatic process, fermentation, and alcalase hydrolysis	Whey, lactose, and protein	Protein and sugars, lactulose, ethanol, β-galactosidase, hyaluronic acid, whey proteins	(Amado et al., 2016; Domingues et al., 2010; Galanakis et al., 2014; Silvério et al., 2016)

are critically explored as value-added materials delivered to meet SDG 3, SDG 7, SDG 8, SDG 11, and SDG 12 is discussed in the figure.

The figure shows the routes to get value out of waste, such as volarization of residues, pretreatment, fractioning, and ultimately value-added products.

Biohydrogen: Acidogenosis is a biological process that is most commonly practiced for the conversion of the waste biomass available in various forms into biobased valuable products viz. hydrogen (H_2), volatile fatty acids, and other trace byproducts (Venkata Mohan, 2016). Biohydrogen production in waste biorefineries has tremendous advantages over conventional biofuels and biogas. The combustion process based on biohydrogen as a main fuel generates cleaner energy along with minimal nitrous oxide (N_2O) and GHG emissions (Sarma et al., 2015). Compressed natural gas (CNG) mixed with 30% biohydrogen can reduce suspended particulate emissions (SPMs), CO_2, and hydrocarbon emissions by 55%, 35%, and 60%, respectively.

Biohythane: The gas mixture composed of CH_4 and H_2 is referred to as biohythane, as presented in Figure 1.4. The composition of CH_4 and H_2 in biohythane varies from 43% to 54% and 46% to 57%, respectively, and it also contains 0.4% of CO_2. Proper utilization of biohythane in the industrial and transportation sector would have a significant impact on the reduction of fossil fuel consumption (Pasupuleti and Venkata Mohan, 2015). The concern for the use of microalgae in biological wastewater processes and the production of bioethanol and biodiesel is steadily increasing due to increased energy demand and an ultra-modern lifestyle. In this context, the use of excessive biogenic waste in AD and the fermentation process would be a potential technological option for the production of biohythane (Ghimire Anish et al., 2017). Various biobased technologies can be designed or engineered especially for the efficient physical-chemical hydrogen and CH_4 production, which will serve the purpose of the promising approach to achieve the SDGs. It has been recognized that the energy sector powered by hydrogen can fundamentally make a revolution in the existing markets, both in the developed and developing economies, as was observed in the 20th century with coal and petroleum products (Ekins and Hughes, 2010)

The composition of CH_4 and H_2 in biohythane varies from 43% to 54% and 46% to 57%, respectively, and it also contains 0.4% of CO_2. Proper utilization of biohythane in the industrial and transportation sector would have a significant impact on the reduction of fossil fuel consumption.

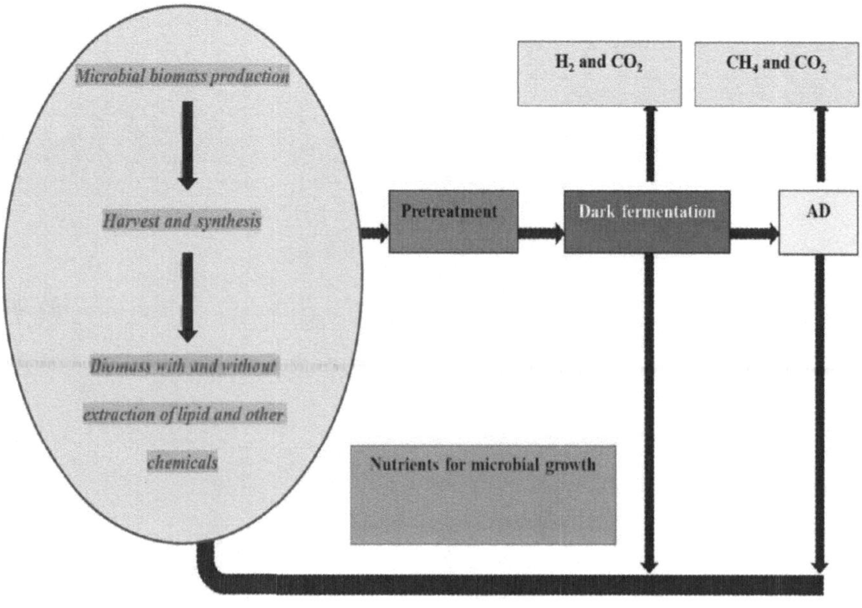

FIGURE 1.4 Biorefinery approach of production of biohythane.

Biorefineries based on the microalgal biomass as a raw material have received significant attention for their proper utilization in the fuel-based energy over since the past decade (Chisti, 2010). Microalgae-based biorefineries offer numerous benefits as compared to conventional plant-based biofuels. The higher abundance of algal biomass and ease in its cultivation make them cost-effective. Microalgal-based biorefineries can produce value-added products like fatty acid methyl esters (FAMEs) and proteins, which have a high capital commercial value. Chisti (2010) concluded that biomass and oil area productivities of algae are higher than that of vascular plants under optimal culture conditions, but the affordability of biofuel production from microalgae with fossil fuels remains a major challenge for the renewable industry. Microalgae-based biorefineries may require larger quantities of nitrogen and phosphorous for the sustained growth of microorganisms, which demands chemical-based fertilizers, and their complex synthesis and extraction process requires a high amount of energy. The concept of a microalgal biorefinery is dependent on the integration of different technologies that are useful to produce value-added products such as biodiesel, ethanol, biohydrogen, biogas, etc. The selection of an appropriate technology and biochemical composition of algal biomass may alter the recovery rate of various biofuels and chemicals. An algal biomass with a lipid content >40% plays a significant role in the recovery of biodiesel, but a lipid content <40% would not be efficient due to the high cost required for the harvesting and synthesis of microalgal biomass (Sialve et al., 2009).

The bioplastics exploited using waste organic components such as vegetables, fruits, food waste litter, wood, corn, etc., are trending rapidly in the markets due to their suitability and bans on the use of plastic bags. Bioplastics are sugar derivatives such as cellulose lactic acid and starch. The major constituents of the bioplastics exploited extensively in the present scenario is starch, i.e., approximately 50% blended with various combinations of degradable polyesters such as ecoflex, polylactic acid, polybutylene adipate-co-terephthalate, or polycaprolactone. The bioplastics derived from cellulosic materials are cellulose esters and nitrocellulose. The combined blend of cellulosic fibers and starch for bioplastics is considered to have better characteristics such as high mechanical strength and gas permeability. Also, the bioplastics made from such components are moisture resistant due to a cellulosic content that is hydrophilic. Several other materials are also known to be exploited towards making bioplastics, i.e., proteins like gluten and casein that are known to have promising characteristics and qualities.

Bioethanol is one of the most extensive biofuels exploited using organic waste residues such as fruits and vegetable sugars (carbohydrates) and agricultural crop residues like corn stalk, sugarcane bagasse, sorghum, etc., through the fermentation process. Other cellulosic feedstocks exploited for the production of economic nonfood sugar high-value biofuels are leaves, twigs, litter, etc., i.e., more often used in its pure form (E100) as an additive towards minimizing vehicular emissions. Similarly, biodiesel is produced through the trans-esterification of oils and fats. The fuel is commonly known to be used extensively in European countries in vehicles to ensure reduced levels of toxic gas emissions.

Biofertilizers are one of the most extensive and highly efficient forms of the value-added product of waste components. For decades the practices of biofertilizers have been well utilized by the farmers and are composted using household organic waste. Various types of compositing practices are being exploited in various countries in the world in terms of soil fertility, high-yielding capacity, and chemical-free edibles. Biogas is one of the highly efficient sources of fuel that is produced through AD of organic waste, such as cattle manure, agricultural residue, kitchen waste, and biomass in different constituents, as presented in Figure 1.5. AD of organic MSW is a well-established treatment process, but recovery of heat and energy requires pretreatment, which further reduces its cost-effectiveness. However, at present research is focused on targeting resource recovery of the high-value chemicals through the application of hybrid technologies like fermentation, pyrolysis, and hydrothermal conversation of waste residue and thus will improve cost-effectiveness.

AD of organic MSW (waste biomass) is a well-established resource recovery process and leads to a high-value chemical product achieved through the application of hybrid technologies like fermentation, pyrolysis, and hydrothermal conversation of waste residue.

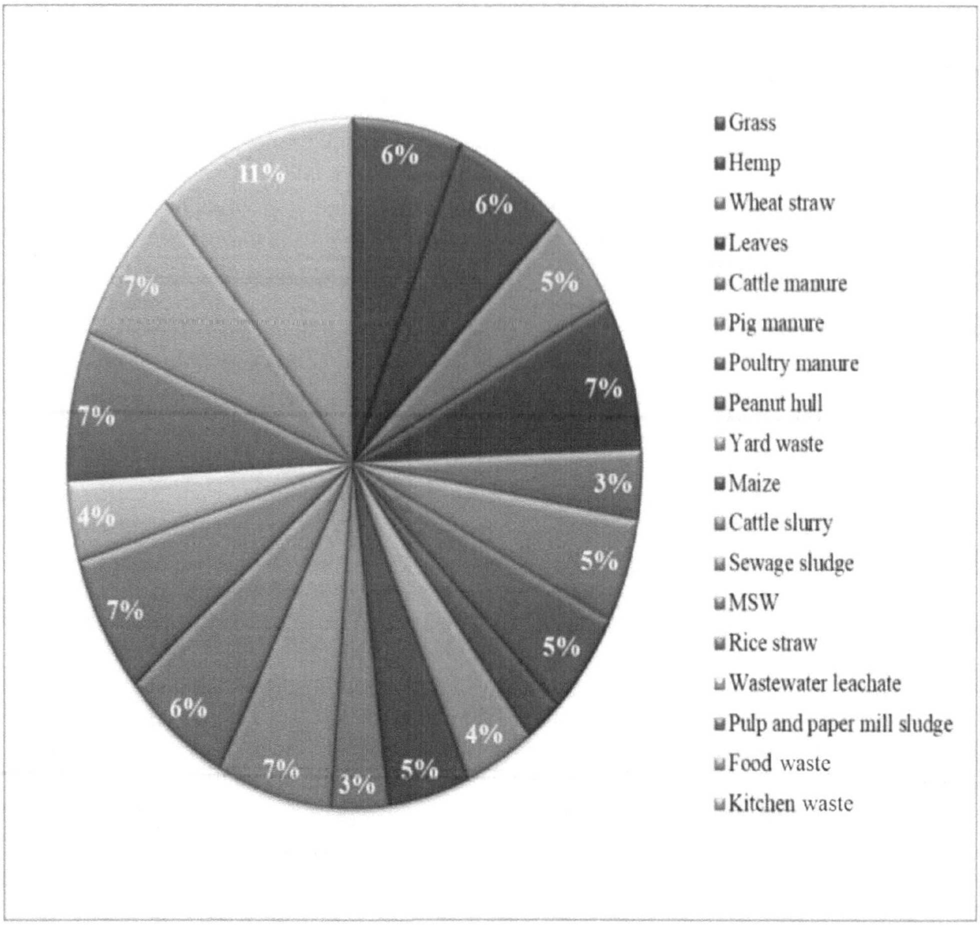

FIGURE 1.5 Different types of the waste biomass that have biogas generation potential using AD process.

1.2.2.1 Nutrient Recovery

Resource recovery from waste biomass is the most significant environmental agenda in developed and developing countries (Zabaleta and Rodic, 2015). Recovery of essential nutrients such as phosphorus (P) might be very relevant, considering its current extraction rate may lead to complete exploitation within the next 300 years (Smil, 2000; Gilbert, 2009; Childers et al., 2011; Cordell et al., 2011; Desmidt et al., 2015). A portion of the organic fraction of MSW, which consists of food and kitchen waste; garden waste; and commercial organic waste from restaurants, caterers, and food processing industries are disposed at open MSW dumpsites and possess a variety of useful substrates like starch, amino sugars, lipids, proteins, lignocellulose biomass, and other compounds. The amount of these substrates in MSW varies depending upon the nutrient content, energy content, and conditions favorable for the growth of the microbial population. Nutrients like carbon and hydrogen and elements such as Mn, Mg, Zn, Fe, S, Cu, Co, N, P, and K are abundantly present in different substrates of MSW (Hanc and Pliva, 2013). Functional aspects of the waste management sector, including practices like waste collection, transportation, treatment and processing, disposal, etc., which are common practices in almost all parts of the world, do not pay attention to the recovery of essential nutrients, which shall be replenished through various extraction and industrial synthetic processes.

1.2.2.2 Healthy Food Supplements

The choice of the microalgal strain *Chlorella vulgaris* is deliberated as value-added materials, as this organism is already sold as a food supplement, eliminating costly and time-consuming approval processes. Furthermore, *C. vulgaris* has been highlighted as a potential candidate for renewable biofuel production in open ponds and showed high productivities when cultivated as biofilms that allow high-yielding capacities under moderate climatic conditions.

1.2.3 Status of Waste Biorefinery in Developed and Developing Countries

Commitments have been made for the protection of the environment and conservation of natural resources by a group of nations, including both developed and developing countries. The agreement was prepared to address the issues, challenges, and opportunities through the SDGs and as per the agreement made during the Conference of Parties (COP21) which was held in Paris for climate change negotiations. The focus is now shifted towards the markets for second-generation biofuels and the design of commercial technologies for the extraction of the value-added products from waste biomass both in developed and developing countries. Promotion of biorefineries for the overall development of the circular economy in rural parts of the developing countries may provide a promising approach to achieve both the short- and long-term SDGs. Waste biomass, like sugarcane bagasse, crop residues, agriculture, and forestry residue, is found abundantly in India and China. Due to high population growth, setting up of waste biorefineries may play a significant role to fulfill the growing demand for value-added products (bioenergy, chemicals, and materials). In Brazil, the consumption of waste biomass in the first-generation biorefineries is fulfilling around 18% of the ethanol requirement of the automotive industry (Hatti-Kaul, 2010). The United States has installed second-generation biofuel refineries and the world's largest cellulosic ethanol-based biorefineries, followed by Canada, China, and the EU. In developed countries, the utilization of waste biomass varies significantly based on the availability of raw materials (feedstock) and their potential for energy or fuel production. Corn stover, leaves, stalks, sugarcane bagasse, MSW, and forestry residues are the most commonly used raw materials as a feedstock in waste biorefineries for the conversion of value-added products in almost all parts of the world. Companies in developed countries of the EU and United States are engaged in partnership projects for the building of advanced ethanol facilities (e.g., Fuyang bioproject started in China is sponsored by the joint cooperation of Guozhen and Italy-based beta renewables). Until now there is no ethanol-based project in the African continent and Latin America except the largest ethanol plant in Brazil. Strategies that have given traction to advanced biofuels in the world are market segmentation of conventional biofuels in the United States and EU and funding through bank loans, which have prompted the overall growth of the biorefinery industry in developed countries (Kaul, 2010). Venture-capital culture and low-interest rates may play a significant role in the growth of waste biorefineries in India and other developing countries.

1.3 POTENTIAL ADVANTAGES OF CONVERSION OF WASTE BIOMASS INTO VALUE-ADDED PRODUCTS

Nature itself plays a very important role in converting value out of waste. The consumption of natural resources is rising tremendously and has influenced human survival. This tendency to deal with the natural process coupled with critical elements, i.e., science and engineering, is emerging tremendously with immense benefits to sustain the limited resources. Waste is one of the biggest resources that can be exploited at its best to attain several associated components limiting the dependency on natural resources. A few of the benefits of the conversion of waste biomass into value-added products are the reduction in pollution load, health, and safety; sustainable resilient environment; green products; extended job opportunities; and the increased economy, as shown in

FIGURE 1.6 Potential benefits of conversion of waste biomass.

Figure 1.6. Therefore, biorefineries for the conversion of waste into valuable products are receiving attention from industrial, scientific, and government communities all over the world (Clark, 2019). Conversion of waste into value-added products is one of the most crucial approaches in reducing the overburden of waste in the landfills. The approach is intended towards making wealth out of waste and at the same time contributing to reducing the adverse effects of toxic gases such as methane, carbon monoxide, carbon dioxide, and leaching of heavy metals. Diversion of waste reaching landfills also contributes to profiling the significant opportunities such as high-value chemical products, reduced transportation cost, and other associated critical issues, and most important is the replacement of fossil fuels and excavation of natural resources. The biomass, i.e., enriched with lignocellulose and cellulosic composition, also available in abundant quantities, represents potential raw material for the production of biofuels, energy, and high-value chemical products. This practice also has a significant impact on human health and a sustainable and clean environment. Recycling the waste into value-added materials opens vast opportunities for marginalized people through employment generation with additional benefits such as increased economy, living standard, respect, health, and hygiene. It also contributes to maintaining a sustainable and ecofriendly environment.

The advancement of existing technologies will enable people to recognize the full potential of different waste biomass materials as a chemical and energy resource, which may further help its implementation and overall growth of the circular economy (Clark, 2019). A country like India has created pressure on natural resources, and to fulfill the increased energy demand, the conversion of waste biomass into gaseous fuel and value-added products is an urgent need (Bhatia et al., 2020).

Waste biorefinery can be considered a key instrument to achieve the bioeconomy goal and for the development of a linear to circular transition of the economy in rural as well as urban areas. The success of the biorefinery projects is based on the circular economy, which depends mainly on the ability to transform the biomass into value-added products and the resulting growth of the economy. The waste management sector may play a vital role in boosting the biorefinery-based circular economy due to the physicochemical properties of the organic waste biodegradable materials composed of carbohydrates, proteins, and lipids, which have high potential for the creation of value-added products (Leong et al. 2021; Víctor et al. 2020). Waste biorefineries offer a viable approach to achieve the SDGs and conservation of natural resources. For the potential benefits from the waste sector for the conversion of value-added products, its functional aspects such as waste generation, collection, treatment, and disposal could be analyzed in the context of rural and urban economies in developing countries to refine and estimate the efficient use of the MSW in biorefineries.

A circular flow diagram shows a few of the benefits of the conversion of waste biomass into value-added products, such as the reduction in pollution load, health, and safety; a sustainable resilient environment; green products; extended job opportunities; and the increased economy.

1.4 CHALLENGES ASSOCIATED WITH CONVERSION OF VALUE OUT OF WASTE

Controlling massive amounts of biomass waste is a major challenge in developing countries due to several factors such as unavailability of feedstock material, poor infrastructure facility, unskilled manpower, poor strategies on waste handling, and a dearth in administration enactment. Also, due to poor quantification of the waste composition, delineation of an appropriate management plan is hindered. Evaluation of technical challenges related to the conversion of a waste biorefinery, i.e., chemical or biological, is crucial, and a few of them are discussed next.

1.4.1 Lack of Availability and Nature of Waste Residue

Diversion of any resource, even though it seems to be unnecessary, is likely to have unintended effects by increasing the price of that resource. It is also important to note that the sheer existence of waste and waste for bioenergy feedstock does not in itself mean that a growing bioenergy industry can use these services preferentially. In the case of the Drax power plant in the United Kingdom, for example, it has been seen that wood to fuel the plant is supplied by felling trees in the United States, despite the existence of the waste and residues. Furthermore, the cost of collecting and shipping these cellulosic materials can, to some degree, restrict the availability of waste material from the development of other products. Although it will be very costly to ship waste on a long-haul basis, policy incentives could likely build a demand for foreign trade in these feedstocks. Finally, there would very likely be competition between biofuels, biopower, and heating for feedstocks in the energy market, so the amounts of cellulosic content should not be considered to be entirely usable for one industry or another. Using trans-esterification, thermochemical, and biochemical paths, biomass residues and waste can be transformed into transportation fuels and bioelectricity. Waste utilization, along with the added advantage of cleaning the atmosphere, is expected to be the most economical method for the development of renewable energy. Significant quantities of biomass residues and waste are eventually created from various industries around the world, and if effective and economically feasible technologies were developed, waste could be a promising bioenergy feedstock. Still the production of biomass residues and waste as an immediate energy supply has many limitations. First and foremost, based on the current technologies developed, waste-to-bioenergy processing is still not as cost-competitive as fossil-based fuels.

1.4.2 Market Identification and Related Barriers

The comparatively high cost of processing and transporting waste material, also referred to as 'the harvesting margin', is the main challenge of using agriculture and forest biomass. Many of these waste sources are now used as low-value inputs for manufacturing and agricultural processes and cannot be diverted to the production of bioenergy without consumer impacts, higher demand for other products, and indirect environmental effects. For comparison, pine pulpwood may provide a strong foundation. In Mississippi, the average value of pine pulpwood was \$17.6/standard cord (shipped wood) for standing timber from 1977 to 2005 and \$51.7/std. The harvesting margin ranged from $23.3 to $48.4, with a mean of $34.1/std cord, described as the differential between stumpage and shipped rates. With an average of 67.1%, the proportion of the harvesting margin in the supplied price varied from 41.2% to 79.6%. Process technology preference depends on the target end product and the feedstocks. In general, as opposed to biochemical techniques for the development of biofuels, thermochemical technology that employs thermal heat may not be susceptible to the composition of biomass waste. Nevertheless, in material processing, transport, and conversion technologies, the production of biofuels from biomass waste is still considered to be more stable relative to conventional editable biofuels based on food crops. However, current scientific projects are committed to resolving the limitations of current technologies and to improving the quality and productivity of the manufacturing technologies used.

1.4.3 Nonavailability of Skilled Manpower and Required Training

To understand the future possibilities that occur in the waste-derived bioeconomy, major steps must be taken by the government, industry, and the scientific community. It is necessary to lobby on waste as a high-value potential across the country with a long-term commercial plan. For a more standardized method of waste disposal through local councils, a long-term approach and purpose should be formulated. Adequate financing is required for the advancement of expertise with demonstration of provision. Successive government funding for the implementation of a long-term strategy for a bioeconomy derived from waste with an opportunity system, political stability, improvement of market distortions, etc., would support the bioeconomy derived from waste. The area of waste biorefinery has considerable room in terms of scope and requires liberal financial assistance to secure its applications. The key goal is to satisfy demand and provide skilled labor, to learn new talents and creative thinking, not just for current employment but also for the jobs that are to be developed in the future. In India, population changes, skills growth programs, and rising levels of schooling seek to expand the size of the pool of trained and skilled employees, generating opportunities for the bioeconomy. Economies with large professional human capital expand jobs as industry openings and monetary expenditure for technological development and thereby increase the nation's per capita income that boosts demand for industrial biotechnologies.

1.4.4 Lack of Awareness for Utilization of Waste as Value-Added Products

The amount of waste is expected to increase about 150 Mt by 2030 from the existing 62 Mt. Based on the present scenario, the landfills in most of the countries are facing immense pressure due to inadequate scientific treatment and indiscriminate disposal each year. However, due to lack of awareness and information, most of the valuable waste is burned by the local people. In waste management, there is a shortage of training, and the number of trained waste management practitioners is limited, and there is a lack of transparency.

1.5 OPPORTUNITIES FOR WASTE BIOREFINING: THE INDIAN PERSPECTIVE

India has great potential to fulfill the current energy requirement of the country through utilizing the concepts and practices of a volarizing waste biorefinery that could be a plan or a concept toward transforming biomass to meet various needs. The availability of paper, pulp, food and yard waste, residues of wood processing, agriculture residues, and residues of forest harvesting from well-established manufacturing chains are essential biomass sources that can provide a large amount of biomass for a wide variety of value-added products to be produced. Most of these wastes play a vital role in protecting the nature of the land, avoiding deforestation, and encouraging biodiversity.

Agricultural waste, food waste, industrial waste, and domestic wastes are mainly prioritized areas in volarizing waste biorefineries. Sustainable processing of agriculture and domestic wastes enables leverage of the bioeconomy. Materializing organic and agricultural biomass into a portfolio of marketable products gives better resources for the rural economy. This section is intended to outlay the current prospects and opportunities in biorefinery and bioconversion of organic and biomass for sustainable development. India is the fastest-growing nation of the developing countries, with a predicted population growth of 1.69 billion people (>500 people km^2). More than 42% of India's rural population lives below the poverty line (USD 1.25 based on purchasing power parity), depends heavily on subsidized kerosene for cooking, or low-income households use traditional resources (biomass, animal waste, and wood) for cooking, heating, and lighting purposes, which holds serious health problems for women and children due to particulate emissions. Several initiatives have been taken to deliver measurable community outcomes that ultimately meet the SDGs, for example, sustainable energy generation from organic waste, biofertilizer from waste residues of AD, and additional co-products such as microalgal food health supplements via recycling nutrient-rich AD effluents. These approaches have been assumed to be directly relevant to present initiatives taken by India such as Swatch Bharat Abhiyan (Clean India mission) and align strongly with research and development priorities in other developing countries. Simultaneously, the approach could be stated as pillars for the state's economy through reduced waste generation and clean energy production for a sustainable future.

1.5.1 Reduction of Waste Quantity

With rapid industrialization and development in India, municipal waste is increasing. Emerging platforms in chemical technologies such as gasification, pyrolysis, liquefaction, and various biochemical conversion pathways became very helpful in limiting waste accumulation and minimizing the environmental impact. Trans-esterification is one of the important techniques in bioconversion of some agricultural wastes and food waste. Trans-esterification aims at biocatalysis of oils to biodiesels. This results in minimizing water pollution and spill pollution mediated by oil waste. There are also various platforms for bioconversion of both agricultural waste residue and food waste into energy to meet the requirements in rural areas. A few of them are biogas, microbial fuel cells, and bioelectricity chambers. Employing such techniques in biorefining organic waste potentially reduces the impact of biomass on the environment. In detail, proper recycling or reutilization of waste materials into value-added products will furthermore decrease the waste quantity from the earth and will provide maximum efficiency of reusability.

1.5.2 Energy Recovery

Volarizing waste biorefinery for energy recovery involves combustion of nonrecyclable materials to generate energy in the forms of heat, bioelectricity, and biofuel. Biomass is converted into energy steam and pressure. In the case of biofuels, microorganisms were employed in the AD of solid and nonrecyclable biomass. Biogas or Gobar gas production is one such example. At the industrial scale, pyrolysis of biomass produces various energy forms such as bio-oil, charcoal, biodiesel, and methane. Recovering energy forms directly from biomass combustion could provide a path for sustainable development and solid waste management.

1.5.3 Reducing GHG Emission from the Waste Sector

In 2006, the Intergovernmental Panel on Climate Change (IPCC) model estimated that GHG emission from waste sectors is expected to rise 1.6 times by 2031. Diversion of solid waste from landfills and drainage gutters into domestic areas is one of the main reasons for GHG emission. Since it is urgent to apply serious actions and plans in controlling the emissions of GHGs, recent technologies employing renewable energy techniques proved to be successful ventures in controlling GHGs. Refrigeration units and air conditioners are being replaced with solar technologies to combat the serious issue.

1.5.4 Waste to Wealth: Value Recovery from Waste

With advancements in today's technologies, bioconversion of agricultural and food industry biomass is successfully recycled and refined into value-added products for the bioeconomy. Value-added products in agriculture and domestic sectors are of more importance. Biocosmetics, ornamental products, biofertilizers, and crop promoters are a few of them. Locally available agricultural/forest waste can be used to manufacture value-added pellets or briquettes in conjunction with other widely available waste products such as paper, wood chips, oil waste, etc. (Tenenbaum, 2008). Any waste materials discarded in landfills may be turned into value-added items to reduce the burden on landfills. Proper content collection, packaging processes, and facilities should be implemented to manufacture the final product with good calorific performance, less pollution, and less residue after combustion to make the whole process economical and economically viable. Scientific waste management and production of value-added goods for industrial uses should be promoted to minimize GHGs and to address the lack of land for dumping purposes. This will not only increase job opportunities in rural areas but will also be a good source of farmers' additional revenue. The ash generated in local industrial units can be used in agricultural fields as manure. The fossil fuels used to move these waste materials to dump sites would also be saved by this operation. In today's background, this necessitates the value of technical solid waste management.

1.6 CONCLUSIONS

The generation of waste has become one of the major challenging implications for the environment and hygienic surroundings. The processes adopted towards volarizing the waste biorefineries are sustainably commercialized for the development of value-added bioproducts, in particular biofuel, bioenergy, biofertilizers, and health food supplements. The uptake of the ecofriendly technologies will be achieved through appropriate dissemination of outcomes and scalability potential (techno-economic-based up-scaling potential for bioproducts) in public and industry forums. Therefore, volarizing the waste biorefinery is receiving significant attention in the current scenario due to its enormous benefits. The practices are acclaimed in almost all of the countries because of its significant advantages in meeting the SDGs. Conversion of waste biomass into value-added products results into reduction in pollution load, health and safety, a sustainable resilient environment, green products, extended job opportunities, and an increased economy. The significant SDGs that are possibly achieved through volarizing waste biorefineries under the umbrella of sustainable waste management are good health and well-being, clean water and sanitation, affordable and clean energy, decent work and economic growth, sustainable cities and communities, and responsible consumption and production.

CONFLICT OF INTEREST

The authors declare no competing financial interest.

ACKNOWLEDGEMENTS

The authors would like to thank the director, CSIR-NEERI, Nagpur India, for facilitating and supporting the activities.

REFERENCES

Ahn, S-J., Graczyk, D., 2012. *Understanding Energy Challenges in India: Policies, Players and Issues.* International Energy Agency, Paris.

Amado, I.R., Vázquez, J.A., Pastrana, L., Teixeira, J.A., 2016. Cheese whey: A cost-effective alternative for hyaluronic acid production by Streptococcus zooepidemicus. *Food Chemistry* 198, 54–61. https://doi.org/10.1016/j.foodchem.2015.11.062

Bayr, S., Rintala, J., 2012. Thermophilic anaerobic digestion of pulp and paper mill primary sludge and co-digestion of primary and secondary sludge. *Water Research* 46, 4713–4720. https://doi.org/10.1016/j.watres.2012.06.033

Bene, C., Barange, M., Roahan, S., Pinstrup-Andersen, P., Merino, G., Hemre, G.-I., Williams, M., 2015. Feeding 9 billion by 2050 putting fish baack on the menu. *Food Security* 7, 261–274.

Bhatia., R.K., Ramadoss, G., Jain, A.K., Dhiman, R.K., Bhatia, S.K., Bhatt, A.K., 2020. Conversion of waste biomass into gaseous fuel: present status and challenges in India. *BioEnergy Research* 13, 1046–1068.

Brunner, P.H., Rechberger H., 2015. Waste to energy – key element for sustainable waste management. *Waste Management* 37, 3–12. doi: 10.1016/j.wasman.2014.02.003. PMID: 24630214.

Cavka, A., Guo, X., Tang, S.J., Winestrand, S., Jönsson, L.J., Hong, F., 2013. Production of bacterial cellulose and enzyme from waste fiber sludge. *Biotechnology for Biofuels* 6, 1–10. https://doi.org/10.1186/1754-6834-6-25

Childers, D.L., Corman, J., Edwards, M., Elser, J.J., 2011. Sustainability challenges of phosphorus and food: Solutions from closing the human phosphorus cycle. *BioScience* 61, 117–124. doi:10.1525/bio.2011.61.2.6

Chisti, Y., 2010. A bioeconomy vision of sustainability. *Biofuels, Bioproducts and Biorefining.* 4(4), 359–361.

Clark, H.J., 2019. Green biorefinery technologies based on waste biomass. *Green Chemistry* 21, 1168–1170. https://doi.org/10.1039/C9GC90021G

Coelho, E., Rocha, M.A.M., Moreira, A.S.P., Domingues, M.R.M., Coimbra, M.A., 2016. Revisiting the structural features of arabinoxylans from brewers' spent grain. *Carbohydrate Polymers* 139, 167–176. https://doi.org/10.1016/j.carbpol.2015.12.006

Conde, E., Hemming, J., Smeds, A., Reinoso, B.D., Moure, A., Willför, S., Domínguez, H., Parajó, J.C., 2013. Extraction of low-molar-mass phenolics and lipophilic compounds from Pinus pinaster wood with compressed CO_2. *The Journal of Supercritical Fluids* 81, 193–199. https://doi.org/10.1016/j.supflu.2013.04.018

Cordell, D., Rosemarin, A., Schröder, J.J., Smit, A.L., 2011. Towards global phosphorus security: A systems framework for phosphorus recovery and reuse options. *Chemosphere* 84, 747–758. doi:10.1016/j.chemosphere.2011.02.032

Cotana, F., Barbanera, M., Foschini, D., Lascaro, E., Buratti, C., 2015. Preliminary optimization of alkaline pretreatment for ethanol production from vineyard pruning. *Energy Procedia* 82, 389–394. https://doi.org/10.1016/j.egypro.2015.11.814

Dávila, I., Gordobil, O., Labidi, J., Gullón, P., 2016. Assessment of suitability of vine shoots for hemicellulosic oligosaccharides production through aqueous processing. *Bioresource Technology* 211, 636–644. https://doi.org/10.1016/j.biortech.2016.03.153

DeLeón-González, G., González-Valdez, J., Mayolo-Deloisa, K., Rito-Palomares, M., 2016. Intensified fractionation of brewery yeast waste for the recovery of invertase using aqueous two-phase systems. *Biotechnology and Applied Biochemistry* 63, 886–894. https://doi.org/10.1002/bab.1435

De Melo, M.M.R., Domingues, R.M.A., Sova, M., Lack, E., Seidlitz, H., Lang, F., Silvestre, A.J.D., Silva, C.M., 2014. Scale-up studies of the supercritical fluid extraction of triterpenic acids from Eucalyptus globulus bark. *The Journal of Supercritical Fluids* 95, 44–50. https://doi.org/10.1016/j.supflu.2014.07.030

Desmidt, E., Ghyselbrecht, K., Zhang, Y., Pinoy, L., Van der Bruggen, B., Verstraete, W., Rabaey, K., Meesschaert, B., 2015. Global phosphorus scarcity and full-scale P-recovery techniques: a review. *Critical Reviews in Environmental Science and Technology* 45, 336–384. doi:10.1080/10643389.2013.866531

Devesa-Rey, R., Vecino, X., Varela-Alende, J.L., Barral, M.T., Cruz, J.M., Moldes, A.B., 2011. Valorization of winery wastes vs. the costs of not recycling. *Waste Management* 31, 2327–2335.

Djukić-Vuković, A., Mladenović, D., Radosavljević, M., Kocić-Tanackov, S., Pejin, J., Mojović, L., 2016. Wastes from bioethanol and beer productions as substrates for l(+) lactic acid production: A comparative study. *Waste Management* 48, 478–482. https://doi.org/10.1016/j.wasman.2015.11.031

Domingues, L., Guimarães, P.M.R., Oliveira, C., 2010. Metabolic engineering of Saccharomyces cerevisiae for lactose/whey fermentation. *Bioengineered Bugs* 1, 164–171. https://doi.org/10.4161/bbug.1.3.10619

Ekins, P., Hughes, N., 2010. The prospects for a hydrogen economy (2): hydrogen transitions. *Technology Analysis and Strategic Management* 22, 1–17. http://doi.org/10.1080/09537320903438013

Galanakis, C.M., Chasiotis, S., Botsaris, G., Gekas, V., 2014. Separation and recovery of proteins and sugars from Halloumi cheese whey. *Food Research International* 65, 477–483. https://doi.org/10.1016/j.foodres.2014.03.060

García-Becerra, L., Mitjans, M., Rivas-Morales, C., Verde-Star, J., Oranday-Cárdenas, A., Vinardell María, P., 2016. Antioxidant comparative effects of two grape pomace Mexican extracts from vineyards on erythrocytes. *Food Chemistry* 194, 1081–1088. https://doi.org/10.1016/j.foodchem.2015.08.131

Ghisellini, P., Cialani, C., Ulgiati, S., 2016. A review in circular economy: the expected transition to a balanced interplay of environmental and economic systems. *Journal of Cleaner Production* 114, 11–32.

Gilbert, N., 2009. Environment: the disappearing nutrient. *Nature* 461, 716–718. doi:10.1038/461716a

Gomes, D., Domingues, L., Gama, M., 2016. Valorizing recycled paper sludge by a bioethanol production process with cellulase recycling. *Bioresource Technology* 216, 637–644. https://doi.org/10.1016/j.biortech.2016.06.004

González-Centeno, M.R., Rosselló, C., Simal, S., Garau, M.C., López, F., Femenia, A., 2010. Physico-chemical properties of cell wall materials obtained from ten grape varieties and their byproducts: Grape pomaces and stems. *LWT – Food Science and Technology* 43, 1580–1586. https://doi.org/10.1016/j.lwt.2010.06.024

Gullón, P., González-Muñoz, M.J., Domínguez, H., Parajó, J.C., 2008. Membrane processing of liquors from Eucalyptus globulus autohydrolysis. *Journal of Food Engineering* 87, 257–265. https://doi.org/10.1016/j.jfoodeng.2007.11.032

Hanc, A., Pliva, P., 2013. Vermicomposting technology as a tool for nutrient recovery from kitchen bio-waste. *Journal of Material Cycles and Waste Management* 15: 431–439.

Hatti-Kaul, R., 2010. Biorefineries: A path to sustainability? *Crop Science* 50, S152–S156. https://doi.org/10.2135/cropsci2009.10.0563.

http://www.sahyog-europa-india.eu/inventories (accessed 06.2022).

https://www.altenergymag.com/article/2009/08/biomass-wastes/530/ (accessed 06.2022).

https://www.hurstboiler.com/biomass_boiler_systems/biomass_fuel_types (accessed 06.2022).

Kumar, S., Smith S.R., Fowler G., Velis C., Kumar S.J., Arya S., Rena, K.R., Cheeseman, C., 2017. Challenges and opportunities associated with waste management in India. *Royal Society Open Science* 4 (3): 160764. https://doi.org/10.1098/rsos.160764

Leong, H.Y., Chang, C.K., Khoo, K.S., Chew, K.W., Chia, S.R., Lim, J.W., Chang, Jo-S., Show, P.K., 2021. Waste biorefinery towards a sustainable circular bioeconomy: A solution to global issues. *Biotechnology for Biofuels* 14 (87): 87. https://doi.org/10.1186/s13068-021-01939-5

Lin, C.S.K., Pfaltzgraff, L.A., Herrero-Davila, L. Mubofu, E.B., Abderrrahim, S., Clark, J.H., Koutinas, A.K., Kopsahelis, N., Mohamed, Z., Brocklesby, R., Luque, R. 2013. Food waste as a valuable resource for the production of chemicals, materials and fuels. Current situation and global perspective. *Environmental Science & Technology* 6, 426–464.

Lin, D., Lopez-Sanchez, P., Li, R., Li, Z., 2014. Production of bacterial cellulose by *Gluconacetobacter hansenii* CGMCC 3917 using only waste beer yeast as nutrient source. *Bioresource Technology* 151, 113–119.

Lou, X.F., Nair, J., Ho, G., 2013. Potential for energy generation from anaerobic digestion of food waste in Australia. *Waste Management & Research* 31, 283–294.

Max, B., Torrado, A.M., Moldes, A.B., Converti, A., Domínguez, J.M., 2009. Ferulic acid and p-coumaric acid solubilization by alkaline hydrolysis of the solid residue obtained after acid prehydrolysis of vine shoot prunings: Effect of the hydroxide and pH. Biochem. *Engineering Journal* 43, 129–134. https://doi.org/10.1016/j.bej.2008.09.015

Mendes, J.A.S., Prozil, S.O., Evtuguin, D.V., Lopes, L.P.C., 2013. Towards comprehensive utilization of winemaking residues: Characterization of grape skins from red grape pomaces of variety Touriga Nacional. *Industrial Crops and Products* 43, 25–32. https://doi.org/10.1016/j.indcrop.2012.06.047

Michelin, M., Teixeira, J.A., 2016. Liquid hot water pretreatment of multi feedstocks and enzymatic hydrolysis of solids obtained thereof. *Bioresource Technology* 216, 862–869. https://doi.org/10.1016/j.biortech.2016.06.018

Nizami, A-S., Mohanakrishna, G., Mishra U., Pant, D. 2016. Chapter- Trends and sustainability criteria for liequid biofuels. Book - *Fuels*. 1st Edition 1–38 pages, CRC Press. eBook ISBN 9781315370743.

Nizami, A.S., Rehan, M., Waqas, M., Naqvi, M., Ouda, O.K.M., Shahzad, K., Miandad, R. Khan, M.Z., Syamsiro, M., Ismail, I.M.I., Pant, D., 2017. Waste biorefineries: Enabling circular economies in developing countries. *Bioresource Technology* 241, 1101–1117. https://doi.org/10.1016/j.biortech.2017.05.097

Obulisamy, P.K., Sanjeet, M., Jonathan, W.C.W., 2017. Bio-refining of food waste for fuel and value products. *Energy Procedia*. 136, 14–21. https://doi.org/10.1016/j.egypro.2017.10.253.

Parfitt, J., Barthel, M., Macnaughton, S., 2010. Food waste within food supply chains: Quantification and potential for change to 2050. *Philosophical Transactions of the Royal Society B: Biological Sciences* 365, 3065–3081.

Pasupuleti, S.B., Venkata Mohan, S., 2015. Single-stage fermentation process for high-value biohythane production with the treatment of distillery spent-wash. *Bioresource Technology* 189, 177–185.

Portilla-Rivera, O., Torrado, A., Domínguez, J.M., Moldes, A.B., 2008. Stability and emulsifying capacity of biosurfactants obtained from lignocellulosic sources using Lactobacillus pentosus. *Journal of Agricultural and Food Chemistry* 56, 8074–8080. https://doi.org/10.1021/jf801428x

Rivera, O.M.P., Moldes, A.M., Domínguez, J.M., 2007. Lactic acid and biosurfactants production from hydrolyzed distilled grape marc. *Process Biochemistry* 42, 1010–1020. https://doi.org/10.1016/j.procbio.2007.03.011

Rodríguez-López, J., Sánchez, A.J., Gómez, D.M., Romaní, A., Parajó, J.C., 2012. Fermentative production of fumaric acid from Eucalyptus globulus wood hydrolyzates. *Journal of Chemical Technology & Biotechnology* 87, 1036–1040. https://doi.org/10.1002/jctb.2729

Romaní, A., Garrote, G., Alonso, J.L., Parajó, J.C., 2010. Experimental assessment on the enzymatic hydrolysis of hydrothermally pretreated eucalyptus globulus wood. *Industrial & Engineering Chemistry Research* 49, 4653–4663. https://doi.org/10.1021/ie100154m

Romaní, A., Michelin, M., Domingues, L., Teixeira, J.A., 2018. Valorization of wastes from agrofood and pulp and paper industries within the biorefinery concept: Southwestern europe scenario. *Waste Biorefinery: Potential and Perspectives*, 487–504, ISBN 978-0-444-63992-9. https://doi.org/10.1016/B978-0-444-63992-9.00016-1

Romaní, A., Yáñez, R., Garrote, G., Alonso, J.L., 2008. SSF production of lactic acid from cellulosic biosludges. *Bioresource Technology* 99, 4247–4254. https://doi.org/10.1016/j.biortech.2007.08.051

Sarma, S.J., Pachapur, V., Brar, S.K., Bihan, Y.L., Buelna, G., 2015. Hydrogen biorefinery: Potential utilization of the liquid waste from fermentative hydrogen production. *Renewable and Sustainable Energy Reviews* 50, 942–951.

Sialve, B., Bernet, N., Bernard, O., 2009. Anaerobic digestion of microalgae as a necessary step to make microalgal biodiesel sustainable. *Biotechnology Advances* 27, 409–416.

Silvério, S.C., Macedo, E.A., Teixeira, J.A., Rodrigues, L.R., 2016. Biocatalytic approaches using lactulose: End product compared with substrate. *Comprehensive Reviews in Food Science and Food Safety* 15, 878–896. https://doi.org/10.1111/1541-4337.12215

Smil, V., 2000. Phosphorus in the environment: Natural flows and human interferences. *Annual Review of Energy and the Environment* 25, 53–88. https://www.annualreviews.org/doi/pdf/10.1146/annurev.energy.25.1.53

Smith, K., O'Farrell, K., Brindley, F., 2011. *Waste and Recycling in Australia 2011: Incorporating a Revised Method for Compiling Waste and Recycling Data.* Department of Sustainability, Environment, Water, Population and Communities.

Spinelli, S., Conte, A., Lecce, L., Padalino, L., Del Nobile, M.A., 2016. Supercritical carbon dioxide extraction of brewer's spent grain. *The Journal of Supercritical Fluids* 107, 69–74. https://doi.org/10.1016/j.supflu.2015.08.017

Tan, S.T., Ho, W.S., Hashim, H., Lee, C.T., Taib, M.R., Ho, C.S., 2015. Energy, economic and environmental (3E) analysis of waste-to-energy (WTE) strategies for municipal solid waste (MSW) management in Malaysia. *Energy Conversion and Management* 102, 111–120.

Tenenbaum, D.J. 2008. Food vs. fuel: Diversion of crops could cause more hunger. *Environmental Health Perspectives* 166(6), A254–A257.

Tozlu, A., Özahi, E., Abusog˘lu, A., 2016. Waste to energy technologies for municipal solid waste management in Gaziantep. *Renewable and Sustainable Energy Reviews* 54, 809–815.

Vecino, X., Devesa-Rey, R., Moldes, A.B., Cruz, J.M., 2014. Formulation of an alginate-vineyard pruning waste composite as a new eco-friendly adsorbent to remove micronutrients from agroindustrial effluents. *Chemosphere* 111, 24–31. https://doi.org/10.1016/j.chemosphere.2014.03.004

Venkata Mohan S., Nikhil G., Patharla Chiranjeevi, P., Nagendranatha Reddy, C., 2016. Waste biorefinery models towards sustainable circular bioeconomy: Critical review and future perspectives. *Bioresource Technology* 215, 2–12.

Víctor, P., Pascual, A., Rodrigo, A., Torreiro, M.G., Latorre-Sánchez, M., Lozano C.C., David-Moreno, A., Oliva-Dominguez, J.M., Serna-Maza, A., García, N.H., Granados, I.G., Roldan-Aguayo, R., Ovejero-Roncero, D., Marin, J.L.M., Smith, M., Musinovic H., Raingué, A., Belard, L., Pascual, C., Lebrero, R., Muñoz, R., 2020. Integrated innovative biorefinery for the transformation of municipal solid waste into biobased products. *Waste Biorefinery*, 41–80, ISBN 9780128182284. https://doi.org/10.1016/B978-0-12-818228-4.00002-2.

Vieira, E.F., Carvalho, J., Pinto, E., Cunha, S., Almeida, A.A., Ferreira, I.M.P.L.V.O., 2016. Nutritive value, antioxidant activity and phenolic compounds profile of brewer's spent yeast extract. *Journal of Food Composition and Analysis* 52, 44–51. https://doi.org/10.1016/j.jfca.2016.07.006

Wysokinska, Z., 2016. The "new" environmental policy of the European Union: A path to development of a circular economy and mitigation of the negative effects of climate change. *Comparative Economic Research* 19, 57–73.

Zabaleta, I., Rodic, L., 2015. Recovery of essential nutrients from municipal solid waste: Impact of waste management infrastructure and governance aspects. *Waste Management* 44, 178–187. http://doi.org/10.1016/j.wasman.2015.07.033

2 Microorganisms Used in Biorefineries

Taner Sar, Sharareh Harirchi, Fatemeh Yazdian, and Mohammad J. Taherzadeh

CONTENTS

DOI: 10.1201/9781003335740-2

2.1 INTRODUCTION

2.1.1 Wastes and Their Compositions

Resource recovery using biorefineries is principally developed to use agricultural residuals and also waste materials. Waste is referred to any substance that is discarded after primary use or is worthless, defective, and of no use. The waste product may become a by-product, co-product, or resource through an invention that increases the value of a waste product to above zero. Examples of wastes are municipal solid wastes (household trash/refuse), hazardous wastes, wastewaters (such as sewage, which contains bodily wastes [feces and urine] and surface runoff), radioactive wastes, food by-products, and others. These types of wastes are to be addressed for the control of environmental pollutants and creating a healthy environment (Ahmed and Ahmaruzzaman 2016).

Wastes can be separated according to their physical properties such as organic wastes, plastics, glass, paper, metals, wood, textiles, nonflammable, etc. (Gidarakos, Havas, and Ntzamilis 2006). The chemical and elemental content of organic wastes and by-products associated with the food and similar industries may vary depending on the content of the raw waste. Information about waste composition is crucial for planning and decision making in waste management. Efficient methods are needed to assess the effect of legislative, logistic, and technical measures on the waste stream. Routine determination of waste composition and trends is essential to assess the effect of such measures. By determining the composition of the wastes, wastes can be recycled into the economy by creating useful products via biological and chemical approaches or producing valuable products by the microbial fermentation.

2.1.2 Microbiology for Biorefinery and Bioeconomy

Plans for a new biobased economy have been developed by almost all countries in the world. Biorefineries are considered a key instrument for achieving the bioeconomy goals and for promoting the linear-to-circular transition. The fundamental contribution of biorefineries to the concept of the circular economy is based on their ability to transform biomass into different final products with high added value.

Biological wastes (wastes from food consumed in homes, restaurants, markets, and commercial premises, as well as biodegradable waste produced in gardens, orchards, and parks) are mainly composed of carbohydrates, proteins, and lipids, all of which represent high potential raw materials for the creation of valuable products. Based on the definition of a biorefinery in the International Energy Agency Bioenergy Task 42, biorefining is the sustainable processing of all kinds of biomass into a spectrum of marketable products and energy (Pandey et al. 2015). In this regard, microorganisms can play a key role in the conversion of biomass, wastes, organic materials, etc., to the final products and fuels. Therefore, it is a prerequisite to cultivate and grow them in the culture media under routine laboratory conditions. There are different types of culture media, including chemically defined, complex, or semi-defined, that can be used for the cultivation and purification of microbial strains. These media can be divided into general, selective, or differential categories based on their type of application. After isolation and purification, microorganisms are identified into various levels based on the polyphasic taxonomy and other classification methods. Sometimes, it is not easy to isolate and purify microbial strains to obtain a pure culture, and instead of a single isolate, a combination of two or more microbial strains may be found. Moreover, over time, the pure culture paradigm has been disclaimed, as more than 98% of microorganisms cannot be cultivated on laboratory media under unnatural conditions and as well in nature—they do not stay alive as a pure species or strain in the dynamic communities of an ecosystem (Woolverton, Sherwood, and Willey 2016; Allen and Banfield 2005; Gunasekaran, Noronha, and Pandey 2016).

Basic microbiology, a scientific discipline initiated by Louis Pasteur, Ferdinand Cohn, and Robert Koch in the late nineteenth century, helped us to practice good laboratory procedures to isolate and purify microorganisms from samples, especially when their niches are the same; determination of the microbial diversity of the samples is a significant step to save time and cost. These procedures make up several series of passages of an initial microbial culture in an appropriate growth culture media under controlled physicochemical conditions (such as trace elements, pH, temperature, or oxygen requirement) that result in a single colony. The term "isolate" can be referred in this step and by discriminating the single colony via molecular techniques such as ribotyping, restriction fragment length polymorphism (RFLP), sodium dodecyl sulfate-polyacrylamide gel electrophoresis (SDS-PAGE) of proteins, pulsed-field gel electrophoresis, randomly amplified polymorphic DNA (RAPD), and phenotypic comparison with validly published type strains, a "strain" can be distinguished (Joanne, Linda, and Christopher 2016; Allen and Banfield 2005; Gunasekaran, Noronha, and Pandey 2016).

2.2 MICROORGANISMS FOR PURE CULTURES

Once the identified single strain is going to be used for the waste treatment, a pure inoculum from a storage seed is required. The common technique for the cultivation of a pure or axenic cell is the clonal growth of a prokaryotic or eukaryotic single cell by streaking on solid plates or suspending in a liquid medium. Therefore, an axenic or pure culture is a culture that involves one type of cell, while a xenic or mixed culture contains several different types of cells (Madigan et al. 2018; Pandey and Teixeira 2016). Moreover, carrying out serial dilution of an unreliable sample can be helpful for the pure culture preparation of an intended strain from that sample. Although it should be considered, the latter method is not a proper method for the isolation and purification of the grown cells on insoluble substrates. Furthermore, pure cultures are obtained from viable and culturable microbial colonies, while there are some microbial cells like *Enterobacteriaceae* members that possess a viable but nonculturable state (VBNC), in which the preparation of a pure culture by routine methods is not possible and requires novel techniques (Oliver 2005; Eilers et al. 2000). Although the pure culture of a microbial strain remains significant for polyphasic taxonomy researches, the absence of competing and no interference with other strains makes it important to be used for various industrial purposes. But there is difficulty in the isolation, purification, preparation, storage, and prevention of the contamination with other microbial strains. Similarly, there are fewer details and understanding about the ecological aspects of microorganisms living as an axenic or pure culture (Okafor 2007; Woolverton, Sherwood, and Willey 2016). However, scientists could develop some successful engineered pure cultures called "superbugs" that simultaneously can perform multiple and desired functions, but these single microorganisms are usually costly and sometimes cannot convert varied substrates proficiently (Woolverton, Sherwood, and Willey 2016).

2.2.1 Ethanol-Producing Microorganisms

Having considered the characteristics of microbial pure cultures, they have been applied for different biorefineries. One of the most conventional biorefineries using pure culture is ethanol production (bioethanol first generation) from starches (maize, cassava, wheat, and other cereals) and sugar crops (sugarcane and sugar beet) (Pandey et al. 2015). The most common microbial pure culture used for the fermentation of sugars to ethanol is *Saccharomyces cerevisiae* (brewer's or baker's yeast) belonging to the phylum Ascomycota. In comparison to the other fungi and prokaryotic microorganisms, this yeast has many physiological advantages in the biorefinery context. It can easily grow both aerobically and anaerobically. *S. cerevisiae* has a broad range of pH for growth, with an optimum at acidic pH, which helps the fermentation not to be infected with the routine bacterial contaminants. Moreover, it can tolerate the elevated temperature up to 40°C, but the optimum growth can occur at 30–35°C. This yeast has a great ability to withstand quite high osmotic pressures and a high concentration of the produced ethanol. By isolating *S. cerevisiae* strains from an industrial environment, considerable abilities may be achieved that are worthy proof of the adaptation of this species to hard conditions. For the biorefinery of sugary feedstocks, the theoretical yield of ethanol production is 0.51 g/g and can be accomplished c.a. 90–95% of this theoretical yield by this yeast. *S. cerevisiae* can be stored on the yeast extract peptone dextrose (YEPD) medium for short-term usage or cryopreserved in skim milk for later purposes. In addition, *S. cerevisiae* has been considered generally recognized as safe (GRAS) and is well known among society and industries (Hahn-Hägerdal et al. 2007; Satyanarayana and Kunze 2009). Sucrose, fructose, galactose, mannose, and glucose can be utilized by this species, but it is not able to ferment some saccharides such as cellobiose, xylobiose, arabinose, or xylose derived from lignocellulosic biomass or starchy wastes (Lynd et al. 2002). Despite the widespread industrial applications of *S. cerevisiae*, some weaknesses have limited its usage. This strain can form a substantial amount of glycerol under anaerobic conditions, and the diauxic growth occurs when glucose is the most abundant sugar in the fermentation media; when the carbon sources are depleted, it starts consuming the final product, ethanol, which affects the economical aspect of the process (Satyanarayana and Kunze 2009; Siegal 2015). An alternative solution is to use nonconventional yeasts such as *Hansenula* spp. (Ryabova, Chmil, and Sibirny 2003) or *Zygosaccharomyces* spp. (Thomas and Davenport 1985) that can consume pentoses, starch, and lignocellulose-derived sugars and have much better resistance to low pH and high temperature than *S. cerevisiae*, but they are not considered for industrial ethanolic fermentation due to their low productivity. However, nonconventional yeasts have served as the gene pools for the metabolic pathways not existing in *S. cerevisiae* to improve its productivity for ethanolic fermentation from various resources (Jacques, Lyons, and Kelsall 2003). Recently, rather than the genus *Saccharomyces*, the genera *Pichia*, *Hanseniaspora*, *Dekkera*, and *Wickerhamomyces* can be used for brewing or winemaking (vinification). These genera can propose novel bioconversion pathways that allow them to produce new flavors, low-calories and low-alcohol beers, and other fermented beverages (Stewart 2016; Okafor 2007).

Members of the filamentous fungi class zygomycetes are also saprophytic organisms that can produce a variety of metabolites, including ethanol. Among them, *Rhizopus, Rhizomucor,* and *Mucor* are good ethanol-producing candidates that use glucose, xylose, and lignocellulose hydrolyzate (Millati, Edebo, and Taherzadeh 2005; Karimi, Emtiazi, and Taherzadeh 2006; Abedinifar et al. 2009). Another filamentous fungus, *Neurospora intermedia*, a member of Ascomycota, has recently been introduced as a potential ethanol producer, along with producing a pigment- and protein-rich fungal biomass (Ferreira, Lennartsson, and Taherzadeh 2014, 2015; Nair et al. 2017). Because these fungi members can naturally synthesize several enzymes, including amylase and xylanase, filamentous fungi can be preferred for efficient ethanol production from lignocellulosic wastes (Shahryari et al. 2019; Kanti and Sudiana 2018).

Bioethanol is a renewable biofuel in the liquid form produced by the fermentation of microorganisms from various biomasses. Lactose-containing cheese whey and whey powder are also the potential substrates for bioethanol production. Some *Kluyveromyces* strains (*K. marxianus* and *K. lactis*) can naturally metabolize lactose, but these strains have a low ethanol tolerance. To be able to produce a high yield of ethanol from a high lactose concentration, some researchers have developed *S. cerevisiae* strains expressing LAC4 (coding for β-galactosidase) and LAC12 (coding for lactose permease) that can ferment lactose directly (Domingues, Lima, and Teixeira 2001; Zou et al. 2013). Guimarães et al. carried out adaptive evolution experiments to obtain recombinant *S. cerevisiae* strains in order to consume lactose rapidly and obtain a high ethanol yield (Guimarães, Teixeira, and Domingues 2008; Guimarães et al. 2008).

Escherichia coli is also a potential ethanol-producing microorganism, as it can naturally ferment several types of sugars, including pentoses and hexoses, but these sugars are converted into ethanol along with lactic acid, acetate, and formate formations. For efficient ethanol production through *E. coli*, first all fermentation pathways are blocked by mutation, and then the genes encoding enzymes (pyruvate decarboxylase [*pdc*] and alcohol dehydrogenase II [*adhB*]) for the ethanol pathway from *Zymomonas mobilis* were transformed, and the *E. coli* FBR5 strain that only provides ethanol production was developed (Ingram et al. 1987; Ingram and Conway 1988; Dien et al. 2000). In subsequent studies, new engineered *E. coli* strains were developed to express the *Vitreoscilla* hemoglobin gene (*vgb*) to improve ethanol production with the FBR5 strain (Sanny et al. 2010). *Vitreoscilla* hemoglobin (VHb) is a bacterial hemoglobin that has been characterized for the first time and is widely used to increase the production of various biological products, promote bioremediation, and improve the growth and survival of engineered organisms (Stark, Dikshit, and Pagilla 2012; Stark, Pagilla, and Dikshit 2015). It has been reported that these new strains are effective in ethanol production using various carbon sources, including pure sugars and corn stover (Sanny et al. 2010), sugar beet molasses (Akbas, Sar, and Ozcelik 2014), cheese whey and whey powder (Akbas, Sar, and Ozcelik 2014), and maize and potato processing wastes (Abanoz, Stark, and Akbas 2012; Sumer, Stark, and Yesilcimen Akbas 2015). Immobilization applications, in conjunction with *Vitreoscilla* hemoglobin technology, could be also a useful strategy to maintain the advantage in ethanol productivity, as they are efficient compared to free cells and can be reused in repeated batch fermentation and after long-term storage (Sar, Stark, and Yesilcimen Akbas 2017; Sar et al. 2017; Sar, Stark, and Akbas 2019). In another study conducted with *E. coli*, an ethanol-producing recombinant strain (KO11) was developed by integrating the *Z. mobilis* genes (*pdc*, *adhB*) into the *E. coli* chromosome within or near the pyruvate formate-lyase gene (*pfl*) (Ohta et al. 1991). Further mutations were done for inactivating succinate production (*frd*) and blocking homologous recombination (*recA*). It has been reported that succinate production is blocked by 95% with these mutations (Ohta et al. 1991).

In the biorefinery concept, it is necessary to establish innovative approaches to treat and refine the produced wastes. Nowadays, bioprocessing of the stillage produced in the conventional and whole crop biorefineries attracted the attention of researchers to use microbial strains with considerable characteristics such as high resistance to low or high pH, temperature, or salts. These kinds of microbial strains have been classified as extremophiles and polyextremophiles that are able to grow under extreme conditions of temperature, acidity, alkalinity or salinity, pressure, and radiation (Di Donato et al. 2019). Then, extremophilic microorganisms could survive in the harsh conditions and produce enzymes (Di Donato et al. 2019). Tomás et al. (2013) reported that the isolated extreme thermophile *Thermoanaerobacter pentosaceus* strain can metabolize lignocellulosic hydrolysate–based inhibitors such as 5-(hydroxymethyl)furfural and furfural and produce ethanol directly from rapeseed straw hydrolysate. Similarly, in another study conducted with thermophilic anaerobic *Thermoanaerobacter* strains by Avci and Dönmez (2006), they investigated ethanol production in different media prepared by molasses or sucrose. However, in this study, it has been suggested that new studies should be developed to produce ethanol from high substrate concentrations through *Thermoanaerobacter* strains for industrial applications Avci and Dönmez (2006). Recently, Harirchi et al. (2020) studied a pure culture of *Aeribacillus pallidus* strain Lhs-10 for the aerobic bioprocessing of the beet vinasse derived from a bioethanol production unit under high temperature. This thermoalkaliphilic strain belonged to the order Bacillales and can tolerate high pH, temperature, and NaCl concentration all together and grows optimally at pH 8, 50°C, in the presence of 25 g/L NaCl (Harirchi, Etemadifar, Mahboubi, et al. 2020). This new example can demonstrate the significance of extremophile or polyextremophile properties for industrial applications under severe conditions and create the opportunity for more investigations concerning these microorganisms and their potentials.

2.2.1.1 Microorganisms for Lignocellulosic Ethanol Production

Refining of lignocellulosic materials, including wood, straw, bagasse, grass, pulp, etc., into their components (cellulose, hemicellulose, and lignin) creates the opportunity for the production of a wide spectrum of products and biofuels. This is not unexpected given that lignocellulosic biomass becomes the future source for food, feed, and fuel production. Enzymatic hydrolysis is the bottleneck step in the conversion of lignocellulosic biomass to fermentable sugars (C5 and C6) that predominately apply for biofuels (ethanol, butanol, or hydrogen) production. Lignin can be used for heat and power production or be sold as a source for high-added-value chemicals (Pandey et al. 2015). Cellulase and xylanase are the most applicable enzymes used for the enzymatic hydrolysis obtained from mold, yeast, protozoa, bacteria, and some archaea. The main cellulase producer belongs to the fungal species *Trichoderma reesei* (Acharya and Chaudhary 2012). Lignocellulosic bacteria are very diverse and widespread among different bacterial phyla. They include *Geobacillus, Cytophaga, Actinomyces, Clostridium, Caldocellum, Erwinia, Cellulomonas, Micrococcus, Brevibacillus, Thermobifida, Thermomonospora, Ruminococcus, Bacteriodes, Acetovibrio, Microbispora, Rhodothermus, Arthrobacter*, and *Paenibacillus* (Maki, Leung, and Qin 2009; Baltz, Demain, and Davies 2010; Menendez, Garcia-Fraile, and Rivas 2015; Blumer-Schuette et al. 2014). Recently, using lignocellulosic thermophiles has increased because these microorganisms produce resistant hydrolyzing enzymes, grow fast, and the possibility of contamination with mesophiles is very limited (Acharya and Chaudhary 2012; Brahmachari 2016). Moreover, thermophiles can be applied for the fermentation of sugars into bioethanol (Olson, Sparling, and Lynd 2015). In addition to lignocellulosic microorganisms used for the enzyme production, yeasts such as *Candida shehatae, Pichia stipites,* and *Pachysolen tannophilus* and bacterial strains such as the species *Z. mobilis* are applied for the fermentation of released sugars during enzymatic hydrolysis, but the most common yeast is *S. cerevisiae*, as has been explained in this chapter (Hari Krishna, Janardhan Reddy, and Chowdary 2001; Hayes 2013).

Many types of yeast strains have been actively used in bioethanol production. One of the most commonly used ethanol-producing microorganisms, *S. cerevisiae*, has several advantages because of having high tolerance against ethanol and inhibitors derived from lignocellulosic biomass hydrolysates. However, the wild type of *S. cerevisiae* cannot use some sugars, including xylose, arabinose, and lactose. Other ethanol-producing *Pichia* and *Candida* strains can naturally metabolize xylose, and *Kluyveromyces* strains also are capable of fermenting lactose to ethanol. The wild-type *Z. mobilis* can ferment narrow sugars (only glucose, fructose, and sucrose), and the bacterium cannot naturally metabolize xylose and arabinose, and is also sensitive to furfural, acetic acid, and other compounds released in the pretreatment process (Shui et al. 2015; Grisales Díaz and Willis 2019).

One of the most common studies in ethanol production with yeast is to obtain recombinant *Saccharomyces* strains that can ferment xylose. For this, recombinant *S. cerevisiae* has been created by expressing xylose reductase (XR—converts xylose to xylitol) and xylitol dehydrogenase (XDH—converts xylitol to xylulose) from *P. stipitis* and xylulokinase (XK) from *S. cerevisiae* (Ho, Chen, and Brainard 1998; Eliasson et al. 2000; Toivari et al. 2001). Den Haan et al. (2007) also developed a recombinant *S. cerevisiae* expressing two cellulase-encoding genes, endoglucanase from *Trichoderma reesei* (*egI*) and the β-glucosidase from *Saccharomycopsis fibuligera* (*bgl*1). This strain can ferment ethanol from amorphous cellulose via co-expression of β-glucosidase and endoglucanase activities (Den Haan et al. 2007). Similarly, there are several studies of ethanol production from cellulose by recombinant *S. cerevisiae* strains expressing cellulolytic enzymes, which are endoglucanase (EG), cellobiohydrolase (CBH), and β-glucosidase (BGL) (Fujita et al. 2004; Tsai et al. 2009; Wen, Sun, and Zhao 2010; Yamada et al. 2011). Zhang et al. (2019) developed *S. cerevisiae* for enhancing growth ability and producing ethanol under stress conditions in the presence of multiple inhibitors, which was achieved by overexpression of ADEnine requiring (*ade*) genes, including *ade1, ade13,* and *ade17.*

Similar studies were conducted with *Z. mobilis* aimed at its ability to ferment pentose sugars. For this, a combination of seven *E. coli* genes, which are xylose isomerase, xylulokinase, L-arabinose isomerase, L-ribulokinase, L-ribulose 5-phosphate 4-epimerase, transaldolase, and transketolase, was transformed into to *Z. mobilis.* With this successful mutant made, an engineered *Z. mobilis* strain catalyst was capable of converting both xylose and arabinose along with glucose to ethanol (Zhang et al. 1998; Mohagheghi et al. 2002). Similarly, mutant variants of *Z. mobilis* strains have been evaluated to increase the xylose consumption and also ferment mixed sugars containing xylose (Viitanen et al. 2009; Dunn and Rao 2014, 2015).

2.2.2 Syngas Biorefineries Microorganisms

Synthetic or syngas (mainly CO and H_2) is a product of thermal gasification of, e.g., biomass that can then be converted to biofuels and chemicals using catalytic conversion or microbial fermentation. There are various fermenting microorganisms such as *Clostridium ljungdahlii*, *Clostridium autoethanogenum*, *Clostridium carboxidivorans*, *Acetobacterium woodii*, and *Peptostreptococcus productus* that have been reported for syngas fermentation to a variety of products, including H_2, formate, acetate, ethanol, butyrate and/or butanol, and other alcohols, bioplastics, and value-added chemicals (Rauch, Hrbek, and Hofbauer 2014; Grimalt-Alemany, Skiadas, and Gavala 2018; Chandolias, Pekgenc, and Taherzadeh 2019). Several investigators have shown the ability of waste gas to be utilized by relevant organisms for protein production from cheap and vast carbon sources, i.e., natural gas. Hence, waste gas as a carbon source could be placed among other potential substrates, which results in saving the world from protein scarcity (Yazdian et al. 2005; Mousavi et al. 2010; Yazdian et al. 2010). Since methane has formed the largest part of natural gas [85–90 (v/v)%] and some countries' own waste gas resource in the world, single cell protein (SCP) production from this substrate could be economically reasonable due to accessibility and cheapness of waste gas resources. This subject, however, in view of increasing the rate of hygienic food consumption and availability of natural gas in some countries, has never been considered technically.

Some feedstocks may contain some pollutants such as inhibitors, sulfur, etc. In particular, sulfur converts to hydrogen sulfide (H_2S) and carbonyl sulfide (Kurella, Bhukya, and Meikap 2017). The presence of pollutants such as CO_2, H_2S, and volatile methyl siloxanes (VMSs) in biogas limits its use as a renewable alternative to natural gas derived from fossils. The purification of raw biogas into biomethane, typically referred to as biogas upgrading, allows the injection of this green gas biofuel into natural gas grids or its use in automation (Task 2018). Today, H_2S removal at a commercial scale is carried out using adsorption-based technologies or in situ chemical precipitation in the digester, while VMS removal is conducted in activated carbon/silica gel filters or by cryogenic condensation (Muñoz et al. 2015; Shen et al. 2018). On the other hand, water scrubbing and membrane separation rank among the most popular technologies for CO_2 removal, along with pressure swing adsorption and chemical scrubbing (Task 2018).

2.2.3 Bacterial Polymer Producing Microorganisms

Bacterial polymers (bacterial cellulose, polyhydroxyalkanoates, bacterial alginate, hyaluronic acid) are part of renewable raw materials that are naturally occurring and accumulated by bacteria cultivated in the presence of carbon media.

2.2.3.1 Bacterial Cellulose

Bacterial cellulose is a major biopolymer produced by several bacteria such as *Komagataeibacter* (mainly known as *Acetobacter* and *Gluconacetobacter*), *Agrobacterium*, *Aerobacter*, *Achromobacter*, *Azotobacter*, *Alcaligenes*, *Rhizobium*, *Sarcina*, and *Salmonella* (Wang, Tavakoli, and Tang 2019; Sar and Yesilcimen Akbas 2022). Bacterial cellulose is applied as a raw material in food and food packaging, as supporting material for immobilization, in the creation of new-generation textile products, or as a biopolymer material in the cosmetics and medical fields (Azeredo et al. 2019; Ullah, Santos, and Khan 2016). Bacterial cellulose and its composites can also be used for different purposes as an anode material or interlayer in rechargeable batteries (Wang et al. 2013; Dursun et al. 2016; Celik et al. 2018). To improve bacterial cellulose production, several kinds of genetic modifications have been applied. Chien et al. (2006) constructed a *vgb*-bearing plasmid under the control of a constitutive *bla* promoter and then transformed it into cellulose-producing *Acetobacter xylinum*. They determined that the *vgb*-expressing *A. xylinum* strain did more produce bacterial cellulose than the host strain (Chien et al. 2006; Liu et al. 2018). During bacterial cellulose production, some of the glucose converted to gluconic acid was expressed by pyrroloquinoline quinone (PQQ) cofactor-dependent glucose dehydrogenase (GDH) located in the cell membrane (Shigematsu et al. 2005). To eliminate the gluconic acid formations, Kuo et al. (2015) knocked out the membrane-bound GDH gene via homologous recombination of a defect GDH gene. It was determined that the obtained mutant strain could generate bacterial cellulose increased with efficient glucose consumption, whereas the cellulose yield by the wild type was decreased (Kuo, Teng, and Lee 2015).

2.2.3.2 Polyhydroxyalkanoates

Polyhydroxyalkanoates (PHAs) are a family of biodegradable polyesters with thermoplastic properties produced by a wide variety of microorganisms as intracellular energy and carbon reserve material (Anderson and Dawes 1990). Industrial interest in PHA relies on its similar properties to common thermoplastics polymers derived from petrol such as polypropylene. However, in contrast with synthetic polymers, PHAs present the advantage of being based on renewable resources and are completely biodegradable by a wide variety of microorganisms and under different environmental conditions (Numata, Abe, and Iwata 2009; Sashiwa et al. 2018). PHA can also be accumulated by microorganisms thanks to removal of methane via biogas desulfurization (Bhaskar et al. 2020). Then, PHA can be extracted by biogas desulfurization (Figure 2.1).

Poly-(R)-3-hydroxybutyrate (PHB) is a kind of PHA synthesized by several bacteria such as *Ralstonia eutropha*, *Bacillus megaterium*, and *Pseudomonas* spp. (Madison and Huisman 1999; Kocharin et al. 2012). Studies with PHB production have generally suggested that cheap carbon sources (such as low-cost agricultural waste materials, dairy wastes, and cheese whey) are the major contributors to the cost of PHB production (Nath et al. 2008; RamKumar Pandian et al. 2010; Getachew and Woldesenbet 2016). Wu et al. (2016) aimed to enhance PHB accumulations by using multiple fission cells by changing cell division patterns. First, engineered *E. coli* deleted fission-related genes *min*C and *min*D were used. Then, *fts*Q, *fts*L, *fts*W, *fts*N, and *fts*Z genes with the cell shape control gene *mre*B were overexpressed in the *E. coli* JM109 strain to improve cell growth and PHB production. With these genetic modifications, more dry cell weight (DCW) and more than 80% of PHB accumulation were observed (Wu et al. 2016).

PHAs formed as polyesters are biologically degradable and compatible, and thus they can replace petroleum-based plastics (Horng et al. 2010). PHA degradation under aerobic conditions is less well described and of a rather complex nature, whereas anaerobic degradation of PHAs from *A. vinelandii* UWD cultures was thoroughly investigated in activated sludge; here, PHA conversion yields to both methane and carbon dioxide were reported between 83% and 96% (Kumagai and Doi 1992). However, biodegradation performance under experimental conditions can strongly differ in terms of outcomes under real environmental conditions due to several complications such as the relative ease of parameter control under experimental conditions compared to real situations, which have to be considered (Swift 1992; Khosravi Darani et al. 2020).

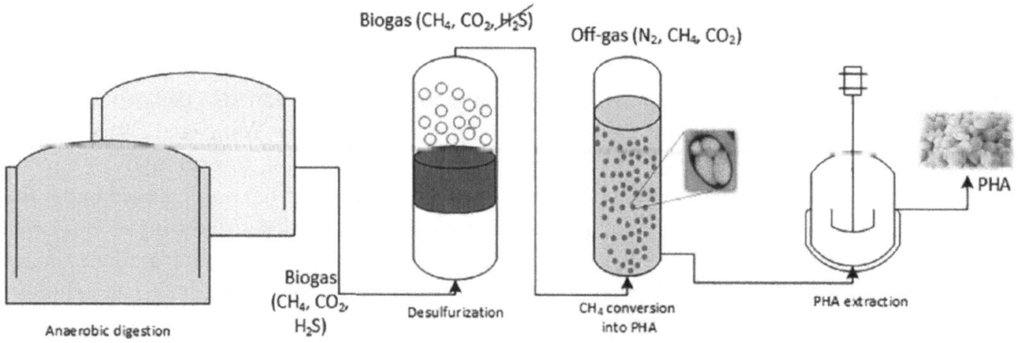

FIGURE 2.1 Schematic diagram showing PHA extraction from the microbial cells through methane bioconversion via biogas desulfurization; modified from Bhaskar et al. (2020).

2.2.3.3 Alginate

Alginate is a polysaccharide compound consisting of β-D-mannuronic acid (M) and α-L-guluronic acid (G) monomers (Hay et al. 2010). Mainly two bacterial genera, *Pseudomonas* and *Azotobacter*, can produce alginate. Concomitantly with the alginate synthesis, PHB can be produced (Hay et al. 2010). Segura et al. (2003) developed *A. vinelandii* mutants to overproduce polymers production, alginate, or PHB. For this, the effects of mutation in *alg*A encoding the first step of alginate biosynthesis or *phb*B were comparatively evaluated. It has been reported that alginate production was inhibited and PHB production was increased up to 5-fold by the *alg*A mutation, and alginate production was also increased 4-fold by the *phb*B mutation (Segura, Guzmán, and Espín 2003).

2.2.3.4 Hyaluronic Acid

Hyaluronic acid (HA), formed by some microorganisms such as the pathogen *Streptococcus,* is a linear glycosaminoglycan polymer comprising D-glucuronic acid and N-acetylglucosamine (Chong et al. 2005; Chien and Lee 2007). The specific genes related to HA biosynthesis are controlled in the *has* operon. The number of genes in the *has* operon may differ, but at least two genes, which are HA synthase (*has*A) and uridine diphosphate (UDP)–glucose dehydrogenase (*has*B), are crucial for HA production (Chong et al. 2005; Widner et al. 2005). Production of HA by fermentation of microbial strains that are recognized as GRAS is highly appreciated (Chien and Lee 2007). For this, Chien and Lee (2007) cloned a nonpathogenic HA-producing *Lactococcus lactis* strain that expresses *has*A and coexpression of *has*A with *has*B of *Streptococcus equi* subsp. *zooepidemicus* in a nisin-controlled expression system. The HA production was detected at a lower concentration by the culture expressing HA synthase alone, while it increased about 8-fold to 0.65 g/L by coexpression of heterologous *has*A with *has*B (Chien and Lee 2007). In another study with *L. lactis,* a strain capable of producing HA was developed by introducing HA synthetic machinery from an operon belonging to the pathogen *Streptococcus zooepidemicus* (Prasad, Jayaraman, and Ramachandran 2010). Prasad et al. (2010) determined the effects of cloning different sets of genes (SJR2, including *has*A and *has*B; SJR3, including *has*A, *has*B, and *has*C; and SJR3 including *has*A, *has*B, *has*C, *has*D, and *has*E) cloned in *L. lactis* for HA production. The maximum yield of HA achieved was 234 mg/L in static flask experiments and 1.8 g/L in batch bioreactor experiments by transformant SJR3 (coexpressing *has*A, *has*B, and *has*C genes) (Prasad, Jayaraman, and Ramachandran 2010).

2.2.4 Oleochemical Biorefinery Microorganisms

An oleochemical biorefinery is considered a different version of a whole-crop biorefinery in which biodiesels (long-chain fatty acid methyl-esters [FAMEs] and fatty acid ethyl-esters [FAEEs]) and vegetable oil–based products such as lubricants, surfactants, solvents, detergents, or functional monomers (e.g., plasticizers) are manufactured. In these kinds of biorefineries, initial feedstocks contain significant amounts of lignocellulosic materials that are subjected to the lignocellulosic feedstock biorefinery (Pandey et al. 2015). Major sources for the oleochemical biorefinery are the plants containing a high level of C12–C18 fatty acids (Moo-Young 2011). Moreover, bacteria and yeast are other sources for the production of oleochemicals from renewable feedstocks that are classified as microbial lipids (Qiao et al. 2017; Easterling et al. 2009). Microbial lipids, namely single cell oils (SCOs), can be produced from carbohydrate feedstockslike starch containing wastes (Xue et al. 2010; Muniraj et al. 2013) by fermentation of microorganisms, and these lipids can be used to obtain biofuels and oleochemicals (Qiao et al. 2017; Easterling et al. 2009). But most of the microbial oleochemicals are not considered end products, and therefore, microbial strains with a high yield of production have not been isolated from nature (Yan and Pfleger 2020). Over the last decade, many studies focused on improving the production of microbial lipids in common strains. The microbial axenic cultures are used for purposes related to microalgae, bacteria, and yeasts. The yeast *Yarrowia lipolytica* is the most used microorganism for microbial oil production and can accumulate lipids at higher levels than 50% of its DCW (Ratledge 2010). The other oleaginous yeasts belong to the genera of *Candida*, *Cryptococcus*, *Rhodotorula*, *Rhizopus*, and *Trichosporon* (Beopoulos et al. 2009). Furthermore, the most important achievements have led to the development of an engineered *S. cerevisiae* as an oleaginous strain, but industrial production of oleochemicals by microorganisms has failed to reach a high volume of production. Qiao et al. (2015) identified a lipogenic enzyme by reverse engineering the mammalian cellular obese phenotypes. Then, delta-9 stearoyl-CoA desaturase (SCD), acetyl-CoA carboxylase (ACC1), and dacylglyceride acyl-transferase (DGA1) were simultaneously overexpressed in *Y. lipolytica*. High yields of lipid conversion (84.7% of theoretical maximal yield) were obtained by the engineered *Y. lipolytica* strain. It was shown that the engineered strain had a 3-fold growth advantage over the wild-type strain (Qiao et al. 2015).

The main obstacles in microbial lipid production include low yield of oleochemicals other than triacylglycerols (TAGs), inability for the selection of desired chain lengths of lipids, low tolerance to oleochemical products, high costs of initial feedstocks for microbial growth, and lack of knowledge in the understanding of metabolic regulation pathways (Yan and Pfleger 2020). But recently, the continuous progress of system biology, metabolic and genome engineering, designing and predicting tools, and new methods for cell identification are the keys to overcome these obstacles.

2.2.5 Marine Biorefinery Microorganisms

Marine sources have a suitable potential that is not yet fully known. Diatoms, microalgae, macroalgae, and cyanobacteria are probably the major biomass on Earth. They can accumulate oil, produce bioactive materials, and preserve starch. Also, they are one of the most important components of the natural CO_2 sequestration. However, the production or harvesting costs of this marine biomass are not still affordable to be a key solution for the bulk production of fuels and chemicals (Pandey et al. 2015). Nevertheless, the main marine microbial strains with potential applications in biorefinery are *Scenedesmus* (green algae), *Chlorella* spp. (green algae), *Gyrosigma* spp. (diatom), *Botryococcus* spp. (green algae), and *Synechocystis* spp. (cyanobacterium) (Govindan et al. 2020; Yoshida et al. 2012; Meixner et al. 2018).

Due to the limited raw material resources used for biofuel production and the shrinkage of agricultural lands, it creates difficulties in the production of renewable energy resources. Biomass created by algae consuming CO_2 can be used as energy feedstocks. Bioethanol, hydrogen, and diesel can be generated with the obtained algae biomass. Many researchers have conducted various studies to increase algae biomass, starch content, or lipid content in order to enhance bioethanol or biodiesel production efficiencies from algae (Sharma et al. 2018; Martín and Grossmann 2013; Dragone et al. 2011). By increasing the lipid content of the algae biomass, it can be utilized not only for biofuel production but also for various applications such as fish and animal feed and fertilizer (Karemore, Pal, and Sen 2013).

2.2.6 Microorganisms Used for Other Product Processing

2.2.6.1 Green Juice Fermentation

An appropriate solution for the sustainable production of various feed, food, biofuels, and chemicals could be green refineries. In this platform, environmentally friendly technologies and systems are integrated to process and use the green biomass, i.e., green corps, legumes, clover, alfalfa, beets, and carrots for the defined targets (Lübeck and Lübeck 2019). After mechanical separation of the green biomass into juice and press cake (solid residues), lactic acid bacteria (LAB) can act on both parts to produce lactic acid. Green juice is a suitable source of proteins, amino acids, fibers, sugars, and minerals that can be fermented to lactic acid or converted to lysine as a feed additive by LAB and *Corynebacterium glutamicum*, respectively (Venus 2006; Kamm 2014). Application of these microorganisms for the fermentation of the green juice is very vital, as the fresh juice is very sensitive to contamination with undesirable microbes that cause unfavorable odors or gas production that can affect the quality of the juice. Later usage of the fresh juice is not possible, and LAB should be preserved for this processing. Furthermore, the “ENLAC” (enzyme-assisted ensiling) method that combines the LAB and functional enzymes together can be used for this purpose (Kamm 2014). Interestingly, LAB due to lactic acid and other inhibitory substance production can extend the shelf life of the products in the green biorefineries (Lübeck and Lübeck 2019). Generally, fresh juice contains indigenous LAB, but when a pure culture of *Lactobacillus salivarius* was added to the juice, the fermentation process could be controlled to ensure the high quality of the final product (Santamaría-Fernández et al. 2017). This result was the same that obtained from a pure culture of *Lactobacillus plantarum* and *Pediococcus cerevisiae* (Ajibola 1984).

2.2.6.2 Lactic Acid

Lactic acid, as an organic acid, is naturally produced by chemical synthesis or fermentation (John et al. 2009). Lactic acid biofermentation by LAB and *Bacillus* strains can be successfully done from a wide range of raw materials such as molasses, starchy waste, lignocellulosic wastes, coffee pulp hydrolysate, orange peel wastes, and other raw materials (Pleissner et al. 2016; de la Torre et al. 2019, Alves de Oliveira et al. 2019; Grewal and Khare 2018; Chen et al. 2020). Lactic acid is a more popular product widely used in food, cosmetic, textile, leather, pharmaceutical, and chemical industries and a precursor for polylactic acid (PLA) synthesis and biodegradable plastic material as an alternative to petrochemical-derived plastics (Murariu and Dubois 2016; Lee et al. 2017). Studies on the effective production of lactic acid aimed to use xylose more effectively, especially in a mixed sugar obtained from the lignocellulosic hydrolysate. Yoshida et al. (2011) developed an engineered *Lactobacillus* strain that successfully produces lactic acid from a sugar mixture containing glucose, xylose, and arabinose. For this, two copies of xylose assimilating the *xylAB* operon from *Lactobacillus pentosus* were integrated into the *L. plantarum* strain which is deficient in the L-lactate dehydrogenase gene (*ldhL1*). This strain carries the transketolase gene (*tkt*) from *L. lactis* instead of the phosphoketolase 1 gene (*xpk1*) and also lacks the phosphoketolase 2 (*xpk2*) gene (Yoshida et al. 2011). Similarly, Zhang et al. (2016) cloned xylose-assimilating genes (encoding xylose isomerase and xylulokinase) into an L-lactate–deficient strain, *L. plantarum*, for cost-effective D-lactic production. It has been reported that the *L. plantarum* NCIMB 8826 *ΔldhL1-pLEM-xylAB* strain could use sugars derived from maize hydrolysate and produce high-purity actic acid (>99%) (Zhang et al. 2016).

2.2.6.3 Citric Acid

Citric acid is one of the largest metabolites used in industrial applications. Citric acid can be microbially produced by some of the fungi, yeast, and bacteria strains. Among them, *Aspergillus niger* is the most common citric acid producer, and it has the ability to ferment different kinds of substrates such as molasses, pumpkin, apple pomace, date syrup, sweet potato starch hydrolysate, and cassava peel substrate (Majumder et al. 2010; Dhillon et al. 2011; Mostafa and Alamri 2012; Betiku and Adesina 2013; Adeoye, Lateef, and Gueguim-Kana 2015). Some traditional irradiation technologies, including ultraviolet (UV) irradiation, γ-rays, and chemical mutagenesis, have been applied to improve metabolite production (Hu et al. 2016). Lotfy et al. (2007) performed random mutation applications using UV irradiation and chemical mutagenizing agents (ethyl methanesulfonate [EMS] and acridine orange [AO]) to increase citric acid production. It was determined that an obtained stable UV mutant strain (W5) was approximately 3.2 times higher than the wild-type strain with the highest yield (60.25%) (Lotfy, Ghanem, and El-Helow 2007). Javed et al. (2010) also investigated the effects of UV radiation and chemical mutagenesis (EMS and ethidium bromide [EB]) and obtained the highest citric acid production by the mutant strain created via EB mutagenesis. Applications of heavy-ion beams tend to increase mutation frequency and have a wide mutation spectrum (Hu et al. 2016). Hu et al. investigated the effects of heavy ion beam irradiations (e.g., carbon $^{12}C^{6+}$) for obtaining different phenotypic *A. niger* mutants to enhance citric acid production (Hu et al. 2016; Hu et al. 2017).

2.2.6.4 Natural Pigments

Pigments obtained from various plants and fruits are used as natural color materials in several applications such as textiles and dyes. Similarly, pigments can be alternatively produced by some filamentous fungi. Among them, *Neurospora* is a good source for obtaining yellow-to-orange carotenoids. Gmoser (Gmoser et al. 2018) reported that light, carbon, pH, and cofactor Zn^{+2} are important parameters in pigment production and that the highest pigment production (1.19 mg carotenoids/g dry weight biomass) was produced in the bubble column reactor. Along with light, incubation temperature and even biomass washing could be also important factors in pigment production (Gmoser, Ferreira, et al. 2019). With these factors, it has been reported that pigment production from *N. intermedia* reached to 1.4 mg/g dry fungal biomass (Gmoser, Ferreira, et al. 2019).

2.2.6.5 Fungal Biomass

Filamentous fungi can naturally synthesize a variety of enzymes (amylase, invertase, cellulase, xylanase, lipase, or protease), and thus can be produced in a variety of substrate sources (Ferreira et al. 2013; Ferreira et al. 2016). Fungal biomass produced by filamentous fungi cultivation can be considered as an animal feed or food source because of its protein content, namely mycoprotein. These fungal products have been traditionally used as red yeast rice produced by *Monascus purpureus*, hamanatto, miso and shoyu produced by *Aspergillus oryzae*, tempeh produced by *Rhizopus oryzae*, and oncom produced by *N. intermedia* in Asia (Moore and Chiu 2001; Souza Filho et al. 2019; Stodolak, Starzyńska-Janiszewska, and Bączkowicz 2020). Currently, a company (Quorn, Marlow Foods, UK) has commercialized and sold a mycoprotein product produced by *Fusarium venenatum* in supermarkets (Marlow Foods 2022Z). This fungal biomass can be produced from dairy industry wastes (Mahboubi et al. 2017a, b; Thunuguntla et al. 2018), fish processing sidestreams (Sar, Ferreira, and Taherzadeh 2020a, b), starch-containing sources (Souza Filho, Zamani, and Taherzadeh 2019; Gmoser, Sintca, et al. 2019; Souza Filho, Zamani, and Taherzadeh 2017; Sar et al. 2022), sugar industry by-products (Nair and Taherzadeh 2016; Karimi et al. 2019), and various fruit processing industry wastes (Lennartsson et al. 2012; Satari et al. 2016; Sar et al. 2020).

2.3 MICROORGANISMS USED IN CO-CULTURES

It is slightly difficult to have a precise definition of co-cultures, because, in the various microbiological resources and literature, terms differing from each other have been used. However, a co-culture can be defined intentionally by the preparation of a mixed culture in the laboratory with known strains. The co-culture may usually contain only two or more strains that might carry different metabolic pathways and, in combination, they can act on some substrate, whereas neither alone could do it. Moreover, a natural co-culture may have nutritional interdependence and form a syntrophic relationship. Using a co-culture can help to prevent contamination in the fermentation systems and reduce the total and operational costs and energy consumption in comparison with pure cultures (Wong, Tyagi, and Pandey 2016; Pandey, Negi, and Soccol 2016). The co-cultivation technique, one of the effective fermentation methods, can be defined as anaerobic or aerobic incubation of different specified microbial strains. It appears to be advantageous compared to pure culture fermentation due to the synergistic utilization potential of the metabolic pathways of all microorganisms in the co-culture system. Therefore, this technique can help to increase yield, improve crop qualities, and utilize substrate efficiency (Li et al. 2013; Bader et al. 2010). Thus, the co-culture is widely used in the production of metabolites, including chemicals, biofuels, biopolymers, and enzymes (Tables 2.1 and 2.2).

TABLE 2.1
Examples of Co-Cultures with Potential Applications in Biofuels Production

Product	Co-Cultures	References
Bioethanol	*Zymomonas mobilis* and *Pichia stipitis*	(Chen 2011; Fu et al. 2009)
	Thermoanaerobacter pentosaceus and *Saccharomyces cerevisiae*	(Tomás et al. 2013)
	Clostridium thermocellum and *Clostridium thermolacticum*	(Xu and Tschirner 2011; Xu and Tschirner 2014)
Hydrogen	*Clostridium thermocellum* and *Thermoanaerobacterium thermosaccharolyticum*	(Liu et al. 2008)
	Clostridium butyricum and *Rhodopseudomonas faecalis*	(Ding et al. 2009)
	Bacillus thermoamylovorans and *Clostridium beijerinckii*	(Chang et al. 2008)
Hydrogen and ethanol	*Escherichia coli* and *Enterobacter* spp.	(Maru et al. 2016)
Butanol	*Clostridium beijerinckii* and *Clostridium tyrobutyricum*	(Li et al. 2013)
	Clostridium thermocellum and *Clostridium saccharoperbutylacetonicum*	(Kiyoshi et al. 2015)
	Clostridium acetobutylicum and *Bacillus subtilis*	(Oliva-Rodríguez et al. 2019)
Microbial fuel cells	*Escherichia coli* and *Geobacter sulfurreducens*	(Qu et al. 2012; Bourdakos, Marsili, and Mahadevan 2014)
	Klebsiella variicola and *Pseudomonas aeruginosa*	(Islam et al. 2018)

TABLE 2.2
Examples of Co-Cultures with Potential Applications in the Biometabolite Production

Metabolite	Product	Co-Cultures	References
Polymer	Bacterial cellulose	*Gluconacetobacter hansenii* and *Escherichia coli*	(Liu and Catchmark 2019b)
	Bacterial cellulose/hyaluronic acid	*Gluconacetobacter hansenii* and *Lactococcus lactis*	(Liu and Catchmark 2020, 2019a)
	Bacterial cellulose/polyhydroxybutyrate	*Gluconacetobacter xylinus* and *Ralstonia eutropha*	(Ding et al. 2021)
Lipid		*Trichosporonoides spathulata* with microalgae	(Kitcha and Cheirsilp 2014)
		Scenedesmus obliquus and *Candida tropicalis*	(Wang et al. 2016)
		Chlamydomonas reinhardtii and *Azotobacter chroococcum*	(Xu, Cheng, and Wang 2018)
Organic acid	Lactic acid	*Lactobacillus delbrueckii* and *Lactococcus lactis*	(Sahoo and Jayaraman 2019)
	Lactic acid	*Bacillus coagulans* and *Lactobacillus rhamnosus*	(Chen et al. 2020)
	Butyric acid	*Bacillus* spp. and *Clostridium tyrobutyricum*	(Dwidar et al. 2013)
	Citric acid	*Aspergillus ornatus* and *Alternaria alternata*	(Ali et al. 2016)
	Citric acid	*Rhizopus stolonifera* and *Aspergillus niger*	(Zafar et al. 2020)
Enzyme	Laccase	*Ganoderma lucidum* and *Trametes versicolor*	(Kuhar, Castiglia, and Levin 2015)
	Xylanase and laccase	*Penicillium oxalicum* and *Pleurotus ostreatus*	(Kuhar, Castiglia, and Levin 2015)
	Cellulase and xylanase	*Saccharomyces cerevisiae* and *Candida tropicalis*	(Qadir et al. 2018)
	Cellulase	*Trichoderma reesei* and *Penicillium citrinum*	(Lodha, Pawar, and Rathod 2020)

Tomás et al. (2013) used *Thermoanaerobacter pentosaceus*, which could ferment rapeseed straw (PRS), directly in ethanol production and obtained the 37% ethanol yield. In addition, it has been reported that when *S. cerevisiae* is cofermented with this strain, the sugar in the hydrolysate is used more effectively, and the theoretical alcohol yield was reached 85% (Tomás et al. 2013). It was thought that *Bacillus thermoamylovorans* could be a good partner for anaerobic *Clostridium* since it can synthesize many different enzymes (lipase, protease, α-amylase, pectinase, and cellulase). It has been determined that when two different type strains are applied as a co-culture, hydrogen production ability increases, and this type of biotechnological application can be improved upon (Chang et al. 2008).

Wastewater discharged from leather industry may contain very high concentrations of organic pollutants, phenols, tannins, and heavy metals (Venugopal et al. 2019). It is possible to obtain higher yields in bioremediation studies by having the advantage of microorganisms with beneficial properties with the co-culturing. Zang et al. (2017) identified that the synergistic effect of yeast and bacteria has been improved by the increase of biomass and enzyme activities in the soil. Sometimes it may be necessary to determine the rate of inoculation of bacterial cultures to determine the effective yield. Tao et al. (2017) determined the biodegradation activity of crude oil was improved when the ratio of inoculation size of *Burkholderiales* and *Bacillus subtilis* was 2:1.

2.4 MIX CULTURE AND MICROBIAL CONSORTIA

Mostly, the pure culture of microbial strains is used by researchers or industrial experts for the production of various products like enzymes, bioactive materials, proteins, solvents, etc., from a certain substrate. However, pure cultures usually have a limitation in terms of the utilization of problematical substrates and may lead to the formation of more toxic and hazardous intermediates, by-products, or final products. Moreover, highly oxidized or reduced processes such as nitrogen cycle processes (nitrogen fixation, nitrification, and denitrification) cannot be performed by a single strain simultaneously (Tony, Goyal, and Khanna 2009; Paerl and Pinckney 1996). In nature, microorganisms do not often live as isolates and usually live in complex communities and consortia that interact in various ways (Bernstein and Carlson 2012). Similarly, it is challenging to use a pure culture for biohydrogen production from organic wastes due to the out-competition of pure culture by other nonhydrogen producers existing in the wastes (Pandey, Negi, and Soccol 2016). Hence, microbiologists prefer to apply a combination of two or more microbial strains (consortium) under typical conditions that can be constructed synthetically or obtained naturally. They may consist of several strains of the same or different species, or even different domains such as bacteria and fungi consortia that have been applied for the biodegradation of plastics (Yamano et al. 2008). Due to the high biodiversity of natural mixed cultures and synergistic metabolic activities among them, various substrates or wastes can be utilized and processed in a way that resuls in nontoxic intermediates or by-products. Moreover, microbial consortia show attractive characteristics such as productivity, stability, and operational robustness (Brahmachari 2016; Pandey, Negi, and Soccol 2016; Jia et al. 2016).

2.4.1 Natural Microbial Consortia

Synergetic or symbiotic relationships of two or more microbial strains lead to effective growth and proliferation than individual strains considered as microbial consortia. The special requirements among microbial groups necessitate the formation of microbial consortia among them, even distinct groups. Undoubtedly, in each ecosystem microbial consortia are present and play an important role in the interactions and associations inside the system. Some of these natural microbial consortia are gut microbiota, methanogenic communities, detrital communities, composts, benthic cyanobacterial mat communities, or planktonic microalgal-bacterial consortia. Among these natural communities, rumen and manure microbiota, wastewater sludge, granular sludge, and methanogenic and acidogenic communities have been mostly applied for the anaerobic digestion of wastes since the 1860s; however, this process has been considered a model for the applications of microbial complex communities in the industry. Generally, due to difficulties in the purification and preservation of anaerobic strains, the in-depth understanding of their microbiological aspects has remained and requires state-of-the-art approaches for further investigations (Paerl and Pinckney 1996; Nakayama et al. 2019). Based on the microbial sources, process conditions, pH, temperature, salts, inhibitors, and feedstock type, the structure, predominant microbial phyla, and functionality of the anaerobic digestion process may be different. For example, Pampillón-González et al. (2017) studied the microbial community structure of pig waste by 16S rRNA gene metagenomics, and the results revealed that bacterial operational taxonomic units (OTUs) were larger than the archaeal ones. Dominant taxa in the entrance of the digester belonged to the families *Clostridiaceae*, *Ruminococcaceae*, and *Lachnospiraceae* and the genera *Methanolobus*, *Methanosarcina*, *Methanosaeta*, and *Methanospirillum*. The main chamber was dominated by OP9 bacterial members and the syntrophic *Candidatus Cloacamonas*, and predominant methanogens were identified as *Methanosaeta*, *Methanoculleus*, and *Methanospirillum* (Pampillón-González et al. 2017). In another study, Sun et al. (Sun et al. 2015) showed responses of bacterial and archaeal communities of the manure to various temperatures and cosubstrates such as straw. The most dominant archaeal community consisted of *Methanobrevibacter*, and increasing the temperature to 52°C resulted in the dominance of *Methanoculleus* and *Methanobacterium*. Among bacterial communities, the members of *Bacteriodetes* and *Firmicutes* were abundant in all samples, and by the digestion, an uncultured class WWE1 dominated.

Recently, a series of -omics approaches in biological science, like metagenomics, proteomics, metabolomics, and novel microbiological methods, have been developed to investigate microbial consortia and reveal their functioning, interactions, and characteristics without cultivating them into pure cultures (Amann, Ludwig, and Schleifer 1995; Handelsman 2004).

2.4.1.1 Gut Microbiota

The gut microbiota is a very complicated microbial community composed of symbiotic, commensal, and opportunistic pathogenic microorganisms associated with the conversion of food to essential vitamins, cofactors, digestible ingredients, and preventing resistant pathogen colonization such that we cannot live without them (Lederberg and McCray 2001; Zengler 2008). However, our knowledge about the human gut microbiota is still limited due to low accessibility with highly invasive procedures. Thus most of the obtained knowledge about it is based on the analysis of stool. Consistent studies about the diversity of gut microbiota, including traditional cultivation approaches and molecular techniques, provide an essential framework for the understanding of its functions, dynamics, and host-microbe interactions (Zengler 2008). Based on the achieved results, the bacterial phyla, including Firmicutes (*Clostridium*, *Enterococcus*, *Lactobacillus*, and *Faecalibacterium*) and Bacteroidetes (*Bacteroides* and *Prevotella*) are generally dominant in a healthy adult. Microorganisms belonging to Actinobacteria (mainly *Bifidobacterium*), Proteobacteria, Verrucomicrobia, and Euryarchaeota are also found at low levels (Matijašić et al. 2014).

Microbial intestinal communities may vary according to the type of diet (e.g., vegan, vegetarian, and omnivores), genotype, eating habits, and age (Senghor et al. 2018); in particular, infants' microbiota contain a high abundance of *Lactobacil*lus. *Bifidobacterium* is also high in the microbiota of breast-fed babies, while the number of these bacteria is low in formula-fed babies (Yu, Chen, and Kling 2012; Bezirtzoglou, Tsiotsias, and Welling 2011). *Bifidobacteria* are responsible for folate production that improves metabolic processing, including DNA synthesis and repair (Pompei et al. 2007). Similarly, other microorganisms can produce hydrolyzing enzymes or metabolites such as antibiotics or molecules with antimicrobial activity and short-chain fatty acids (SCFAs; acetate, propionate, and butyrate) (Wong et al. 2006).

2.4.1.2 Methanogenic Communities

Similar to the gut microbiota, the rumen also consists of a complex community, including bacteria, protozoa, fungi, and archaea. Some genera from rumen microbiota (*Fibrobacter, Ruminococcus, Ruminobacter, Lachnospira, Succinomonas*, etc.) can degrade complex materials such as xylan, pectin, and lignocellulose. Additionally, some of the ruminal microorganisms can metabolize the degraded materials and generate SCFAs and methane (22–29%) (Van Soest 2018; Barragán-Trinidad, Carrillo-Reyes, and Buitrón 2017). Methane-producing bacteria (*Methanobrevibacter ruminantium, Methanomicrobium mobile, Methanosarcina mazei, Methanosarcina barkeri*, and *Methanobacterium formicicum*) can be isolated from rumen (Mitsumori, Tajima, and Itabashi, 2002). Deng et al. (2017) performed metagenomic analysis on buffalo rumen and achieved high cellulolytic activity by using *Methanobrevibacter* and *Ruminococcus* in coinoculated fermentation. Moreover, the gut microbiota of other animals has the potential to be used in the biorefinery platforms. For example, land pulmonate's gut microbiota (an agricultural pest) can break down the lignocellulosic materials into an extensive range of degraded components such that it is considered a potential source of the novel hydrolyzing enzymes (Joynson et al. 2017).

2.4.1.3 Detrital Communities

Detrital communities are associated with the terrestrial and marine detritus that includes the fragments, bodies, or fecal materials of dead organisms (animals, plants, or other domains). Microbial detrital communities can colonize these materials to decompose or remineralize them and play a key role in the food chain and web and element cycles (Getz 2011; Hoffmann et al. 2017). The environmental factors such as seasonal changes, temperature, soil texture, water depth, detritus types, or particle size can affect the microbial decomposer communities' structure and dynamics (Yeager and Sinsabaugh 1998; Hoffmann et al. 2017). The bacteria, algae, protozoa, or fungi can form these kinds of communities, and their functions on the detritus may decrease the toxicity of the recalcitrant compounds (Pokarzhevskii et al. 2003). To better understand detrital communities, Abell and Bowman studied the function and activities of the bacterial communities on diatom detritus *(Nitzschia closterium* Ehrenberg) in seawater mesocosms using real-time polymerase chain reaction (PCR), fluorescent in situ hybridization (FISH), and denaturing gradient gel electrophoresis (DGGE). Their findings indicated the quick bacterial colonization on the diatom detritus. Predominant bacteria belonged to the class α-Proteobacteira, γ-Proteobacteria, and Flavobacteria. The members of the class Flavobacteria contributed to the initial colonization that could be considered an essential factor in the bacterial colonization on the detritus. However, other physicochemical and biological factors may result in different responses in these communities. Consequently, the determination of these responses would be significant for the understanding of secondary metabolite production (Abell and Bowman 2005). In addition to diatom detritus, algal species blooms provide important carbon input to the anoxic layers of the aquatic ecosystems. Algal species such as *Chlorella vulgaris* have been remarked on as promising resources for biofuel production; therefore, microbial communities involved in their detritus turnover have considerable importance (Morrison et al. 2017).

2.4.1.4 Composting Communities

Composting is the transformation of organic wastes into a humus-rich complex mixture with the enzymes secreted by microbial populations for the disposal of solid organic wastes (Federici et al. 2011; Zang et al. 2017). In an aerobic-composting microbial consortium culture, different types of the bacterial genera (*Bacillus, Halobacillus, Staphylococcus, Microbacterium*, and *Streptomyces*) can be found (Chander et al. 2018). Along with aerobic bacteria, some anaerobic (*Clostridium* and *Thermoanaerobacterium*) and facultative anaerobic bacteria (*Rhodocyclaceae*) can be found and isolated (Wongwilaiwalin et al. 2010). Some special microorganisms with capability to degrade recalcitrant compounds such as tannin (e.g., *Lysinibacillus fusiformis, Kocuria palustris, Tetrathiobacter kashmirensis, Rhodococcus rhodochrous, Pantoea, and Serratia*) and phenol (*Diaphorobacter, Pseudomonas, Thauera,* and *Comamonas*) can be isolated through the composting of some wastewater such as olive oil mill water and petroleum refinery wastewater (Silva et al. 2012; Federici et al. 2011; Pepi et al. 2010).

2.4.1.5 Benthic Cyanobacterial Mat Communities

Cyanobacteria usually exist in the form of thick microbial benthic mats. They can consist of distinct microbial groups with vertical stratification. The order *Oscillatoriales* is a member of cyanobacteria mostly located in the top layer. Cyanobacterial mats are both a dominant feature of Antarctic ecosystems and also extreme environments such as hot springs and desert crusts (Jungblut et al. 2005; Prieto-Barajas, Valencia-Cantero, and Santoyo 2018). Cyanobacteria can be detected with non-cyanobacteria members that are mainly bacteria and others (diatoms, archaea, fungi, and viruses) (Cissell and McCoy 2020). Dekas et al. (2009) identified the archaeal/bacterial consortia in benthic ecosystems. Here, methane-oxidizing archaea could fix N_2 and also share the products with sulfate-reducing bacteria symbionts (Dekas, Poretsky, and Orphan 2009). Similarly, the effects of marine microbial communities on anaerobic oxidation of methane in in vitro conditions have been investigated (Holler et al. 2011; Orcutt and Meile 2008; Nauhaus et al. 2002).

2.4.1.6 Planktonic Microalgal-Bacterial Consortia

Microalgal-bacterial consortia include nitrifying, heterotrophic, and photosynthetic microorganisms that can be considered as an alternative technology for the improvement of wastewater treatment (Petrini et al. 2020; Vargas et al. 2016). In addition to improving ammonia removal in wastewater treatment (Vargas et al. 2016), bacteria can also promote the biomass production of microalgae. Consequently, the obtained biomass can be used for biofuel production (Zhang et al. 2020). Furthermore, microalgal-bacterial consortia can be used in various biotechnological applications such as reducing CO_2 emissions and generating electricity (Zhang et al. 2020).

2.4.2 Engineered or Synthetic Microbial Consortia

Regarding the technical difficulties of multistep metabolic engineering, use of engineered microbial co-cultures or consortia can be a promising solution (Gunasekaran, Noronha, and Pandey 2016; Zhou et al. 2015). These mixed cultures have been developed recently via genetic and metabolic engineering, system biology, and synthetic biology. They represent exclusive metabolic capabilities, novel enzymatic cascades, and interactions such as synergism, mutualism, or commensalism that support their growth to survive under extreme or toxic conditions. Even though preadaptation is a significant process to improve the ability of engineered microbial consortia, the choosing of appropriate strains is essential to achieve an effective system (Liu, Monroe, and Davis 2018; Varjani et al. 2020; Park et al. 2014). One of the challenges facing microbial consortia is the variability of the feedstocks that affect the functions and yield that the process needs to be reoptimized, which cause noteworthy delay and costs in the biorefinery plants (Kenney et al. 2013). A suggested solution for this issue is to mix various feedstocks to help microbial consortia adaptation (Langholtz, Stokes, and Eaton 2016). In addition, the design of synthetic consortia is significant for effectiveness of the whole system to have a successful industrial achievement (Liu, Monroe, and Davis 2018; Bernstein and Carlson 2012).

2.4.3 Artificial Microbial Consortia

As the name of these consortia indicates, artificial microbial consortia (AMC) include two or more wild microbial populations and their interactions do not naturally take place. The design of such systems requires a fundamental knowledge of microbial eco-physiology and the software that simplifies the potential pairing of microbial communities. Industrial AMCs have been used for renewable energy production, consolidated bioprocessing (CBP), food processing, and remediation of recalcitrant wastes or compounds.

2.4.4 Bioproducts through Microbial Consortia

Mixed cultures can supply valuable combinations of various metabolic pathways for the consumption of complex materials due to their hydrolytic and degradative enzymes. Subsequently, the production of high value-added materials like lactic acid, by using the wastes such as corn stover, date juice, or cassava bagasse, has occurred as by-products or the final product of the treatment process (Cui, Li, and Wan 2011; Nancib, Nancib, and Boudrant 2009; John, Nampoothiri, and Pandey 2006).

In early times, researchers have been focused on natural microbial consortia (NMA) for producing biohydrogen or biofuel (ethanol, butanol) from lignocellulosic substrates (Cheng et al. 2012; Ronan et al. 2013; Pérez-Rangel et al. 2015). When using natural microbial communities, some kinds of problems, such as losses in primary production, decreases in product yield, prolongation of incubation time, and increased costs, may be encountered (Valdez-Vazquez et al. 2015). For this reason, synthetic microbial communities have been formed by using two or more known microorganisms, and bioproduction studies have been started to be implemented with these communities (Bernstein and Carlson 2012).

For biometabolite production from the lignocellulosic biomasses, a microorganism that can hydrolyze the substrate and another microorganism that produces metabolites from releasing fermentable sugars are needed. Many bacteria can produce the hydrogen, which is a special mechanism to dispose of excess electrons through hydrogenase activity. For the most part, some strict anaerobes (*Clostridia*, methanogenic bacteria, archaea), facultative anaerobes (*E. coli*, *Enterobacter*, *Citrobacter*), and aerobes (*Alcaligenes*, *Bacillus*) could be used for hydrogen production (Li and Fang 2007). Extreme thermophiles *Caldicellulosiruptor saccharolyticus* with *Caldicellulosiruptor kristjanssonii* are also used to improve H_2 production (Zeidan and Van Niel 2009). Rossi et al. (2011) also identified the bacterial strains related to H_2 production through DNA techniques for identifying the bacteria genera in bacterial consortia and announced that the genera *Klebsiella* and *Pantoea* were present in the consortia.

Another example for energy generation from microbial consortia is microbial fuel cells (MFCs). MFCs can be generated through microbial consortia consuming biomass belonging with biorefineries or via treatment of wastewater. MFCs can consist of a pure culture or mixed communities. Exoelectrogenic microorganisms can be isolated from freshwater and marine sediments or sludge and belong to different microbial phyla (Feng, Song, and Bae 2016; Nandy, Kumar, and Kundu 2013). Despite many reports about the axenic MFCs such as *Geobacter* spp., *Shewenlla* spp., *Haloferax volcanii*, and *Natrialba magadii*, several researchers have showed that the mixed communities have more advantages over single-species MFCs (Prokhorova et al. 2017, Li et al. 2016). The bacterial and archaeal MFCs are promising approaches that have been used to treat some agro-food industry wastewater, including dairy, brewery, winery, and nitrogen containing wastes, to produce clean energy (Abrevaya et al. 2011; Logan et al. 2006; Cecconet et al. 2018; Lu et al. 2019).

Biogas production is becoming widespread every day, as well as bioethanol production, for reducing greenhouse gas emissions for sustainable energy. Biogas production is a typical example for microbial consortia applications. Methane production mainly consists of a four-step process, which are hydrolysis, acidogenesis, acetogenesis/dehydrogenation, and methanogenesis (Figure 2.2) (Weiland 2010). In the hydrolyzing step, the microorganisms (*Bacteroides*, *Clostridium*, and *Streptococcus*) hydrolyze the polymers into the monomers. Afterwards, the acidogens convert the monomers into propionate, butyrate, valerate, alcohols, and lactate. Thereafter, those acids are converted to acetate, hydrogen, and carbon dioxide. Members of methanogenic bacteria (*Methanosarcina*, *Metanonococcus*, and *Methanotrix*) produce methane from acetate or hydrogen and carbon dioxide (Weiland 2010; Lukitawesa 2020).

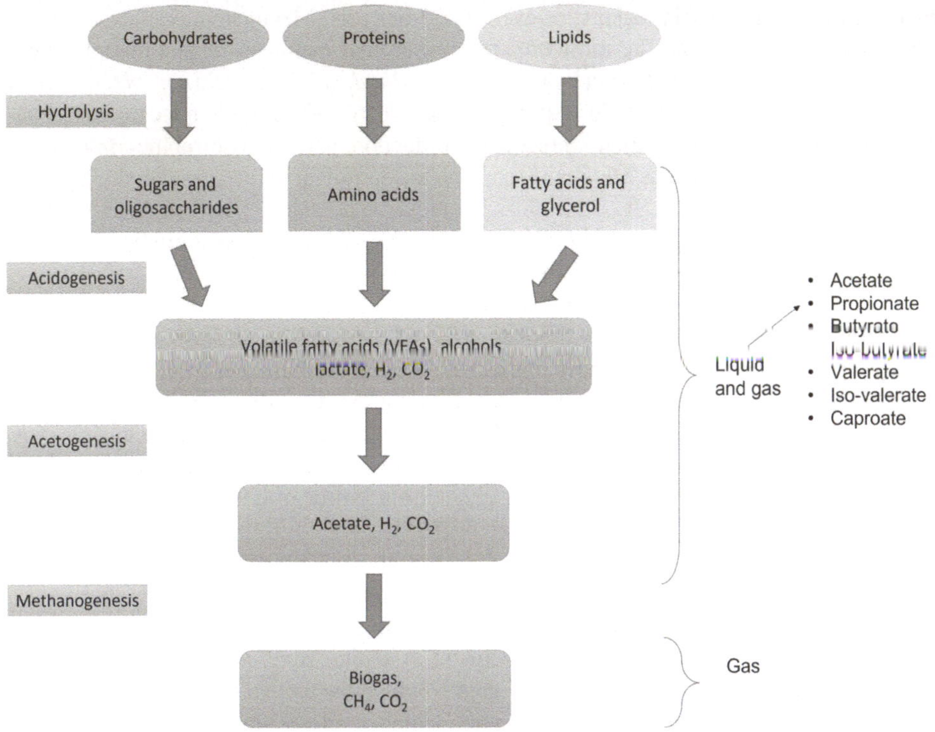

FIGURE 2.2 Overview of biogas production steps in anaerobic digestion; modified from Lukitawesa (2020) and Harirchi et al. (2022).

2.4.5 Wastewater Treatment

Undoubtedly, the efficacious operation of wastewater treatment plants depends on the normal function of microbial communities in the bioreactors, which are identified by modern molecular approaches and environmental genomic tools and include fermentative bacteria, acetogens, methanogens, and other related taxa. These communities consist of many members from various microbial phyla, and some of them are the dominant populations and key players of biodegradation pathways. Hence, it is significant to have a better understanding of all microbial members and their growth requirements for successful nutrient removal, bulking, foaming, and other processing (Russell 2019; Angenent et al. 2016; Dahiya and Venkata Mohan 2016). However, the global view of microbial diversity in the wastewater treatment plants remains hard to define. In this regard, the Global Water Microbiome Consortium (GWMC; http://gwmc.ou.edu/, [Wu et al. 2019]) was formed in 2014, and its overall mission is to provide a platform to facilitate international collaboration and communication on research and education on global water and wastewater microbiome studies (Chowdappa et al. 2014).

The wastewater composition, including heavy metals, inhibitors, or other organic materials, can affect the microbial communities' structure. Microbial consortia have the potential to convert organic wastewaters into valuable biomaterials in the renewable platforms. The carboxylate platform is one of the interesting platforms in which anaerobic microbial consortia have been used for this purpose. Under typical conditions, anaerobic microbial communities in the bioreactors produce methane as the final product from various organic materials, ranging from fats, carbohydrates, to proteins. But these communities can produce other final products, such as medium-chain carboxylate (MCC), under certain conditions. The most attractive MCC is *n*-Caproate (*n*-hexanoic acid) that can be extracted easily and applied for the manufacturing of fragrances, additives, dyes, pharmaceuticals, antimicrobial agents, or even liquid biofuels. Metagenomic analysis and microbial diversity studies revealed the most important microbial members in this platform belonged to *Clostridium kluyveri*, *Eubacterium pyruvativorans*, *Rhodospirillum rubrum*, *Megasphaera elsdenii*, and *Clostridia*-related bacteria (Angenent et al. 2016), while methanogenic consortia such as *Methanomethylovorans* contribute to methane production, and methylotrophic *Methanosarcina* species associated with the genus *Methanobacterium* are dominant in the granular sludge of anaerobic digesters (Chen et al. 2017). It is significant to preserve a balance between different microbial consortia, particularly hydrogen-consuming methanogens, so that hydrogen production without its removal may result in acetate accumulation (Jiang 2006).

In addition to anaerobic bacteria, many strains of the genus *Bacillus* have been known as protein, fat, and starchy material degraders. Moreover, some of the *Bacillus* strains can remove nitrogen and phosphorus as effectively as organic materials. Another predominant genus in the wastewater microbial communities is *Pseudomonas*, with substantial biodegrading abilities that can rapidly reduce biological oxygen demand (BOD). Similarly, the species of *Achromobacter* and *Comamonas* can take part in domestic wastewater bioprocessing. These kinds of microbial communities can participate in the removal of phenolic compounds, recalcitrant hydrocarbons, and other toxic chemicals (Dahiya and Venkata Mohan 2016; Mehrotra et al. 2019).

In the comprehensive study performed by Zhang et al. (2017), activated sludge collected from different wastewater treatment units and their microbial community structures have been analyzed. The results showed the genus *Nitrospira* and the families Caldilineaceae and Anaerolineaceae were associated with domestic biological treatment systems, while *Thauera*, which is a Gram-negative aromatic hydrocarbon decomposer genus belonging to the family Zoogloeaceae, was related to industrial wastewater treatment units. Some contaminating heavy compounds (oil absorbers, chlorinated products) were removed by the microbial consortia (Gertler et al. 2009; Gallego et al. 2007; Pham et al. 2009; Bacosa and Inoue 2015). Similarly, dechlorination of tetrachloroethene (PCE) to ethene was successfully achieved by a mixed culture of species containing *Dehalococcoides ethenogenes* strains (Löffler et al. 2013). Furthermore, it was found out that the genus *Thiobacillus*—an obligate chemolithotroph—along with *Thauera* are functional microorganisms in chemical wastewater treatment (Zhang, Xu, and Zhu 2017). Moreover, they revealed that the genera *Desulfomicrobium* and *Macellibacteroides* play a key role in the treatment of textile dyeing wastewater. There are many examples of microbial mixed cultures, such as a report related to the consortium of *Sphingobacterium*, *Pseudomonas*, and *Bacillus* capable of decolorizing textile wastewater more promptly compared to the pure cultures (Tamboli et al. 2010).

Mazzucotelli et al. (2014) used an artificial consortium including *Lactococcus garvieae*, *Bacillus thuringiensis*, *E. coli*, and *Stenotrophomonas* spp. for the treatment of dairy wastewater. Under optimal conditions, an 83.44% degradation percentage was achieved after 48h that was higher than by using the pure culture of each strain (Mazzucotelli et al. 2014). Most of these AMCs include two microbial cultures, and a few researchers reported successful interactions with more than two strains, such as a consortium of *A. vinelandii* ATCC 12837, *Bacillus licheniformis* ATCC 25972, and *Paenibacillus curdlanolyticus* ATCC 51899, for syntrophic biodegradation of an antibiotic (Kim et al. 2008).

As Johnson reported in 2008, acidophilic microbial consortia are more robust than the axenic cultures and play a key role in biomining processes (Johnson 2008).

Wu et al. (2019) showed a global scale of bacterial communities in the wastewater treatment plants. Based on the results, most of the core bacterial community members belonged to the phylum *Proteobacteria*. The most abundant OTUs were related to *Dokdonella kunshanensis* DC-3 and the genus *Zoogloea*. In addition to these OTUs, *Nitrospira* and *Nitrosomonas* are important OTUs for ammonium nitrogen removal in the activated sludge. The polyphosphate-accumulating microorganisms identified as *Candidatus Accumulibacter* and *Candidatus Accumulimonas* are known as core taxa with metabolic flexibility. Remarkably, it was revealed that the core community of the activated sludge did not have any overlap with the other environments (Wu et al. 2019).

2.5 CONCLUSION AND FUTURE PERSPECTIVES

This chapter has provided a review of the various types of microorganisms used in the biorefinery platforms. While most microorganisms are not isolated as pure cultures, they can be applied for biorefinery purposes in different forms. It is known that microbial strains never live as the axenic form in natural environments and have significant interactions and communicate with other members of an ecosystem. This concept allows us to engineer their metabolic pathways and processes to protect the environment and support the production of high-value-added materials from inexpensive resources such as various wastes. Therefore, further investigations and development are required to improve our knowledge about microbial interactions, functions, and their communities' structures. The main areas for research include 1) developing new designs for co-cultures or mixed cultures, 2) reducing the cost of metabolic and genetic engineering, 3) improving the conditions for the use of microbial strains in industries, and 4) researching how to increase the productivity of the microbial producers. Furthermore, biological converting of the wastes to value-added materials is also of interest. Hence, finding novel technologies for these conversions at large scales provides an opportunity to do more shifts from the petroleum refinery to the biorefinery. However, possible obstacles should not be underestimated because early companies and wastewater treatment plants will be expected to face challenges and take time to achieve high production. Optimistically, these kinds of efforts can be warranted to become profitable and are a green step toward the eventual aim of sustainable clean environments and an economy.

ACKNOWLEDGEMENT

The authors are grateful to Sweden's innovation agency (VINNOVA) and Swedish Research council for their financial support.

REFERENCES

Abanoz, K., B. C. Stark, and M. Y. Akbas. 2012. "Enhancement of ethanol production from potato-processing wastewater by engineering Escherichia coli using Vitreoscilla haemoglobin." *Letters in Applied Microbiology* no. 55 (6):436–443. https://doi.org/10.1111/lam.12000.

Abedinifar, Sorahi, Keikhosro Karimi, Morteza Khanahmadi, and Mohammad J. Taherzadeh. 2009. "Ethanol production by Mucor indicus and Rhizopus oryzae from rice straw by separate hydrolysis and fermentation." *Biomass and Bioenergy* no. 33 (5):828–833. https://doi.org/10.1016/j.biombioe.2009.01.003.

Abell, G. C. J., and John P. Bowman. 2005. "Colonization and community dynamics of class Flavobacteria on diatom detritus in experimental mesocosms based on Southern Ocean seawater." *FEMS Microbiology Ecology* no. 53 (3):379–391. https://doi.org/10.1016/j.femsec.2005.01.008.

Abrevaya, Ximena C., Natalia Sacco, Pablo J. D. Mauas, and Eduardo Cortón. 2011. "Archaea-based microbial fuel cell operating at high ionic strength conditions." *Extremophiles* no. 15 (6):633. https://doi.org/10.1007/s00792-011-0394-z.

Acharya, Somen, and Anita Chaudhary. 2012. "Bioprospecting thermophiles for cellulase production: a review." *Brazilian Journal of Microbiology* no. 43:844–856.

Adeoye, A. O., A. Lateef, and E. B. Gueguim-Kana. 2015. "Optimization of citric acid production using a mutant strain of Aspergillus niger on cassava peel substrate." *Biocatalysis and Agricultural Biotechnology* no. 4 (4):568–574. https://doi.org/10.1016/j.bcab.2015.08.004.

Ahmed, Md Juned K., and M. Ahmaruzzaman. 2016. "A review on potential usage of industrial waste materials for binding heavy metal ions from aqueous solutions." *Journal of Water Process Engineering* no. 10:39–47. https://doi.org/10.1016/j.jwpe.2016.01.014.

Ajibola, O. O. 1984. "Physical, chemical and microbiological changes during green plant juice fermentation." *Food Chemistry* no. 13 (3):181–192. https://doi.org/10.1016/0308-8146(84)90072-4.

Akbas, Meltem Yesilcimen, Taner Sar, and Busra Ozcelik. 2014. "Improved ethanol production from cheese whey, whey powder, and sugar beet molasses by 'Vitreoscilla hemoglobin expressing' Escherichia coli." *Bioscience, Biotechnology, and Biochemistry* no. 78 (4):687–694. https://doi.org/10.1080/09168451.2014.896734.

Ali, Sohaib Rafaqat, Zahid Anwar, Muhammad Irshad, Saima Mukhtar, and Nabeela Tariq Warraich. 2016. "Bio-synthesis of citric acid from single and co-culture-based fermentation technology using agro-wastes." *Journal of Radiation Research and Applied Sciences* no. 9 (1):57–62. https://doi.org/10.1016/j.jrras.2015.09.003.

Allen, E. E., and J. F. Banfield. 2005. "Community genomics in microbial ecology and evolution." *Nature Reviews Microbiology* no. 3 (6):489–498. https://doi.org/10.1038/nrmicro1157.

Alves de Oliveira, Regiane, Roland Schneider, Carlos Eduardo Vaz Rossell, Rubens Maciel Filho, and Joachim Venus. 2019. "Polymer grade l-lactic acid production from sugarcane bagasse hemicellulosic hydrolysate using Bacillus coagulans." *Bioresource Technology Reports* no. 6:26–31. https://doi.org/10.1016/j.biteb.2019.02.003.

Amann, R. I., W. Ludwig, and K. H. Schleifer. 1995. "Phylogenetic identification and in situ detection of individual microbial cells without cultivation." *Microbiological Reviews* no. 59 (1):143–169.

Anderson, A. J., and E. A. Dawes. 1990. "Occurrence, metabolism, metabolic role, and industrial uses of bacterial polyhydroxyalkanoates." *Microbiological Reviews* no. 54 (4):450–472.

Angenent, L. T., H. Richter, W. Buckel, C. M. Spirito, K. J. Steinbusch, C. M. Plugge, D. P. Strik, T. I. Grootscholten, C. J. Buisman, and H. V. Hamelers. 2016. "Chain elongation with reactor microbiomes: open-culture biotechnology to produce biochemicals." *Environmental Science & Technology* no. 50 (6):2796–810. https://doi.org/10.1021/acs.est.5b04847.

Avci, Ayşe, and Sedat Dönmez. 2006. "Effect of zinc on ethanol production by two Thermoanaerobacter strains." *Process Biochemistry* no. 41 (4):984–989. https://doi.org/10.1016/j.procbio.2005.11.007.

Azeredo, Henriette M. C., Hernane Barud, Cristiane S. Farinas, Vanessa M. Vasconcellos, and Amanda M. Claro. 2019. "Bacterial cellulose as a raw material for food and food packaging applications." *Frontiers in Sustainable Food Systems* no. 3 (7). https://doi.org/10.3389/fsufs.2019.00007.

Bacosa, Hernando Pactao, and Chihiro Inoue. 2015. "Polycyclic aromatic hydrocarbons (PAHs) biodegradation potential and diversity of microbial consortia enriched from tsunami sediments in Miyagi, Japan." *Journal of Hazardous Materials* no. 283:689–697. https://doi.org/10.1016/j.jhazmat.2014.09.068.

Bader, J., E. Mast-Gerlach, M. K. Popović, R. Bajpai, and U. Stahl. 2010. "Relevance of microbial coculture fermentations in biotechnology." *Journal of Applied Microbiology* no. 109 (2):371–387. https://doi.org/10.1111/j.1365-2672.2009.04659.x.

Baltz, R. H., A. L. Demain, and J. E. Davies. 2010. *Manual of Industrial Microbiology and Biotechnology*: ASM Press.

Barragán-Trinidad, Martín, Julián Carrillo-Reyes, and Germán Buitrón. 2017. "Hydrolysis of microalgal biomass using ruminal microorganisms as a pretreatment to increase methane recovery." *Bioresource Technology* no. 244:100–107. https://doi.org/10.1016/j.biortech.2017.07.117.

Beopoulos, Athanasios, Julien Cescut, Ramdane Haddouche, Jean-Louis Uribelarrea, Carole Molina-Jouve, and Jean-Marc Nicaud. 2009. "Yarrowia lipolytica as a model for bio-oil production." *Progress in Lipid Research* no. 48 (6):375–387. https://doi.org/10.1016/j.plipres.2009.08.005.

Bernstein, Hans C., and Ross P. Carlson. 2012. "Microbial Consortia Engineering for Cellular Factories: in vitro to in silico systems." *Computational and Structural Biotechnology Journal* no. 3:e201210017–e201210017. https://doi.org/10.5936/csbj.201210017.

Betiku, Eriola, and Olusola A. Adesina. 2013. "Statistical approach to the optimization of citric acid production using filamentous fungus Aspergillus niger grown on sweet potato starch hydrolyzate." *Biomass and Bioenergy* no. 55:350–354. https://doi.org/10.1016/j.biombioe.2013.02.034.

Bezirtzoglou, Eugenia, Arsenis Tsiotsias, and Gjalt W. Welling. 2011. "Microbiota profile in feces of breast- and formula-fed newborns by using fluorescence in situ hybridization (FISH)." *Anaerobe* no. 17 (6):478–482. https://doi.org/10.1016/j.anaerobe.2011.03.009.

Bhaskar, T., A. Pandey, E. R. Rene, and D. C. W. Tsang. 2020. *Waste Biorefinery: Integrating Biorefineries for Waste Valorisation*: Elsevier Science.

Blumer-Schuette, Sara E., Steven D. Brown, Kyle B. Sander, Edward A. Bayer, Irina Kataeva, Jeffrey V. Zurawski, Jonathan M. Conway, Michael W. W. Adams, and Robert M. Kelly. 2014. "Thermophilic lignocellulose deconstruction." *FEMS Microbiology Reviews* no. 38 (3):393–448. https://doi.org/10.1111/1574-6976.12044.

Bourdakos, Nicholas, Enrico Marsili, and Radhakrishnan Mahadevan. 2014. "A defined co-culture of Geobacter sulfurreducens and Escherichia coli in a membrane-less microbial fuel cell." *Biotechnology and Bioengineering* no. 111 (4):709–718. https://doi.org/10.1002/bit.25137.

Brahmachari, G. 2016. *Biotechnology of Microbial Enzymes: Production, Biocatalysis and Industrial Applications*: Elsevier Science.

Cecconet, Daniele, Daniele Molognoni, Arianna Callegari, and Andrea G. Capodaglio. 2018. "Agro-food industry wastewater treatment with microbial fuel cells: energetic recovery issues." *International Journal of Hydrogen Energy* no. 43 (1):500–511. https://doi.org/10.1016/j.ijhydene.2017.07.231.

Celik, Kamile Burcu, Elif Ceylan Cengiz, Taner Sar, Burcu Dursun, Osman Ozturk, Meltem Yesilcimen Akbas, and Rezan Demir-Cakan. 2018. "In-situ wrapping of tin oxide nanoparticles by bacterial cellulose derived carbon nanofibers and its application as freestanding interlayer in lithium sulfide based lithium-sulfur batteries." *Journal of Colloid and Interface Science* no. 530:137–145. https://doi.org/10.1016/j.jcis.2018.06.054.

Chander, Girish, Suhas P. Wani, S. Gopalakrishnan, Ankita Mahapatra, Swati Chaudhury, C. S. Pawar, Manoj Kaushal, and A. V. R. Kesava Rao. 2018. "Microbial consortium culture and vermi-composting technologies for recycling on-farm wastes and food production." *International Journal of Recycling of Organic Waste in Agriculture* no. 7 (2):99–108. https://doi.org/10.1007/s40093-018-0195-9.

Chandolias, Konstantinos, Enise Pekgenc, and Mohammad J. Taherzadeh. 2019. "Floating membrane bioreactors with high gas hold-up for syngas-to-biomethane conversion." *Energies* no. 12 (6):1046.

Chang, Jui-Jen, Chia-Hung Chou, Cheng-Yu Ho, Wei-En Chen, Jiunn-Jyi Lay, and Chieh-Chen Huang. 2008. "Syntrophic co-culture of aerobic Bacillus and anaerobic Clostridium for bio-fuels and bio-hydrogen production." *International Journal of Hydrogen Energy* no. 33 (19):5137–5146. https://doi.org/10.1016/j.ijhydene.2008.05.021.

Chen, Hao, Bo Chen, Zihang Su, Kang Wang, Boxuan Wang, Yong Wang, Zhihao Si, Yilu Wu, Di Cai, and Peiyong Qin. 2020. "Efficient lactic acid production from cassava bagasse by mixed culture of Bacillus coagulans and lactobacillus rhamnosus using stepwise pH controlled simultaneous saccharification and co-fermentation." *Industrial Crops and Products* no. 146:112175. https://doi.org/10.1016/j.indcrop.2020.112175.

Chen, Wei-Yu, Lucia Kraková, Jer-Horng Wu, Domenico Pangallo, Lenka Jeszeová, Bing Liu, and Hidenari Yasui. 2017. "Community and proteomic analysis of anaerobic consortia converting tetramethylammonium to methane." *Archaea* no. 2017:2170535. https://doi.org/10.1155/2017/2170535.

Chen, Yanli. 2011. "Development and application of co-culture for ethanol production by co-fermentation of glucose and xylose: a systematic review." *Journal of Industrial Microbiology & Biotechnology* no. 38 (5):581–597. https://doi.org/10.1007/s10295-010-0894-3.

Cheng, Chieh-Lun, Pei-Yi Che, Bor-Yann Chen, Wen-Jhy Lee, Chiu-Yue Lin, and Jo-Shu Chang. 2012. "Biobutanol production from agricultural waste by an acclimated mixed bacterial microflora." *Applied Energy* no. 100:3–9. https://doi.org/10.1016/j.apenergy.2012.05.042.

Chien, Liang-Jung, Hsiao-Ting Chen, Pei-Fen Yang, and Cheng-Kang Lee. 2006. "Enhancement of cellulose pellicle production by constitutively expressing Vitreoscilla hemoglobin in Acetobacter xylinum." *Biotechnology Progress* no. 22 (6):1598–1603.

Chien, Liang-Jung, and Cheng-Kang Lee. 2007. "Hyaluronic acid production by recombinant Lactococcus lactis." *Applied Microbiology and Biotechnology* no. 77 (2):339–346. https://doi.org/10.1007/s00253-007-1153-z.

Chong, Barrie Fong, Lars M. Blank, Richard McLaughlin, and Lars K. Nielsen. 2005. "Microbial hyaluronic acid production." *Applied Microbiology and Biotechnology* no. 66 (4):341–351. https://doi.org/10.1007/s00253-004-1774-4.

Chowdappa, P., Shivakumar Gowda, C. S. Chethana, and S. Madhura. 2014. "Antifungal activity of chitosan-silver nanoparticle composite against Colletotrichum gloeosporioides associated with mango anthracnose." *African Journal of Microbiology Research* no. 8 (17):1803–1812.

Cissell, Ethan C., and Sophie J. McCoy. 2020. "Shotgun metagenomic sequencing reveals the full taxonomic, trophic, and functional diversity of a coral reef benthic cyanobacterial mat from Bonaire, Caribbean Netherlands." *Science of The Total Environment* no. 755:142719. https://doi.org/10.1016/j.scitotenv.2020.142719.

Cui, Fengjie, Yebo Li, and Caixia Wan. 2011. "Lactic acid production from corn stover using mixed cultures of Lactobacillus rhamnosus and Lactobacillus brevis." *Bioresource Technology* no. 102 (2):1831–1836. https://doi.org/10.1016/j.biortech.2010.09.063.

Dahiya, Shikha, and S. Venkata Mohan. 2016. "Strategic design of synthetic consortium with embedded wastewater treatment potential: deciphering the competence of isolates from diverse microbiome." *Frontiers in Environmental Science* no. 4 (30). https://doi.org/10.3389/fenvs.2016.00030.

de la Torre, I., M. G. Acedos, M. Ladero, and V. E. Santos. 2019. "On the use of resting L. delbrueckii spp. delbrueckii cells for D-lactic acid production from orange peel wastes hydrolysates." *Biochemical Engineering Journal* no. 145:162–169. https://doi.org/10.1016/j.bej.2019.02.012.

Dekas, Anne E., Rachel S. Poretsky, and Victoria J. Orphan. 2009. "Deep-sea archaea fix and share nitrogen in methane-consuming microbial consortia." *Science* no. 326 (5951):422. https://doi.org/10.1126/science.1178223.

Deng, Yuying, Zhenxing Huang, Wenquan Ruan, Mingxing Zhao, Hengfeng Miao, and Hongyan Ren. 2017. "Co-inoculation of cellulolytic rumen bacteria with methanogenic sludge to enhance methanogenesis of rice straw." *International Biodeterioration & Biodegradation* no. 117:224–235. https://doi.org/10.1016/j.ibiod.2017.01.017.

Den Haan, Riaan, Shaunita H. Rose, Lee R. Lynd, and Willem H. van Zyl. 2007. "Hydrolysis and fermentation of amorphous cellulose by recombinant Saccharomyces cerevisiae." *Metabolic Engineering* no. 9 (1):87–94. https://doi.org/10.1016/j.ymben.2006.08.005.

Dhillon, Gurpreet Singh, Satinder Kaur Brar, Mausam Verma, and Rajeshwar Dayal Tyagi. 2011. "Utilization of different agro-industrial wastes for sustainable bioproduction of citric acid by Aspergillus niger." *Biochemical Engineering Journal* no. 54 (2):83–92. https://doi.org/10.1016/j.bej.2011.02.002.

Di Donato, Paola, Ilaria Finore, Annarita Poli, Barbara Nicolaus, and Licia Lama. 2019. "The production of second generation bioethanol: the biotechnology potential of thermophilic bacteria." *Journal of Cleaner Production* no. 233:1410–1417. https://doi.org/10.1016/j.jclepro.2019.06.152.

Dien, B. S., N. N. Nichols, P. J. O'Bryan, and R. J. Bothast. 2000. "Development of new ethanologenic Escherichia coli strains for fermentation of lignocellulosic biomass." *Applied Biochemistry and Biotechnology* no. 84–86:181–196.

Ding, Jie, Bing-Feng Liu, Nan-Qi Ren, De-Feng Xing, Wan-Qian Guo, Ji-Fei Xu, and Guo-Jun Xie. 2009. "Hydrogen production from glucose by co-culture of Clostridium Butyricum and immobilized Rhodopseudomonas faecalis RLD-53." *International Journal of Hydrogen Energy* no. 34 (9):3647–3652. https://doi.org/10.1016/j.ijhydene.2009.02.078.

Ding, Ran, Shengjun Hu, Mengying Xu, Qiang Hu, Shan Jiang, Kai Xu, Pier-Luc Tremblay, and Tian Zhang. 2021. "The facile and controllable synthesis of a bacterial cellulose/polyhydroxybutyrate composite by co-culturing Gluconacetobacter xylinus and Ralstonia eutropha." *Carbohydrate Polymers* no. 252:117137. https://doi.org/10.1016/j.carbpol.2020.117137.

Domingues, Lucília, Nelson Lima, and José A. Teixeira. 2001. "Alcohol production from cheese whey permeate using genetically modified flocculent yeast cells." *Biotechnology and Bioengineering* no. 72 (5):507–514. https://doi.org/10.1002/1097-0290(20010305)72:5<507::AID-BIT1014>3.0.CO;2-U.

Dragone, Giuliano, Bruno D. Fernandes, Ana P. Abreu, António A. Vicente, and José A. Teixeira. 2011. "Nutrient limitation as a strategy for increasing starch accumulation in microalgae." *Applied Energy* no. 88 (10):3331–3335. https://doi.org/10.1016/j.apenergy.2011.03.012.

Dunn, Kori L., and Christopher V. Rao. 2014. "Expression of a xylose-specific transporter improves ethanol production by metabolically engineered Zymomonas mobilis." *Applied Microbiology and Biotechnology* no. 98 (15):6897–6905. https://doi.org/10.1007/s00253-014-5812-6.

Dunn, Kori L., and Christopher V. Rao. 2015. "High-throughput sequencing reveals adaptation-induced mutations in pentose-fermenting strains of Zymomonas mobilis." *Biotechnology and Bioengineering* no. 112 (11):2228–2240. https://doi.org/10.1002/bit.25631.

Dursun, Burcu, Taner Sar, Ali Ata, Mathieu Morcrette, Meltem Yesilcimen Akbas, and Rezan Demir-Cakan. 2016. "Pyrolyzed bacterial cellulose-supported SnO2 nanocomposites as high-capacity anode materials for sodium-ion batteries." *Cellulose* no. 23 (4):2597–2607. https://doi.org/10.1007/s10570-016-0966-2.

Dwidar, Mohammed, Seil Kim, Byoung Seung Jeon, Youngsoon Um, Robert J. Mitchell, and Byoung-In Sang. 2013. "Co-culturing a novel Bacillus strain with Clostridium tyrobutyricum ATCC 25755 to produce butyric acid from sucrose." *Biotechnology for Biofuels* no. 6 (1):35. https://doi.org/10.1186/1754-6834-6-35.

Easterling, Emily R., W. Todd French, Rafael Hernandez, and Margarita Licha. 2009. "The effect of glycerol as a sole and secondary substrate on the growth and fatty acid composition of Rhodotorula glutinis." *Bioresource Technology* no. 100 (1):356–361. https://doi.org/10.1016/j.biortech.2008.05.030.

Eilers, Heike, Jakob Pernthaler, Frank Oliver Glöckner, and Rudolf Amann. 2000. "Culturability and in situ abundance of pelagic bacteria from the North Sea." *Applied and Environmental Microbiology* no. 66 (7):3044–3051. https://doi.org/10.1128/aem.66.7.3044-3051.2000.

Eliasson, Anna, Camilla Christensson, C. Fredrik Wahlbom, and Bärbel Hahn-Hägerdal. 2000. "Anaerobic xylose fermentation by recombinant Saccharomyces cerevisiae carrying XYL1, XYL2, andXKS1 in mineral medium chemostat cultures." *Applied and Environmental Microbiology* no. 66 (8):3381. https://doi.org/10.1128/AEM.66.8.3381-3386.2000.

Federici, Ermanno, Milva Pepi, Alessandro Esposito, Silvia Scargetta, Laura Fidati, Simone Gasperini, Giovanni Cenci, and Roberto Altieri. 2011. "Two-phase olive mill waste composting: community dynamics and functional role of the resident microbiota." *Bioresource Technology* no. 102 (23):10965–10972. https://doi.org/10.1016/j.biortech.2011.09.062.

Feng, Qing, Young-Chae Song, and Byung-Uk Bae. 2016. "Influence of applied voltage on the performance of bioelectrochemical anaerobic digestion of sewage sludge and planktonic microbial communities at ambient temperature." *Bioresource Technology* no. 220:500–508. https://doi.org/10.1016/j.biortech.2016.08.085.

Ferreira, J. A., P. R. Lennartsson, L. Edebo, and M. J. Taherzadeh. 2013. "Zygomycetes-based biorefinery: present status and future prospects." *Bioresource Technology* no. 135:523–532. https://doi.org/10.1016/j.biortech.2012.09.064.

Ferreira, Jorge A., Amir Mahboubi, Patrik R. Lennartsson, and Mohammad J. Taherzadeh. 2016. "Waste biorefineries using filamentous ascomycetes fungi: present status and future prospects." *Bioresource Technology* no. 215:334–345. https://doi.org/10.1016/j.biortech.2016.03.018.

Ferreira, Jorge A., Patrik R. Lennartsson, and Mohammad J. Taherzadeh. 2014. "Production of ethanol and biomass from thin stillage using food-grade zygomycetes and ascomycetes filamentous fungi." *Energies* no. 7 (6). https://doi.org/10.3390/en7063872.

Ferreira, Jorge A., Patrik R. Lennartsson, and Mohammad J. Taherzadeh. 2015. "Production of ethanol and biomass from thin stillage by Neurospora intermedia: a pilot study for process diversification." *Engineering in Life Sciences* no. 15 (8):751–759. https://doi.org/10.1002/elsc.201400213.

Fu, Nan, Paul Peiris, Julie Markham, and John Bavor. 2009. "A novel co-culture process with Zymomonas mobilis and Pichia stipitis for efficient ethanol production on glucose/xylose mixtures." *Enzyme and Microbial Technology* no. 45 (3):210–217. https://doi.org/10.1016/j.enzmictec.2009.04.006.

Fujita, Y., J. Ito, M. Ueda, H. Fukuda, and A. Kondo. 2004. "Synergistic saccharification, and direct fermentation to ethanol, of amorphous cellulose by use of an engineered yeast strain codisplaying three types of cellulolytic enzyme." *Applied and Environmental Microbiology* no. 70 (2):1207–1212. https://doi.org/10.1128/aem.70.2.1207-1212.2004.

Gallego, José Luis R., María Jesús García-Martínez, Juan F. Llamas, Carmen Belloch, Ana I. Peláez, and Jesús Sánchez. 2007. "Biodegradation of oil tank bottom sludge using microbial consortia." *Biodegradation* no. 18 (3):269–281.

Gertler, Christoph, Gunnar Gerdts, Kenneth N. Timmis, and Peter N. Golyshin. 2009. "Microbial consortia in mesocosm bioremediation trial using oil sorbents, slow-release fertilizer and bioaugmentation." *FEMS Microbiology Ecology* no. 69 (2):288–300. https://doi.org/10.1111/j.1574-6941.2009.00693.x.

Getachew, Anteneh, and Fantahun Woldesenbet. 2016. "Production of biodegradable plastic by polyhydroxybutyrate (PHB) accumulating bacteria using low cost agricultural waste material." *BMC Research Notes* no. 9 (1):509. https://doi.org/10.1186/s13104-016-2321-y.

Getz, Wayne M. 2011. "Biomass transformation webs provide a unified approach to consumer–resource modelling." *Ecology Letters* no. 14 (2):113–124. https://doi.org/10.1111/j.1461-0248.2010.01566.x.

Gidarakos, E., G. Havas, and P. Ntzamilis. 2006. "Municipal solid waste composition determination supporting the integrated solid waste management system in the island of Crete." *Waste Management* no. 26 (6):668–679. https://doi.org/10.1016/j.wasman.2005.07.018.

Gmoser, Rebecca, Jorge A. Ferreira, Magnus Lundin, Mohammad J. Taherzadeh, and Patrik R. Lennartsson. 2018. "Pigment production by the edible filamentous fungus Neurospora intermedia." *Fermentation* no. 4 (1):11.

Gmoser, Rebecca, Jorge A. Ferreira, Mohammad J. Taherzadeh, and Patrik R. Lennartsson. 2019. "Post-treatment of fungal biomass to enhance pigment production." *Applied Biochemistry and Biotechnology* no. 189 (1):160–174. https://doi.org/10.1007/s12010-019-02961-y.

Gmoser, Rebecca, Carissa Sintca, Mohammad J. Taherzadeh, and Patrik R. Lennartsson. 2019. "Combining submerged and solid state fermentation to convert waste bread into protein and pigment using the edible filamentous fungus N. intermedia." *Waste Management* no. 97:63–70. https://doi.org/10.1016/j.wasman.2019.07.039.

Govindan, Natanamurugaraj, Gaanty Pragas Maniam, Mashitah M. Yusoff, Mohd Hasbi Ab. Rahim, Tawan Chatsungnoen, Rameshprabu Ramaraj, and Yusuf Chisti. 2020. "Statistical optimization of lipid production by the diatom Gyrosigma sp. grown in industrial wastewater." *Journal of Applied Phycology* no. 32 (1):375–387. https://doi.org/10.1007/s10811-019-01971-x.

Grewal, Jasneet, and S. K. Khare. 2018. "One-pot bioprocess for lactic acid production from lignocellulosic agro-wastes by using ionic liquid stable Lactobacillus brevis." *Bioresource Technology* no. 251:268–273. https://doi.org/10.1016/j.biortech.2017.12.056.

Grimalt-Alemany, Antonio, Ioannis V. Skiadas, and Hariklia N. Gavala. 2018. "Syngas biomethanation: state-of-the-art review and perspectives." *Biofuels, Bioproducts and Biorefining* no. 12 (1):139–158. https://doi.org/10.1002/bbb.1826.

Grisales Díaz, Víctor Hugo, and Mark J. Willis. 2019. "Ethanol production using Zymomonas mobilis: development of a kinetic model describing glucose and xylose co-fermentation." *Biomass and Bioenergy* no. 123:41–50. https://doi.org/10.1016/j.biombioe.2019.02.004.

Guimarães, Pedro M. R., Jean François, Jean Luc Parrou, José A. Teixeira, and Lucília Domingues. 2008. "Adaptive evolution of a lactose-consuming Saccharomyces cerevisiae recombinant." *Applied and Environmental Microbiology* no. 74 (6):1748. https://doi.org/10.1128/AEM.00186-08.

Guimarães, P. M., J. A. Teixeira, and L. Domingues. 2008. "Fermentation of high concentrations of lactose to ethanol by engineered flocculent Saccharomyces cerevisiae." *Biotechnology Letters* no. 30 (11):1953–1958. https://doi.org/10.1007/s10529-008-9779-1.

Gunasekaran, Paramasamy, Santosh Noronha, and Ashok Pandey. 2016. *Current Developments in Biotechnology and Bioengineering: Functional Genomics and Metabolic Engineering*: Elsevier.

Hahn-Hägerdal, Bärbel, Kaisa Karhumaa, César Fonseca, Isabel Spencer-Martins, and Marie F. Gorwa-Grauslund. 2007. "Towards industrial pentose-fermenting yeast strains." *Applied Microbiology and Biotechnology* no. 74 (5):937–953. https://doi.org/10.1007/s00253-006-0827-2.

Handelsman, J. 2004. "Metagenomics: application of genomics to uncultured microorganisms." *Microbiology and Molecular Biology Reviews* no. 68 (4):669–685. https://doi.org/10.1128/mmbr.68.4.669-685.2004.

Hari Krishna, S., T. Janardhan Reddy, and G. V. Chowdary. 2001. "Simultaneous saccharification and fermentation of lignocellulosic wastes to ethanol using a thermotolerant yeast." *Bioresource Technology* no. 77 (2):193–196. https://doi.org/10.1016/S0960-8524(00)00151-6.

Harirchi, Sharareh, Zahra Etemadifar, Amir Mahboubi, Fatemeh Yazdian, and Mohammad J. Taherzadeh. 2020. "The effect of calcium/magnesium ratio on the biomass production of a novel Thermoalkaliphilic Aeribacillus pallidus strain with highly heat-resistant spores." *Current Microbiology* no. 77 (10):2565–2574. https://doi.org/10.1007/s00284-020-02010-6.

Harirchi, Sharareh, Zahra Etemadifar, Fatemeh Yazdian, and Mohammad J. Taherzadeh. 2020. "Efficacy of polyextremophilic Aeribacillus pallidus on bioprocessing of beet vinasse derived from ethanol industries." *Bioresource Technology* no. 313:123662. https://doi.org/10.1016/j.biortech.2020.123662.

Harirchi, Sharareh, Steven Wainaina, Taner Sar, Seyed Ali Nojoumi, Milad Parchami, Mohsen Parchami, Sunita Varjani, Samir Kumar Khanal, Jonathan Wong, Mukesh Kumar Awasthi, and Mohammad J. Taherzadeh. 2022. "Microbiological insights into anaerobic digestion for biogas, hydrogen or volatile fatty acids (VFAs): a review." *Bioengineered* no. 13 (3):6521–6557. https://doi.org/10.1080/21655979.2022.2035986.

Hay, Iain D., Zahid Ur Rehman, Aamir Ghafoor, and Bernd H. A. Rehm. 2010. "Bacterial biosynthesis of alginates." *Journal of Chemical Technology & Biotechnology* no. 85 (6):752–759. https://doi.org/10.1002/jctb.2372.

Hayes, Daniel J. M. 2013. "Second-generation biofuels: why they are taking so long." *WIREs Energy and Environment* no. 2 (3):304–334. https://doi.org/10.1002/wene.59.

Ho, Nancy W. Y., Zhengdao Chen, and Adam P. Brainard. 1998. "Genetically engineered Saccharomycesyeast capable of effective cofermentation of glucose and xylose." *Applied and Environmental Microbiology* no. 64 (5):1852. https://doi.org/10.1128/AEM.64.5.1852-1859.1998.

Hoffmann, Katy, Christiane Hassenrück, Verena Salman-Carvalho, Moritz Holtappels, and Christina Bienhold. 2017. "Response of bacterial communities to different detritus compositions in Arctic deep-sea sediments." *Frontiers in Microbiology* no. 8 (266). https://doi.org/10.3389/fmicb.2017.00266.

Holler, Thomas, Friedrich Widdel, Katrin Knittel, Rudolf Amann, Matthias Y. Kellermann, Kai-Uwe Hinrichs, Andreas Teske, Antje Boetius, and Gunter Wegener. 2011. "Thermophilic anaerobic oxidation of methane by marine microbial consortia." *The ISME Journal* no. 5 (12):1946–1956. https://doi.org/10.1038/ismej.2011.77.

Horng, Y. T., K. C. Chang, C. C. Chien, Y. H. Wei, Y. M. Sun, and P. C. Soo. 2010. "Enhanced polyhydroxybutyrate (PHB) production via the coexpressed phaCAB and vgb genes controlled by arabinose PBAD promoter in Escherichia coli." *Letters in Applied Microbiology* no. 50 (2):158–167. https://doi.org/10.1111/j.1472-765X.2009.02772.x.

Hu, Wei, Ji-hong Chen, Shu-yang Wang, Jing Liu, Yuan Song, Qing-feng Wu, and Wen-jian Li. 2016. "Changes in the physiological properties and kinetics of citric acid accumulation via carbon ion irradiation mutagenesis of Aspergillus niger." *Journal of Zhejiang University-SCIENCE B* no. 17 (4):262–270. https://doi.org/10.1631/jzus.B1500120.

Hu, Wei, Wenjian Li, Hao Chen, Jing Liu, Shuyang Wang, and Jihong Chen. 2017. "Changes in transcript levels of starch hydrolysis genes and raising citric acid production via carbon ion irradiation mutagenesis of Aspergillus niger." *PLoS ONE* no. 12 (6):e0180120. https://doi.org/10.1371/journal.pone.0180120.

Ingram, L. O., and T. Conway. 1988. "Expression of different levels of ethanologenic enzymes from Zymomonas mobilis in recombinant strains of Escherichia coli." *Applied and Environmental Microbiology* no. 54 (2):397–404. https://doi.org/10.1128/aem.54.2.397-404.1988.

Ingram, L. O., T. Conway, D. P. Clark, G. W. Sewell, and J. F. Preston. 1987. "Genetic engineering of ethanol production in Escherichia coli." *Applied and Environmental Microbiology* no. 53 (10):2420–2425. https://doi.org/10.1128/aem.53.10.2420-2425.1987.

Islam, M. Amirul, Huei Ruey Ong, Baranitharan Ethiraj, Chin Kui Cheng, and Md Maksudur Rahman Khan. 2018. "Optimization of co-culture inoculated microbial fuel cell performance using response surface methodology." *Journal of Environmental Management* no. 225:242–251. https://doi.org/10.1016/j.jenvman.2018.08.002.

Jacques, K. A., T. P. Lyons, and D. R. Kelsall. 2003. *The Alcohol Textbook: A Reference for the Beverage, Fuel and Industrial Alcohol Industries*: Nottingham University Press.

Javed, Sadia, Muhammad Asgher, Munir A. Sheikh, and Haq Nawaz. 2010. "Strain improvement through UV and chemical mutagenesis for enhanced citric acid production in molasses-based solid state fermentation." *Food Biotechnology* no. 24 (2):165–179. https://doi.org/10.1080/08905436.2010.482014.

Jia, Xiaoqiang, Chang Liu, Hao Song, Mingzhu Ding, Jin Du, Qian Ma, and Yingjin Yuan. 2016. "Design, analysis and application of synthetic microbial consortia." *Synthetic and Systems Biotechnology* no. 1 (2):109–117. https://doi.org/10.1016/j.synbio.2016.02.001.

Jiang, B. 2006. *The effect of trace elements on the metabolism of methanogenic consortia*, S.n., [S.l.].

Joanne, M. Willey, M. Sherwood Linda, and J. Woolverton Christopher. 2016. *Prescott's Microbiology*: McGraw Hill.

John, Rojan P., G. S. Anisha, K. Madhavan Nampoothiri, and Ashok Pandey. 2009. "Direct lactic acid fermentation: focus on simultaneous saccharification and lactic acid production." *Biotechnology Advances* no. 27 (2):145–152. https://doi.org/10.1016/j.biotechadv.2008.10.004.

John, Rojan P., K. Madhavan Nampoothiri, and Ashok Pandey. 2006. "Solid-state fermentation for l-lactic acid production from agro wastes using Lactobacillus delbrueckii." *Process Biochemistry* no. 41 (4):759–763. https://doi.org/10.1016/j.procbio.2005.09.013.

Johnson, D. B. 2008. "Biodiversity and interactions of acidophiles: key to understanding and optimizing microbial processing of ores and concentrates." *Transactions of Nonferrous Metals Society of China* no. 18 (6):1367–1373. https://doi.org/10.1016/S1003-6326(09)60010-8.

Joynson, Ryan, Leighton Pritchard, Ekenakema Osemwekha, and Natalie Ferry. 2017. "Metagenomic analysis of the gut microbiome of the common black slug arion ater in search of novel lignocellulose degrading enzymes." *Frontiers in Microbiology* no. 8 (2181). https://doi.org/10.3389/fmicb.2017.02181.

Jungblut, Anne-Dorothee, Ian Hawes, Doug Mountfort, Bettina Hitzfeld, Daniel R. Dietrich, Brendan P. Burns, and Brett A. Neilan. 2005. "Diversity within cyanobacterial mat communities in variable salinity meltwater ponds of McMurdo Ice Shelf, Antarctica." *Environmental Microbiology* no. 7 (4):519–529. https://doi.org/10.1111/j.1462-2920.2005.00717.x.

Kamm, B. 2014. *Microorganisms in Biorefineries*: Springer Berlin Heidelberg.

Kanti, A., and I. M. Sudiana. 2018. *Production of phytase, amylase and cellulase by Aspergillus, Rhizophus and Neurospora on mixed rice rtraw powder and soybean curd residue*. Paper read at Proceedings of the IOP Conference Series: Earth and Environmental Science.

Karemore, Ankush, Ruma Pal, and Ramkrishna Sen. 2013. "Strategic enhancement of algal biomass and lipid in Chlorococcum infusionum as bioenergy feedstock." *Algal Research* no. 2 (2):113–121. https://doi.org/10.1016/j.algal.2013.01.005.

Karimi, Keikhosro, Giti Emtiazi, and Mohammad J. Taherzadeh. 2006. "Ethanol production from dilute-acid pretreated rice straw by simultaneous saccharification and fermentation with Mucor indicus, Rhizopus oryzae, and Saccharomyces cerevisiae." *Enzyme and Microbial Technology* no. 40 (1):138–144. https://doi.org/10.1016/j.enzmictec.2005.10.046.

Karimi, Sajjad, Nasrollah Mahboobi Soofiani, Torbjörn Lundh, Amir Mahboubi, Anders Kiessling, and Mohammad J. Taherzadeh. 2019. "Evaluation of filamentous fungal biomass cultivated on vinasse as an alternative nutrient source of fish feed: protein, lipid, and mineral composition." *Fermentation* no. 5 (4). https://doi.org/10.3390/fermentation5040099.

Kenney, Kevin L., William A. Smith, Garold L. Gresham, and Tyler L. Westover. 2013. "Understanding biomass feedstock variability." *Biofuels* no. 4 (1):111–127. https://doi.org/10.4155/bfs.12.83.

Khosravi Darani, Kianoosh, Fatemeh Yazdian, Hamid Rashedi, Neda Madadian Bozorg, Mohsen Moradi, Soheil Rezazadeh Mofradnia, and Martin Koller. 2020. "Simulation of bioreactors for PHB production from natural gas." *Iranian Journal of Chemistry and Chemical Engineering (IJCCE)* no. 39 (1):313–336. https://doi.org/10.30492/ijcce.2020.33133.

Kim, Hyun Jung, James Q. Boedicker, Jang Wook Choi, and Rustem F. Ismagilov. 2008. "Defined spatial structure stabilizes a synthetic multispecies bacterial community." *Proceedings of the National Academy of Sciences* no. 105 (47):18188–18193. https://doi.org/10.1073/pnas.0807935105.

Kitcha, Suleeporn, and Benjamas Cheirsilp. 2014. "Enhanced lipid production by co-cultivation and co-encapsulation of oleaginous yeast Trichosporonoides spathulata with microalgae in alginate gel beads." *Applied Biochemistry and Biotechnology* no. 173 (2):522–534. https://doi.org/10.1007/s12010-014-0859-5.

Kiyoshi, Keiji, Masataka Furukawa, Tomoko Seyama, Toshimori Kadokura, Atsumi Nakazato, and Shunichi Nakayama. 2015. "Butanol production from alkali-pretreated rice straw by co-culture of Clostridium thermocellum and Clostridium saccharoperbutylacetonicum." *Bioresource Technology* no. 186:325–328. https://doi.org/10.1016/j.biortech.2015.03.061.

Kocharin, Kanokarn, Yun Chen, Verena Siewers, and Jens Nielsen. 2012. "Engineering of acetyl-CoA metabolism for the improved production of polyhydroxybutyrate in Saccharomyces cerevisiae." *AMB Express* no. 2 (1):52. https://doi.org/10.1186/2191-0855-2-52.

Kuhar, Francisco, Valeria Castiglia, and Laura Levin. 2015. "Enhancement of laccase production and malachite green decolorization by co-culturing Ganoderma lucidum and Trametes versicolor in solid-state fermentation." *International Biodeterioration & Biodegradation* no. 104:238–243. https://doi.org/10.1016/j.ibiod.2015.06.017.

Kumagai, Yoshiharu, and Yoshiharu Doi. 1992. "Enzymatic degradation of binary blends of microbial poly(3-hydroxybutyrate) with enzymatically active polymers." *Polymer Degradation and Stability* no. 37 (3):253–256. https://doi.org/10.1016/0141-3910(92)90167-4.

Kuo, Chia-Hung, Hsin-Yin Teng, and Cheng-Kang Lee. 2015. "Knock-out of glucose dehydrogenase gene in Gluconacetobacter xylinus for bacterial cellulose production enhancement." *Biotechnology and Bioprocess Engineering* no. 20 (1):18–25. https://doi.org/10.1007/s12257-014-0316-x.

Kurella, Swamy, Pawan Kishan Bhukya, and B. C. Meikap. 2017. "Removal of H2S pollutant from gasifier syngas by a multistage dual-flow sieve plate column wet scrubber." *Journal of Environmental Science and Health, Part A* no. 52 (6):515–523. https://doi.org/10.1080/10934529.2017.1281690.

Langholtz, Matthew H, Bryce J Stokes, and Laurence M Eaton. 2016. "2016 Billion-ton report: advancing domestic resources for a thriving bioeconomy, Volume 1: economic availability of feedstock." *Oak Ridge National Laboratory, Oak Ridge, Tennessee, managed by UT-Battelle, LLC for the US Department of Energy* no. 2016:1–411.

Lederberg, Joshua, and Alexa T. McCray. 2001. "Ome SweetOmics: a genealogical treasury of words." *The Scientist* no. 15 (7):8–8.

Lee, Jae Won, Jung Hoon In, Joon-Bum Park, Jonghyeok Shin, Jin Hwan Park, Bong Hyun Sung, Jung-Hoon Sohn, Jin-Ho Seo, Jin-Byoung Park, Soo Rin Kim, and Dae-Hyuk Kweon. 2017. "Co-expression of two heterologous lactate dehydrogenases genes in Kluyveromyces marxianus for l-lactic acid production." *Journal of Biotechnology* no. 241:81–86. https://doi.org/10.1016/j.jbiotec.2016.11.015.

Lennartsson, Patrik R., Päivi Ylitervo, Christer Larsson, Lars Edebo, and Mohammad J. Taherzadeh. 2012. "Growth tolerance of Zygomycetes Mucor indicus in orange peel hydrolysate without detoxification." *Process Biochemistry* no. 47 (5):836–842. https://doi.org/10.1016/j.procbio.2012.02.019.

Li, Cheng, Keaton Larson Lesnik, Yanzhen Fan, and Hong Liu. 2016. "Millimeter scale electron conduction through exoelectrogenic mixed species biofilms." *FEMS Microbiology Letters* no. 363 (15). https://doi.org/10.1093/femsle/fnw153.

Li, Chenlin, and Herbert H. P. Fang. 2007. "Fermentative hydrogen production from wastewater and solid wastes by mixed cultures." *Critical Reviews in Environmental Science and Technology* no. 37 (1):1–39. https://doi.org/10.1080/10643380600729071.

Li, Lin, Hongxia Ai, Shexi Zhang, Shuang Li, Zexin Liang, Zhen-Qiang Wu, Shang-Tian Yang, and Ju-Fang Wang. 2013. "Enhanced butanol production by coculture of Clostridium beijerinckii and Clostridium tyrobutyricum." *Bioresource Technology* no. 143:397–404. https://doi.org/10.1016/j.biortech.2013.06.023.

Liu, Fang, Eric Monroe, and Ryan W Davis. 2018. "Engineering microbial consortia for bioconversion of multisubstrate biomass streams to biofuels." In *Biofuels-Challenges and opportunities*: IntechOpen.

Liu, Ke, and Jeffrey M. Catchmark. 2019a. "Bacterial cellulose/hyaluronic acid nanocomposites production through co-culturing Gluconacetobacter hansenii and Lactococcus lactis in a two-vessel circulating system." *Bioresource Technology* no. 290:121715. https://doi.org/10.1016/j.biortech.2019.121715.

Liu, Ke, and Jeffrey M. Catchmark. 2019b. "Enhanced mechanical properties of bacterial cellulose nanocomposites produced by co-culturing Gluconacetobacter hansenii and Escherichia coli under static conditions." *Carbohydrate Polymers* no. 219:12–20. https://doi.org/10.1016/j.carbpol.2019.04.071.

Liu, Ke, and Jeffrey M. Catchmark. 2020. "Bacterial cellulose/hyaluronic acid nanocomposites production through co-culturing Gluconacetobacter hansenii and Lactococcus lactis under different initial pH values of fermentation media." *Cellulose* no. 27 (5):2529–2540. https://doi.org/10.1007/s10570-019-02924-w.

Liu, Miao, Siqi Li, Yongzhen Xie, Shiru Jia, Ying Hou, Yang Zou, and Cheng Zhong. 2018. "Enhanced bacterial cellulose production by Gluconacetobacter xylinus via expression of Vitreoscilla hemoglobin and oxygen tension regulation." *Applied Microbiology and Biotechnology* no. 102 (3):1155–1165. https://doi.org/10.1007/s00253-017-8680-z.

Liu, Yan, Peng Yu, Xin Song, and Yinbo Qu. 2008. "Hydrogen production from cellulose by co-culture of Clostridium thermocellum JN4 and Thermoanaerobacterium thermosaccharolyticum GD17." *International Journal of Hydrogen Energy* no. 33 (12):2927–2933. https://doi.org/10.1016/j.ijhydene.2008.04.004.

Lodha, Ankit, Shweta Pawar, and Virendra Rathod. 2020. "Optimised cellulase production from fungal co-culture of Trichoderma reesei NCIM 1186 and Penicillium citrinum NCIM 768 under solid state fermentation." *Journal of Environmental Chemical Engineering* no. 8 (5):103958. https://doi.org/10.1016/j.jece.2020.103958.

Löffler, F. E., J. Yan, K. M. Ritalahti, L. Adrian, E. A. Edwards, K. T. Konstantinidis, J. A. Müller, H. Fullerton, S. H. Zinder, and A. M. Spormann. 2013. "Dehalococcoides mccartyi gen. nov., sp. nov., obligately organohalide-respiring anaerobic bacteria relevant to halogen cycling and bioremediation, belong to a novel bacterial class, Dehalococcoidia classis nov., order Dehalococcoidales ord. nov. and family Dehalococcoidaceae fam. nov., within the phylum Chloroflexi." *International Journal of Systematic and Evolutionary Microbiology* no. 63 (Pt 2):625–635. https://doi.org/10.1099/ijs.0.034926-0.

Logan, Bruce E., Bert Hamelers, René Rozendal, Uwe Schröder, Jürg Keller, Stefano Freguia, Peter Aelterman, Willy Verstraete, and Korneel Rabaey. 2006. "Microbial fuel cells: methodology and technology." *Environmental Science & Technology* no. 40 (17):5181–5192. https://doi.org/10.1021/es0605016.

Lotfy, Walid A., Khaled M. Ghanem, and Ehab R. El-Helow. 2007. "Citric acid production by a novel Aspergillus niger isolate: I. Mutagenesis and cost reduction studies." *Bioresource Technology* no. 98 (18):3464–3469. https://doi.org/10.1016/j.biortech.2006.11.007.

Lu, Sidan, Hongna Li, Guangcai Tan, Fang Wen, Michael T. Flynn, and Xiuping Zhu. 2019. "Resource recovery microbial fuel cells for urine-containing wastewater treatment without external energy consumption." *Chemical Engineering Journal* no. 373:1072–1080. https://doi.org/10.1016/j.cej.2019.05.130.

Lübeck, Mette, and Peter Stephensen Lübeck. 2019. "Application of lactic acid bacteria in green biorefineries." *FEMS Microbiology Letters* no. 366 (3). https://doi.org/10.1093/femsle/fnz024.

Lukitawesa. 2020. *Methane and Volatile Fatty Acids Production from Toxic Substrate*. Doctoral thesis, comprehensive summary, Skrifter från Högskolan i Borås, Högskolan i Borås, Borås.

Lynd, Lee R., Paul J. Weimer, Willem H. van Zyl, and Isak S. Pretorius. 2002. "Microbial cellulose utilization: fundamentals and biotechnology." *Microbiology and Molecular Biology Reviews* no. 66 (3):506. https://doi.org/10.1128/MMBR.66.3.506-577.2002.

Madigan, M. T., K. S. Bender, D. H. Buckley, W. M. Sattley, and D. A. Stahl. 2018. *Brock Biology of Microorganisms*: Pearson.

Madison, L. L., and G. W. Huisman. 1999. "Metabolic engineering of poly(3-hydroxyalkanoates): from DNA to plastic." *Microbiology and Molecular Biology Reviews: MMBR* no. 63 (1):21–53.

Mahboubi, Amir, Jorge A. Ferreira, Mohammad J. Taherzadeh, and Patrik R. Lennartsson. 2017a. "Production of fungal biomass for feed, fatty acids, and glycerol by Aspergillus oryzae from fat-rich dairy substrates." *Fermentation* no. 3 (4). https://doi.org/10.3390/fermentation3040048.

Mahboubi, Amir, Jorge A. Ferreira, Mohammad J. Taherzadeh, and Patrik R. Lennartsson. 2017b. "Value-added products from dairy waste using edible fungi." *Waste Management* no. 59:518–525. https://doi.org/10.1016/j.wasman.2016.11.017.

Majumder, Laboni, Ibrahim Khalil, M Kamruzzaman Munshi, Khorshed Alam, HO Rashid, R Begum, and N Alam. 2010. "Citric acid production by Aspergillus niger using molasses and pumpkin as substrates." *European Journal of Biological Sciences* no. 2 (1):1–8.

Maki, Miranda, Kam Tin Leung, and Wensheng Qin. 2009. "The prospects of cellulase-producing bacteria for the bioconversion of lignocellulosic biomass." *International Journal of Biological Sciences* no. 5 (5):500–516. https://doi.org/10.7150/ijbs.5.500.

Marlow Foods. 2022 [cited 06.06.2022Z. Available from https://www.quorn.co.uk/recipes.

Martín, Mariano, and Ignacio E. Grossmann. 2013. "Optimal engineered algae composition for the integrated simultaneous production of bioethanol and biodiesel." *AIChE Journal* no. 59 (8):2872–2883. https://doi.org/10.1002/aic.14071.

Maru, B. T., F. López, S. W. M. Kengen, M. Constantí, and F. Medina. 2016. "Dark fermentative hydrogen and ethanol production from biodiesel waste glycerol using a co-culture of Escherichia coli and Enterobacter sp." *Fuel* no. 186:375–384. https://doi.org/10.1016/j.fuel.2016.08.043.

Matijašić, Bojana Bogovič, Tanja Obermajer, Luka Lipoglavšek, Iztok Grabnar, Gorazd Avguštin, and Irena Rogelj. 2014. "Association of dietary type with fecal microbiota in vegetarians and omnivores in Slovenia." *European Journal of Nutrition* no. 53 (4):1051–1064. https://doi.org/10.1007/s00394-013-0607-6.

Mazzucotelli, C. A., I. Durruty, C. E. Kotlar, M. R. Moreira, A. G. Ponce, and S. I. Roura. 2014. "Development of a microbial consortium for dairy wastewater treatment." *Biotechnology and Bioprocess Engineering* no. 19 (2):221–230. https://doi.org/10.1007/s12257-013-0517-8.

Mehrotra, Tithi, Abhinav Srivastava, R Pragadeeshwara Rao, and Rachana Singh. 2019. "A novel immobilized bacterial consortium bioaugmented in a bioreactor for sustainable wastewater treatment." *Journal of Pure and Applied Microbiology* no. 13 (1):371–383.

Meixner, K., A. Kovalcik, E. Sykacek, M. Gruber-Brunhumer, W. Zeilinger, K. Markl, C. Haas, I. Fritz, N. Mundigler, F. Stelzer, M. Neureiter, W. Fuchs, and B. Drosg. 2018. "Cyanobacteria biorefinery: production of poly(3-hydroxybutyrate) with Synechocystis salina and utilisation of residual biomass." *Journal of Biotechnology* no. 265:46–53. https://doi.org/10.1016/j.jbiotec.2017.10.020.

Menendez, Esther, Paula Garcia-Fraile, and Raul Rivas. 2015. "Biotechnological applications of bacterial cellulases." *AIMS Bioengineering* no. 2 (3):163.

Millati, Ria, Lars Edebo, and Mohammad J. Taherzadeh. 2005. "Performance of Rhizopus, Rhizomucor, and Mucor in ethanol production from glucose, xylose, and wood hydrolyzates." *Enzyme and Microbial Technology* no. 36 (2):294–300. https://doi.org/10.1016/j.enzmictec.2004.09.007.

Mitsumori Michael, Tajima Kiyoshi, Itabashi Hisao. 2002. Detection of methanogens from the rumen by PCR-based techniques. In: Takahashi J, Young BA (eds), *Greenhouse Gases and Animal Agriculture,* pp. 125–128. Elsevier, Amsterdam.

Mohagheghi, Ali, Kent Evans, Yat-Chen Chou, and Min Zhang. 2002. "Cofermentation of glucose, xylose, and arabinose by genomic DNA-integrated xylose/arabinose fermenting strain of Zymomonas mobilis AX101." *Applied Biochemistry and Biotechnology* no. 98 (1):885–898. https://doi.org/10.1385/ABAB:98-100:1-9:885.

Moore, David, and Siu Wai Chiu. 2001. "Fungal products as food." In *Bio-Exploitation of Filamentous Fungi,* edited by S. B. Pointing and K. D. Hyde, 223–251: Fungal Diversity Press.

Moo-Young, M. 2011. *Comprehensive Biotechnology*: Elsevier Science.

Morrison, Jessica M, Chelsea L Murphy, Kristina Baker, Richard M Zamor, Steve J Nikolai, Shawn Wilder, Mostafa S Elshahed, and Noha H Youssef. 2017. "Microbial communities mediating algal detritus turnover under anaerobic conditions." *PeerJ* no. 5:e2803.

Mostafa, Yasser S., and Saad A. Alamri. 2012. "Optimization of date syrup for enhancement of the production of citric acid using immobilized cells of Aspergillus niger." *Saudi Journal of Biological Sciences* no. 19 (2):241–246. https://doi.org/10.1016/j.sjbs.2012.01.004.

Mousavi, S. M., S. A. Shojaosadati, J. Golestani, and F. Yazdian. 2010. "CFD simulation and optimization of effective parameters for biomass production in a horizontal tubular loop bioreactor." *Chemical Engineering and Processing: Process Intensification* no. 49 (12):1249–1258. https://doi.org/10.1016/j.cep.2010.09.013.

Muniraj, Iniya Kumar, Liwen Xiao, Zhenhu Hu, Xinmin Zhan, and Jianghong Shi. 2013. "Microbial lipid production from potato processing wastewater using oleaginous filamentous fungi Aspergillus oryzae." *Water Research* no. 47 (10):3477–3483. https://doi.org/10.1016/j.watres.2013.03.046.

Muñoz, Raúl, Leslie Meier, Israel Diaz, and David Jeison. 2015. "A review on the state-of-the-art of physical/chemical and biological technologies for biogas upgrading." *Reviews in Environmental Science and Bio/Technology* no. 14 (4):727–759. https://doi.org/10.1007/s11157-015-9379-1.

Murariu, Marius, and Philippe Dubois. 2016. "PLA composites: from production to properties." *Advanced Drug Delivery Reviews* no. 107:17–46. https://doi.org/10.1016/j.addr.2016.04.003.

Nair, Ramkumar B., Magnus Lundin, Patrik R. Lennartsson, and Mohammad J. Taherzadeh. 2017. "Optimizing dilute phosphoric acid pretreatment of wheat straw in the laboratory and in a demonstration plant for ethanol and edible fungal biomass production using Neurospora intermedia." *Journal of Chemical Technology & Biotechnology* no. 92 (6):1256–1265. https://doi.org/10.1002/jctb.5119.

Nair, Ramkumar B., and Mohammad J. Taherzadeh. 2016. "Valorization of sugar-to-ethanol process waste vinasse: a novel biorefinery approach using edible ascomycetes filamentous fungi." *Bioresource Technology* no. 221:469–476. https://doi.org/10.1016/j.biortech.2016.09.074.

Nakayama, Cristina Rossi, Eduardo Dellosso Penteado, Rubens Tadeu Delgado Duarte, Admir José Giachini, and Flávia Talarico Saia. 2019. "Improved methanogenic communities for biogas production." In *Improving Biogas Production: Technological Challenges, Alternative Sources, Future Developments*, edited by Helen Treichel and Gislaine Fongaro, 69–98: Springer International Publishing.

Nancib, Aicha, Nabil Nancib, and Joseph Boudrant. 2009. "Production of lactic acid from date juice extract with free cells of single and mixed cultures of Lactobacillus casei and Lactococcus lactis." *World Journal of Microbiology and Biotechnology* no. 25 (8):1423–1429. https://doi.org/10.1007/s11274-009-0029-z.

Nandy, Arpita, Vikash Kumar, and Patit P. Kundu. 2013. "Utilization of proteinaceous materials for power generation in a mediatorless microbial fuel cell by a new electrogenic bacteria Lysinibacillus sphaericus VA5." *Enzyme and Microbial Technology* no. 53 (5):339–344. https://doi.org/10.1016/j.enzmictec.2013.07.006.

Nath, A., M. Dixit, A. Bandiya, S. Chavda, and A. J. Desai. 2008. "Enhanced PHB production and scale up studies using cheese whey in fed batch culture of Methylobacterium sp. ZP24." *Bioresource Technology* no. 99 (13):5749–5755. https://doi.org/10.1016/j.biortech.2007.10.017.

Nauhaus, Katja, Antje Boetius, Martin Krüger, and Friedrich Widdel. 2002. "In vitro demonstration of anaerobic oxidation of methane coupled to sulphate reduction in sediment from a marine gas hydrate area." *Environmental Microbiology* no. 4 (5):296–305. https://doi.org/10.1046/j.1462-2920.2002.00299.x.

Numata, Keiji, Hideki Abe, and Tadahisa Iwata. 2009. "Biodegradability of Poly(hydroxyalkanoate) Materials." *Materials* no. 2 (3):1104–1126. https://doi.org/10.3390/ma2031104.

Ohta, K., D. S. Beall, J. P. Mejia, K. T. Shanmugam, and L. O. Ingram. 1991. "Genetic improvement of Escherichia coli for ethanol production: chromosomal integration of Zymomonas mobilis genes encoding pyruvate decarboxylase and alcohol dehydrogenase II." *Applied and Environmental Microbiology* no. 57 (4):893.

Okafor, N. 2007. *Modern Industrial Microbiology and Biotechnology*: Science Publishers.

Oliva-Rodríguez, Alejandra G., Julián Quintero, Miguel A. Medina-Morales, Thelma K. Morales-Martínez, José A. Rodríguez-De la Garza, Mayela Moreno-Dávila, Germán Aroca, and Leopoldo J. Rios González. 2019. "Clostridium strain selection for co-culture with Bacillus subtilis for butanol production from agave hydrolysates." *Bioresource Technology* no. 275:410–415. https://doi.org/10.1016/j.biortech.2018.12.085.

Oliver, J. D. 2005. "The viable but nonculturable state in bacteria." *Journal of Microbiology* no. 43 Spec No:93–100.

Olson, Daniel G., Richard Sparling, and Lee R. Lynd. 2015. "Ethanol production by engineered thermophiles." *Current Opinion in Biotechnology* no. 33:130–141. https://doi.org/10.1016/j.copbio.2015.02.006.

Orcutt, B., and C. Meile. 2008. "Constraints on mechanisms and rates of anaerobic oxidation of methane by microbial consortia: process-based modeling of ANME-2 archaea and sulfate reducing bacteria interactions." *Biogeosciences Discussions* no. 5 (3):1933–1967.

Paerl, H. W., and J. L. Pinckney. 1996. "A mini-review of microbial consortia: their roles in aquatic production and biogeochemical cycling." *Microbial Ecology* no. 31 (3):225–247. https://doi.org/10.1007/BF00171569.

Pampillón-González, L., N. L. Ortiz-Cornejo, M. Luna-Guido, L. Dendooven, and Y. E. Navarro-Noya. 2017. "Archaeal and bacterial community structure in an anaerobic digestion reactor (lagoon type) used for biogas production at a pig farm." *Microbial Physiology* no. 27 (5):306–317. https://doi.org/10.1159/000479108.

Pandey, A., R. Höfer, M. Taherzadeh, M. Nampoothiri, and C. Larroche. 2015. *Industrial Biorefineries and White Biotechnology*: Elsevier Science.

Pandey, A., S. Negi, and C. R. Soccol. 2016. *Current Developments in Biotechnology and Bioengineering: Production, Isolation and Purification of Industrial Products*: Elsevier Science.

Pandey, A., and J. A. C. Teixeira. 2016. *Current Developments in Biotechnology and Bioengineering: Foundations of Biotechnology and Bioengineering*: Elsevier Science.

Park, Miso, Qing Sun, Fang Liu, Matthew P. DeLisa, and Wilfred Chen. 2014. "Positional assembly of enzymes on bacterial outer membrane vesicles for cascade reactions." *PLoS ONE* no. 9 (5):e97103. https://doi.org/10.1371/journal.pone.0097103.

Pepi, Milva, Lucia R. Lampariello, Roberto Altieri, Alessandro Esposito, Guido Perra, Monia Renzi, Arianna Lobianco, Antonio Feola, Simone Gasperini, and Silvano E. Focardi. 2010. "Tannic acid degradation by bacterial strains Serratia spp. and Pantoea sp. isolated from olive mill waste mixtures." *International Biodeterioration & Biodegradation* no. 64 (1):73–80. https://doi.org/10.1016/j.ibiod.2009.10.009.

Pérez-Rangel, Marisol, Francisco R. Quiroz-Figueroa, Jaquelina González-Castañeda, and Idania Valdez-Vazquez. 2015. "Microscopic analysis of wheat straw cell wall degradation by microbial consortia for hydrogen production." *International Journal of Hydrogen Energy* no. 40 (1):151–160. https://doi.org/10.1016/j.ijhydene.2014.10.050.

Petrini, Serena, Paola Foladori, Lorenza Donati, and Gianni Andreottola. 2020. "Comprehensive respirometric approach to assess photosynthetic, heterotrophic and nitrifying activity in microalgal-bacterial consortia treating real municipal wastewater." *Biochemical Engineering Journal* no. 161:107697. https://doi.org/10.1016/j.bej.2020.107697.

Pham, Hai, Nico Boon, Massimo Marzorati, and Willy Verstraete. 2009. "Enhanced removal of 1,2-dichloroethane by anodophilic microbial consortia." *Water Research* no. 43 (11):2936–2946. https://doi.org/10.1016/j.watres.2009.04.004.

Pleissner, Daniel, Anna-Katrin Neu, Kerstin Mehlmann, Roland Schneider, Gloria Inés Puerta-Quintero, and Joachim Venus. 2016. "Fermentative lactic acid production from coffee pulp hydrolysate using Bacillus coagulans at laboratory and pilot scales." *Bioresource Technology* no. 218:167–173. https://doi.org/10.1016/j.biortech.2016.06.078.

Pokarzhevskii, Andrei D., Nico M. van Straalen, Dmitrii P. Zaboev, and Andrei S. Zaitsev. 2003. "Microbial links and element flows in nested detrital food-webs." *Pedobiologia* no. 47 (3):213–224. https://doi.org/10.1078/0031-4056-00185.

Pompei, Anna, Lisa Cordisco, Alberto Amaretti, Simona Zanoni, Diego Matteuzzi, and Maddalena Rossi. 2007. "Folate production by bifidobacteria as a potential probiotic property." *Applied and Environmental Microbiology* no. 73 (1):179. https://doi.org/10.1128/AEM.01763-06.

Prasad, Shashi Bala, Guhan Jayaraman, and K. B. Ramachandran. 2010. "Hyaluronic acid production is enhanced by the additional co-expression of UDP-glucose pyrophosphorylase in Lactococcus lactis." *Applied Microbiology and Biotechnology* no. 86 (1):273–283. https://doi.org/10.1007/s00253-009-2293-0.

Prieto-Barajas, Cristina M., Eduardo Valencia-Cantero, and Gustavo Santoyo. 2018. "Microbial mat ecosystems: structure types, functional diversity, and biotechnological application." *Electronic Journal of Biotechnology* no. 31:48–56. https://doi.org/10.1016/j.ejbt.2017.11.001.

Prokhorova, Anna, Katrin Sturm-Richter, Andreas Doetsch, and Johannes Gescher. 2017. "Resilience, dynamics, and interactions within a model multispecies exoelectrogenic-biofilm community." *Applied and Environmental Microbiology* no. 83 (6):e03033–16. https://doi.org/10.1128/aem.03033-16.

Qadir, Faizan, Maria Shariq, Asia Ahmed, and Muhammad Sohail. 2018. "Evaluation of a yeast co-culture for cellulase and xylanase production under solid state fermentation of sugarcane bagasse using multivariate approach." *Industrial Crops and Products* no. 123:407–415. https://doi.org/10.1016/j.indcrop.2018.07.021.

Qiao, Kangjian, Syed Hussain Imam Abidi, Hongjuan Liu, Haoran Zhang, Sagar Chakraborty, Nicki Watson, Parayil Kumaran Ajikumar, and Gregory Stephanopoulos. 2015. "Engineering lipid overproduction in the oleaginous yeast Yarrowia lipolytica." *Metabolic Engineering* no. 29:56–65. https://doi.org/10.1016/j.ymben.2015.02.005.

Qiao, Kangjian, Thomas M. Wasylenko, Kang Zhou, Peng Xu, and Gregory Stephanopoulos. 2017. "Lipid production in Yarrowia lipolytica is maximized by engineering cytosolic redox metabolism." *Nature Biotechnology* no. 35 (2):173–177. https://doi.org/10.1038/nbt.3763.

Qu, Youpeng, Yujie Feng, Xin Wang, and Bruce E. Logan. 2012. "Use of a coculture to enable current production by Geobacter sulfurreducens." *Applied and Environmental Microbiology* no. 78 (9):3484–3487. https://doi.org/10.1128/aem.00073-12.

RamKumar Pandian, Sureshbabu, Venkatraman Deepak, Kalimuthu Kalishwaralal, Neelamegam Rameshkumar, Muniyandi Jeyaraj, and Sangiliyandi Gurunathan. 2010. "Optimization and fed-batch production of PHB utilizing dairy waste and sea water as nutrient sources by Bacillus megaterium SRKP-3." *Bioresource Technology* no. 101 (2):705–711. https://doi.org/10.1016/j.biortech.2009.08.040.

Ratledge, Colin. 2010. "Single cell oils for the 21st century." In *Single Cell Oils* (Second Edition), edited by Zvi Cohen and Colin Ratledge, 3–26: AOCS Press.

Rauch, Reinhard, Jitka Hrbek, and Hermann Hofbauer. 2014. "Biomass gasification for synthesis gas production and applications of the syngas." *WIREs Energy and Environment* no. 3 (4):343–362. https://doi.org/10.1002/wene.97.

Ronan, Patrick, C. William Yeung, John Schellenberg, Richard Sparling, Gideon M. Wolfaardt, and Martina Hausner. 2013. "A versatile and robust aerotolerant microbial community capable of cellulosic ethanol production." *Bioresource Technology* no. 129:156–163. https://doi.org/10.1016/j.biortech.2012.10.164.

Rossi, Daniele Misturini, Janaína Berne da Costa, Elisangela Aquino de Souza, Maria do Carmo Ruaro Peralba, Dimitrios Samios, and Marco Antônio Záchia Ayub. 2011. "Comparison of different pretreatment methods for hydrogen production using environmental microbial consortia on residual glycerol from biodiesel." *International Journal of Hydrogen Energy* no. 36 (8):4814–4819. https://doi.org/10.1016/j.ijhydene.2011.01.005.

Russell, D. L. 2019. *Practical Wastewater Treatment*: Wiley.

Ryabova, Olena B., Oksana M. Chmil, and Andrii A. Sibirny. 2003. "Xylose and cellobiose fermentation to ethanol by the thermotolerant methylotrophic yeast Hansenula polymorpha." *FEMS Yeast Research* no. 4 (2):157–164. https://doi.org/10.1016/S1567-1356(03)00146-6.

Sahoo, Tridweep K., and Guhan Jayaraman. 2019. "Co-culture of Lactobacillus delbrueckii and engineered Lactococcus lactis enhances stoichiometric yield of d-lactic acid from whey permeate." *Applied Microbiology and Biotechnology* no. 103 (14):5653–5662. https://doi.org/10.1007/s00253-019-09819-7.

Sanny, Tony, Marina Arnaldos, Stephanie A. Kunkel, Krishna R. Pagilla, and Benjamin C. Stark. 2010. "Engineering of ethanolic E. coli with the Vitreoscilla hemoglobin gene enhances ethanol production from both glucose and xylose." *Applied Microbiology and Biotechnology* no. 88 (5):1103–1112. https://doi.org/10.1007/s00253-010-2817-7.

Santamaría-Fernández, M., B. Molinuevo-Salces, P. Kiel, S. Steenfeldt, H. Uellendahl, and M. Lübeck. 2017. "Lactic acid fermentation for refining proteins from green crops and obtaining a high quality feed product for monogastric animals." *Journal of Cleaner Production* no. 162:875–881. https://doi.org/10.1016/j.jclepro.2017.06.115.

Sar, Taner, and Meltem Yesilcimen Akbas. 2022. "Potential use of olive oil mill wastewater for bacterial cellulose production." *Bioengineered* no. 13 (3):7659–7669. https://doi.org/10.1080/21655979.2022Z.2050492.

Sar, Taner, Jorge A. Ferreira, and Mohammad J. Taherzadeh. 2020a. "Bioprocessing strategies to increase the protein fraction of Rhizopus oryzae biomass using fish industry sidestreams." *Waste Management* no. 113:261–269. https://doi.org/10.1016/j.wasman.2020.06.005.

Sar, Taner, Jorge A. Ferreira, and Mohammad J. Taherzadeh. 2020b. "Conversion of fish processing wastewater into fish feed ingredients through submerged cultivation of Aspergillus oryzae." *Systems Microbiology and Biomanufacturing.* https://doi.org/10.1007/s43393-020-00009-5.

Sar, Taner, Karin Larsson, Rikard Fristedt, Ingrid Undeland, and Mohammad J. Taherzadeh. 2022. "Demo-scale production of protein-rich fungal biomass from potato protein liquor for use as innovative food and feed products." *Food Bioscience* no. 47:101637.

Sar, Taner, Murat Ozturk, Mohammad J. Taherzadeh, and Jorge A. Ferreira. 2020. "New insights on protein recovery from olive oil mill wastewater through bioconversion with edible filamentous fungi." *Processes* no. 8 (10). https://doi.org/10.3390/pr8101210.

Sar, Taner, Gamze Seker, Ayse Gokce Erman, Benjamin C. Stark, and Meltem Yesilcimen Akbas. 2017. "Repeated batch fermentation of immobilized E. coli expressing Vitreoscilla hemoglobin for long-term use." *Bioengineered* no. 8 (5):651–660. https://doi.org/10.1080/21655979.2017.1303024.

Sar, Taner, Benjamin C. Stark, and Meltem Yesilcimen Akbas. 2017. "Effective ethanol production from whey powder through immobilized E. coli expressing Vitreoscilla hemoglobin." *Bioengineered* no. 8 (2):171–181. https://doi.org/10.1080/21655979.2016.1218581.

Sar, Taner, Benjamin C. Stark, and Meltem Yesilcimen Akbas. 2019. "Bioethanol production from whey powder by immobilized E. coli expressing Vitreoscilla hemoglobin: optimization of sugar concentration and inoculum size." *Biofuels*.1–6. https://doi.org/10.1080/17597269.2019.1583716.

Sashiwa, H., R. Fukuda, T. Okura, S. Sato, and A. Nakayama. 2018. "Microbial degradation behavior in seawater of polyester blends containing poly(3-hydroxybutyrate-co-3-hydroxyhexanoate) (PHBHHx)." *Marine Drugs* no. 16 (1). https://doi.org/10.3390/md16010034.

Satari, Behzad, Keikhosro Karimi, Mohammad J. Taherzadeh, and Akram Zamani. 2016. "Co-production of fungal biomass derived constituents and ethanol from citrus wastes free sugars without auxiliary nutrients in airlift bioreactor." *International Journal of Molecular Sciences* no. 17 (3). https://doi.org/10.3390/ijms17030302.

Satyanarayana, T., and G. Kunze. 2009. *Yeast Biotechnology: Diversity and Applications*: Springer Netherlands.

Segura, D., J. Guzmán, and G. Espín. 2003. "Azotobacter vinelandii mutants that overproduce poly-β-hydroxybutyrate or alginate." *Applied Microbiology and Biotechnology* no. 63 (2):159–163. https://doi.org/10.1007/s00253-003-1397-1.

Senghor, Bruno, Cheikh Sokhna, Raymond Ruimy, and Jean-Christophe Lagier. 2018. "Gut microbiota diversity according to dietary habits and geographical provenance." *Human Microbiome Journal* no. 7–8:1–9. https://doi.org/10.1016/j.humic.2018.01.001.

Shahryari, Zohre, Mohammad H. Fazaelipoor, Younes Ghasemi, Patrik R. Lennartsson, and Mohammad J. Taherzadeh. 2019. "Amylase and xylanase from edible fungus neurospora intermedia: production and characterization." *Molecules* no. 24 (4):721.

Sharma, Jyoti, Smita S. Kumar, Narsi R. Bishnoi, and Arivalagan Pugazhendhi. 2018. "Enhancement of lipid production from algal biomass through various growth parameters." *Journal of Molecular Liquids* no. 269:712–720. https://doi.org/10.1016/j.molliq.2018.08.103.

Shen, Maocai, Yaxin Zhang, Duofei Hu, Jinshi Fan, and Guangming Zeng. 2018. "A review on removal of siloxanes from biogas: with a special focus on volatile methylsiloxanes." *Environmental Science and Pollution Research* no. 25 (31):30847–30862. https://doi.org/10.1007/s11356-018-3000-4.

Shigematsu, Toru, Kazunori Takamine, Masaya Kitazato, Tetsuya Morita, Takaaki Naritomi, Shigeru Morimura, and Kenji Kida. 2005. "Cellulose production from glucose using a glucose dehydrogenase gene (gdh)-deficient mutant of Gluconacetobacter xylinus and its use for bioconversion of sweet potato pulp." *Journal of Bioscience and Bioengineering* no. 99 (4):415–422. https://doi.org/10.1263/jbb.99.415.

Shui, Zong-Xia, Han Qin, Bo Wu, Zhi-yong Ruan, Lu-shang Wang, Fu-Rong Tan, Jing-Li Wang, Xiao-Yu Tang, Li-Chun Dai, Guo-Quan Hu, and Ming-Xiong He. 2015. "Adaptive laboratory evolution of ethanologenic Zymomonas mobilis strain tolerant to furfural and acetic acid inhibitors." *Applied Microbiology and Biotechnology* no. 99 (13):5739–5748. https://doi.org/10.1007/s00253-015-6616-z.

Siegal, Mark L. 2015. "Shifting sugars and shifting paradigms." *PLOS Biology* no. 13 (2):e1002068. https://doi.org/10.1371/journal.pbio.1002068.

Silva, Cynthia C., Helen Hayden, Tim Sawbridge, Pauline Mele, Ricardo H. Kruger, Marili V. N. Rodrigues, Gustavo G. L. Costa, Ramon O. Vidal, Maíra P. Sousa, Ana Paula R. Torres, Vânia M. J. Santiago, and Valéria M. Oliveira. 2012. "Phylogenetic and functional diversity of metagenomic libraries of phenol degrading sludge from petroleum refinery wastewater treatment system." *AMB Express* no. 2 (1):18. https://doi.org/10.1186/2191-0855-2-18.

Souza Filho, Pedro F., Dan Andersson, Jorge A. Ferreira, and Mohammad J. Taherzadeh. 2019. "Mycoprotein: environmental impact and health aspects." *World Journal of Microbiology and Biotechnology* no. 35 (10):147. https://doi.org/10.1007/s11274-019-2723-9.

Souza Filho, Pedro F., Akram Zamani, and Mohammad J. Taherzadeh. 2017. "Production of edible fungi from potato protein liquor (PPL) in airlift bioreactor." *Fermentation* no. 3 (1):12.

Souza Filho, Pedro F., Akram Zamani, and Mohammad J. Taherzadeh. 2019. "Edible protein production by filamentous fungi using starch plant wastewater." *Waste and Biomass Valorization* no. 10 (9):2487–2496. https://doi.org/10.1007/s12649-018-0265-2.

Stark, Benjamin C., Kanak L. Dikshit, and Krishna R. Pagilla. 2012. "The biochemistry of vitreoscilla hemoglobin." *Computational and Structural Biotechnology Journal* no. 3 (4):e201210002. https://doi.org/10.5936/csbj.201210002.

Stark, Benjamin C., Krishna R. Pagilla, and Kanak L. Dikshit. 2015. "Recent applications of Vitreoscilla hemoglobin technology in bioproduct synthesis and bioremediation." *Applied Microbiology and Biotechnology* no. 99 (4):1627–1636. https://doi.org/10.1007/s00253-014-6350-y.

Stewart, Graham G. 2016. "Saccharomyces species in the Production of Beer." *Beverages* no. 2 (4). https://doi.org/10.3390/beverages2040034.

Stodolak, Bożena, Anna Starzyńska-Janiszewska, and Małgorzata Bączkowicz. 2020. "Aspergillus oryzae (Koji Mold) and Neurospora intermedia (Oncom Mold) application for flaxseed oil cake processing." *LWT* no. 131:109651. https://doi.org/10.1016/j.lwt.2020.109651.

Sumer, Fatma, Benjamin C. Stark, and Meltem Yesilcimen Akbas. 2015. "Efficient ethanol production from potato and corn processing industry waste using E. coli engineered to express Vitreoscilla haemoglobin." *Environmental Technology* no. 36 (18):2319–2327. https://doi.org/10.1080/09593330.2015.1026846.

Sun, Li, Phillip B. Pope, Vincent G. H. Eijsink, and Anna Schnürer. 2015. "Characterization of microbial community structure during continuous anaerobic digestion of straw and cow manure." *Microbial biotechnology* no. 8 (5):815–827. https://doi.org/10.1111/1751-7915.12298.

Swift, Graham. 1992. "Biodegradability of polymers in the environment: complexities and significance of definitions and measurements." *FEMS Microbiology Letters* no. 103 (2):339–345. https://doi.org/10.1016/0378-1097(92)90328-L.

Tamboli, Dhawal P., Mayur B. Kurade, Tatoba R. Waghmode, Swati M. Joshi, and Sanjay P. Govindwar. 2010. "Exploring the ability of Sphingobacterium sp. ATM to degrade textile dye Direct Blue GLL, mixture of dyes and textile effluent and production of polyhydroxyhexadecanoic acid using waste biomass generated after dye degradation." *Journal of Hazardous Materials* no. 182 (1):169–176. https://doi.org/10.1016/j.jhazmat.2010.06.011.

Tao, Kaiyun, Xiaoyan Liu, Xueping Chen, Xiaoxin Hu, Liya Cao, and Xiaoyu Yuan. 2017. "Biodegradation of crude oil by a defined co-culture of indigenous bacterial consortium and exogenous Bacillus subtilis." *Bioresource Technology* no. 224:327–332. https://doi.org/10.1016/j.biortech.2016.10.073.

Task, IEA Bioenergy. 2018. "37. Green Gas-Facilitating a future green gas grid through the production of renewable gas." *International Energy Agency.*

Thomas, D. S., and R. R. Davenport. 1985. "Zygosaccharomyces bailii: a profile of characteristics and spoilage activities." *Food Microbiology* no. 2 (2):157–169. https://doi.org/10.1016/S0740-0020(85)80008-3.

Thunuguntla, Rahul, Amir Mahboubi, Jorge A. Ferreira, and Mohammad J. Taherzadeh. 2018. "Integration of membrane bioreactors with edible filamentous fungi for valorization of expired milk." *Sustainability* no. 10 (6). https://doi.org/10.3390/su10061940.

Toivari, Mervi H., Aristos Aristidou, Laura Ruohonen, and Merja Penttilä. 2001. "Conversion of xylose to ethanol by recombinant saccharomyces cerevisiae: importance of xylulokinase (XKS1) and oxygen availability." *Metabolic Engineering* no. 3 (3):236–249. https://doi.org/10.1006/mben.2000.0191.

Tomás, Ana Faria, Pınar Karagöz, Dimitar Karakashev, and Irini Angelidaki. 2013. "Extreme thermophilic ethanol production from rapeseed straw: using the newly isolated Thermoanaerobacter pentosaceus and combining it with Saccharomyces cerevisiae in a two-step process." *Biotechnology and Bioengineering* no. 110 (6):1574–1582. https://doi.org/10.1002/bit.24813.

Tony, Bella Devassy, Dinesh Goyal, and Sunil Khanna. 2009. "Decolorization of textile azo dyes by aerobic bacterial consortium." *International Biodeterioration & Biodegradation* no. 63 (4):462–469. https://doi.org/10.1016/j.ibiod.2009.01.003.

Tsai, Shen-Long, Jeongseok Oh, Shailendra Singh, Ruizhen Chen, and Wilfred Chen. 2009. "Functional assembly of minicellulosomes on the Saccharomyces cerevisiae cell surface for cellulose hydrolysis and ethanol production." *Applied and Environmental Microbiology* no. 75 (19):6087. https://doi.org/10.1128/AEM.01538-09.

Ullah, Hanif, Hélder A. Santos, and Taous Khan. 2016. "Applications of bacterial cellulose in food, cosmetics and drug delivery." *Cellulose* no. 23 (4):2291–2314. https://doi.org/10.1007/s10570-016-0986-y.

Valdez-Vazquez, Idania, Marisol Pérez-Rangel, Adán Tapia, Germán Buitrón, Carlos Molina, Gustavo Hernández, and Lorena Amaya-Delgado. 2015. "Hydrogen and butanol production from native wheat straw by synthetic microbial consortia integrated by species of Enterococcus and Clostridium." *Fuel* no. 159:214–222. https://doi.org/10.1016/j.fuel.2015.06.052.

Van Soest, Peter J. 2018. *Nutritional Ecology of the Ruminant*: Cornell University Press.

Vargas, Gustavo, Andrés Donoso-Bravo, Christian Vergara, and Gonzalo Ruiz-Filippi. 2016. "Assessment of microalgae and nitrifiers activity in a consortium in a continuous operation and the effect of oxygen depletion." *Electronic Journal of Biotechnology* no. 23:63–68. https://doi.org/10.1016/j.ejbt.2016.08.002.

Varjani, S., A. Pandey, E. Gnansounou, S. K. Khanal, and S. Raveendran. 2020. *Current Developments in Biotechnology and Bioengineering: Resource Recovery from Wastes*: Elsevier Science.

Venugopal, Sivaranjani, Lawrance Irudayarajan, Yasmin Khambhaty, and Saravanan Palanivel. 2019. "Batch experiments towards remediation of phenolic syntan using individual as well as co-culture of Bacillus cereus and Pseudomonas aeruginosa." *World Journal of Microbiology and Biotechnology* no. 35 (9):137. https://doi.org/10.1007/s11274-019-2683-0.

Venus, J. 2006. "Utilization of renewables for lactic acid fermentation." *Biotechnology Journal* no. 1 (12):1428–1432. https://doi.org/10.1002/biot.200600180.

Viitanen, Paul V., Carol M. Mc Cutchen, Xu Li, Mark Emptage, Perry G. Caimi, Min Zhang, Yat-Chen Chou, and Mary Ann Franden. 2009. *Ethanol production in fermentation of mixed sugars containing xylose*. Google Patents.

Wang, Bin, Xianglong Li, Bin Luo, Jingxuan Yang, Xiangjun Wang, Qi Song, Shiyan Chen, and Linjie Zhi. 2013. "Pyrolyzed bacterial cellulose: a versatile support for lithium ion battery anode materials." *Small* no. 9 (14):2399–2404. https://doi.org/10.1002/smll.201300692.

Wang, Jing, Javad Tavakoli, and Youhong Tang. 2019. "Bacterial cellulose production, properties and applications with different culture methods: a review." *Carbohydrate Polymers* no. 219:63–76. https://doi.org/10.1016/j.carbpol.2019.05.008.

Wang, Ruimin, Yuan Tian, Shengzhang Xue, Dongmei Zhang, Qinghua Zhang, Xia Wu, Dezhu Kong, and Wei Cong. 2016. "Enhanced microalgal biomass and lipid production via co-culture of Scenedesmus obliquus and Candida tropicalis in an autotrophic system." *Journal of Chemical Technology & Biotechnology* no. 91 (5):1387–1396. https://doi.org/10.1002/jctb.4735.

Weiland, Peter. 2010. "Biogas production: current state and perspectives." *Applied Microbiology and Biotechnology* no. 85 (4):849–860. https://doi.org/10.1007/s00253-009-2246-7.

Wen, Fei, Jie Sun, and Huimin Zhao. 2010. "Yeast surface display of trifunctional minicellulosomes for simultaneous saccharification and fermentation of cellulose to ethanol." *Applied and Environmental Microbiology* no. 76 (4):1251. https://doi.org/10.1128/AEM.01687-09.

Widner, Bill, Régine Behr, Steve Von Dollen, Maria Tang, Tia Heu, Alan Sloma, Dave Sternberg, Paul L. DeAngelis, Paul H. Weigel, and Steve Brown. 2005. "Hyaluronic acid production in Bacillus subtilis." *Applied and Environmental Microbiology* no. 71 (7):3747. https://doi.org/10.1128/AEM.71.7.3747-3752.2005.

Wong, J. M., R. de Souza, C. W. Kendall, A. Emam, and D. J. Jenkins. 2006. "Colonic health: fermentation and short chain fatty acids." *Journal of Clinical Gastroenterology* no. 40 (3):235–243. https://doi.org/10.1097/00004836-200603000-00015.

Wong, J. W. C., R. D. Tyagi, and A. Pandey. 2016. *Current Developments in Biotechnology and Bioengineering: Solid Waste Management*: Elsevier Science.

Wongwilaiwalin, Sarunyou, Ukrit Rattanachomsri, Thanaporn Laothanachareon, Lily Eurwilaichitr, Yasuo Igarashi, and Verawat Champreda. 2010. "Analysis of a thermophilic lignocellulose degrading microbial consortium and multi-species lignocellulolytic enzyme system." *Enzyme and Microbial Technology* no. 47 (6):283–290. https://doi.org/10.1016/j.enzmictec.2010.07.013.

Woolverton, Christopher J., L. Sherwood, and J. Willey. 2016. *Prescott's Microbiology*: McGraw-Hill Education.

Wu, Hong, Zhongyun Fan, Xiaoran Jiang, Jinchun Chen, and Guo-Qiang Chen. 2016. "Enhanced production of polyhydroxybutyrate by multiple dividing E. coli." *Microbial Cell Factories* no. 15 (1):128. https://doi.org/10.1186/s12934-016-0531-6.

Wu, Linwei, Daliang Ning, Bing Zhang, Yong Li, Ping Zhang, Xiaoyu Shan, Qiuting Zhang, and et al. 2019. "Global diversity and biogeography of bacterial communities in wastewater treatment plants." *Nature Microbiology* no. 4 (7):1183–1195. https://doi.org/10.1038/s41564-019-0426-5.

Xu, L., and U. Tschirner. 2011. "Improved ethanol production from various carbohydrates through anaerobic thermophilic co-culture." *Bioresources Technology* no. 102 (21):10065–10071. https://doi.org/10.1016/j.biortech.2011.08.067.

Xu, Lei, and Ulrike Tschirner. 2014. "Immobilized anaerobic fermentation for bio-fuel production by Clostridium co-culture." *Bioprocess and Biosystems Engineering* no. 37 (8):1551–1559. https://doi.org/10.1007/s00449-014-1127-3.

Xu, Lili, Xianglong Cheng, and Quanxi Wang. 2018. "Enhanced lipid production in chlamydomonas reinhardtii by co-culturing with Azotobacter chroococcum." *Frontiers in Plant Science* no. 9 (741). https://doi.org/10.3389/fpls.2018.00741.

Xue, Feiyan, Bin Gao, Yongqiang Zhu, Xu Zhang, Wei Feng, and Tianwei Tan. 2010. "Pilot-scale production of microbial lipid using starch wastewater as raw material." *Bioresource Technology* no. 101 (15):6092–6095. https://doi.org/10.1016/j.biortech.2010.01.124.

Yamada, Ryosuke, Naho Taniguchi, Tsutomu Tanaka, Chiaki Ogino, Hideki Fukuda, and Akihiko Kondo. 2011. "Direct ethanol production from cellulosic materials using a diploid strain of Saccharomyces cerevisiae with optimized cellulase expression." *Biotechnology for Biofuels* no. 4 (1):8. https://doi.org/10.1186/1754-6834-4-8.

Yamano, Naoko, Atsuyoshi Nakayama, Norioki Kawasaki, Noboru Yamamoto, and Seiichi Aiba. 2008. "Mechanism and characterization of polyamide 4 degradation by Pseudomonas sp." *Journal of Polymers and the Environment* no. 16 (2):141–146. https://doi.org/10.1007/s10924-008-0090-y.

Yan, Qiang, and Brian F. Pfleger. 2020. "Revisiting metabolic engineering strategies for microbial synthesis of oleochemicals." *Metabolic Engineering* no. 58:35–46. https://doi.org/10.1016/j.ymben.2019.04.009.

Yazdian, F., M. Pesaran Hajiabbas, S. A. Shojaosadati, M. Nosrati, E. Vasheghani-Farahani, and M. R. Mehrnia. 2010. "Study of hydrodynamics, mass transfer, energy consumption, and biomass production from natural gas in a forced-liquid vertical tubular loop bioreactor." *Biochemical Engineering Journal* no. 49 (2):192–200. https://doi.org/10.1016/j.bej.2009.12.013.

Yazdian, Fatemeh, Solmaz Hajizadeh, Seyed Abbas Shojaosadati, Rasoul Khalilzadeh, Mohsen Jahanshahi, and Mohsen Nosrati. 2005. "Production of single cell protein from natural gas: parameter optimization and RNA evaluation." *Iranian Journal of Biotechnology* no. 3 (4):235–242.

Yeager, P. E., and R. L. Sinsabaugh. 1998. "Microbial diversity along a sediment detrital particle size gradient." *Aquatic Ecology* no. 32 (4):281–289. https://doi.org/10.1023/A:1009944328443.

Yoshida, Masaki, Yuuhiko Tanabe, Natsuki Yonezawa, and Makoto M. Watanabe. 2012. "Energy innovation potential of oleaginous microalgae." *Biofuels* no. 3 (6):761–781. https://doi.org/10.4155/bfs.12.63.

Yoshida, Shogo, Kenji Okano, Tsutomu Tanaka, Chiaki Ogino, and Akihiko Kondo. 2011. "Homo-d-lactic acid production from mixed sugars using xylose-assimilating operon-integrated Lactobacillus plantarum." *Applied Microbiology and Biotechnology* no. 92 (1):67–76. https://doi.org/10.1007/s00253-011-3356-6.

Yu, Z. T., C. Chen, and D. E. Kling. 2012. "The principal fucosylated oligosaccharides of human milk oligosaccharides of human milk exhibit prebiotic properties on cultured infant microbiota." *Glycobiology* no. 23:169–177.

Zafar, Muddassar, Fatima Arshad, Seemab Faizi, Zahid Anwar, Muhammad Imran, and Raja Tahir Mehmood. 2020. "HPLC based characterization of citric acid produced from indigenous fungal strain through single and co-culture fermentation." *Biocatalysis and Agricultural Biotechnology* no. 29:101796. https://doi.org/10.1016/j.bcab.2020.101796.

Zang, X., M. Liu, H. Wang, Y. Fan, H. Zhang, J. Liu, E. Xing, X. Xu, and H. Li. 2017. "The distribution of active β-glucosidase-producing microbial communities in composting." *Canadian Journal of Microbiology* no. 63 (12):998–1008. https://doi.org/10.1139/cjm-2017-0368.

Zeidan, Ahmad A., and Ed W. J. Van Niel. 2009. "Developing a thermophilic hydrogen-producing co-culture for efficient utilization of mixed sugars." *International Journal of Hydrogen Energy* no. 34 (10):4524–4528. https://doi.org/10.1016/j.ijhydene.2008.07.092.

Zengler, K. 2008. *Accessing Uncultivated Microorganisms: From the Environment to Organisms and Genomes and Back*: ASM Press.

Zhang, Bing, Wei Li, Yuan Guo, Zhiqiang Zhang, Wenxin Shi, Fuyi Cui, Piet N. L. Lens, and Joo Hwa Tay. 2020. "Microalgal-bacterial consortia: from interspecies interactions to biotechnological applications." *Renewable and Sustainable Energy Reviews* no. 118:109563. https://doi.org/10.1016/j.rser.2019.109563.

Zhang, Bo, Xiangyang Xu, and Liang Zhu. 2017. "Structure and function of the microbial consortia of activated sludge in typical municipal wastewater treatment plants in winter." *Scientific Reports* no. 7 (1):17930. https://doi.org/10.1038/s41598-017-17743-x.

Zhang, Min, Yat-Chen Chou, Stephen K. Picataggio, and Mark Finkelstein. 1998. *Single Zymomonas mobilis strain for xylose and arabinose fermentation*. Google Patents.

Zhang, Ming-Ming, Liang Xiong, Ya-Jie Tang, Muhammad Aamer Mehmood, Zongbao Kent Zhao, Feng-Wu Bai, and Xin-Qing Zhao. 2019. "Enhanced acetic acid stress tolerance and ethanol production in Saccharomyces cerevisiae by modulating expression of the de novo purine biosynthesis genes." *Biotechnology for Biofuels* no. 12 (1):116. https://doi.org/10.1186/s13068-019-1456-1.

Zhang, Yixing, Praveen V. Vadlani, Amit Kumar, Philip R. Hardwidge, Revathi Govind, Tsutomu Tanaka, and Akihiko Kondo. 2016. "Enhanced D-lactic acid production from renewable resources using engineered Lactobacillus plantarum." *Applied Microbiology and Biotechnology* no. 100 (1):279–288.

Zhou, Kang, Kangjian Qiao, Steven Edgar, and Gregory Stephanopoulos. 2015. "Distributing a metabolic pathway among a microbial consortium enhances production of natural products." *Nature Biotechnology* no. 33 (4):377–383. https://doi.org/10.1038/nbt.3095.

Zou, J., X. Guo, T. Shen, J. Dong, C. Zhang, and D. Xiao. 2013. "Construction of lactose-consuming Saccharomyces cerevisiae for lactose fermentation into ethanol fuel." *Journal of Industrial Microbiology and Biotechnology* no. 40 (3–4):353–363. https://doi.org/10.1007/s10295-012-1227-5.

3 Biorefining Lignocellulosic Feedstocks for Microbial Production of Organic Acids

Huma Fatima, Jasneet Grewal,
Sumit Kumar, and Sunil Kumar Khare

CONTENTS

DOI: 10.1201/9781003335740-3

3.1 INTRODUCTION

Energy production by combustion of fossil fuels is the key contributor to the release of greenhouse gases, specifically carbon dioxide (CO_2). The continuous consumption of fossil fuels contaminates the environment, leading to anthropogenic climate change. The elevated level of pollutants poses a major threat to human civilization. On the other hand, with an accelerated depletion of fossil fuels, meeting the growing energy demand of the population has become a great challenge. The development of renewable and environmentally friendly sources of energy (as an alternative for fossil fuel) is extremely important for the survival of the human race. In terms of renewable raw materials, biomass presents itself as a promising potential candidate to replace fossil fuels for energy production.

A biorefinery is an industry with a collection of facilities required for the conversion of biomass to valuable and marketable products (Takkellapati et al. 2018; Komesu et al. 2022Z). According to the Department of Energy in the United States, the biorefinery processes the biomass feedstocks for extraction or conversion of value-added products (Kamm et al 2006). The National Renewable Energy Laboratory (NREL) defines a biorefinery as the integration of biomass conversion processes and equipment that produces power, fuels, and chemicals (Takkellapati et al. 2018).

Various types of biomasses are used in biorefineries to produce biofuels. The biofuels thus produced can be classified into three generations:

- First generation: Includes the biofuels produced from edible crops, such as wheat, rice, corn, vegetable oils, sugar cane, whey, barley, potato wastes, and sugar beets.
- Second generation: Derived from the biomass, which is nonedible feedstock or side stream wastes. It includes three types of biomasses, namely a) homogeneous biomass, for example, white wood chips; b) quasi-homogeneous, which includes agricultural and forest residues; and c) non-homogeneous, including low-value feedstock as municipal solid wastes (Lee and Lavoie 2013).
- Third generation: Uses algal biomass, which shows higher growth yield as compared with lignocellulosic biomass, but geographical and technical challenges are associated with algal biomass production (Lee and Lavoie 2013). Fourth-generation biofuels are the extension of third-generation biofuels, where genetically modified algae are used as a biomass feedstock.

Based on the composition of the feedstock, biorefineries are categorized into three categories, namely:

- Triglyceride biorefinery uses animal fats, oil from algae, vegetable oils, waste cooking oil, and sugar.
- Starchy biorefinery utilizes sugar beet, wheat, sugar cane, corn, maize, etc.
- Lignocellulosic biorefinery utilizes lignocellulosic feedstocks, such as crop straws, wood, grasses, husks, etc. (Ingle et al. 2020; Ginni et al. 2021).

The lignocellulosic biorefineries have gained much attention from all around the world as they utilize wastes to produce valuable products. It is in such biorefineries that the agricultural wastes have been effectively utilized to produce commercially important products such as biofuels (bioethanol), organic chemicals, platform chemicals, and polymers (Chaturvedi et al. 2018a; Chaturvedi et al. 2018b; Takkellapati et al. 2018; Ingle et al. 2020).

Platform chemicals (also popularly known as blocks) are defined by the Department of Energy in the United States as the "molecules with multiple functional groups that possess the potential to be transformed into new families of useful molecules" (Werpy and Petersen 2004). In other words, platform chemicals are carbon compounds (made of two to six carbons) that are utilized to make a variety of industrially relevant chemicals for commercial usage. The chapter discusses the processes in a lignocellulosic biorefinery used for biomass conversion to platform chemicals, i.e., a) pretreatment, b) saccharification, c) microbial fermentation, and d) downstream processing, followed by the platform chemicals (specifically carboxylic acids), chemical and biological routes of their production, their derivatives, and applications.

3.2 LIGNOCELLULOSIC FEEDSTOCK

India being an agricultural economy, a large amount of crop residue is generated every year. Inadequate sustainable agricultural and waste management practices have led to approximately 92 metric tons of crop waste being burned every year in India. This significantly contributes to air pollution leading to health problems and global warming (Bhuvaneshwari et al. 2019). Biomass waste utilization has a very high significance in resource recovery and simultaneous waste management. Lignocellulosic biomass can be the best alternative to reduce the global dependency on fossil fuels, as it is the most abundant renewable source in nature (Özdenkçi et al. 2017). The key source of lignocellulosic biomass is plant cell walls, which mainly are made up of three polymers, i.e., cellulose (35–50%), hemicellulose (20–35%), and lignin (10–25%). It also consists of some other constituents such as pectin, proteins, ash, lipids, wax, and additional extractives (Khare et al. 2015; Fang et al. 2017). Cellulose is a linear polymer made up of β-glucose (monomers) that are linked by β-1,4-glycosidic bonds and form a chain-like structure also known as β-1,4-glucan chains or cellulose microfibrils. Hemicellulose, on the other hand, is a branched polymer consisting of C5 sugars, i.e., arabinose, xylose, and rhamnose; C6 sugars, i.e., glucose, galactose, and mannose; and uronic acids as its components. The cellulose microfibrils (β-1,4-glucan chains) are covered and cross-linked with each other by hemicellulose and lignin. Further, lignin also cross-links with hemicellulose to form lignin-carbohydrate complexes (Balan et al. 2013; Fang et al. 2017). The complex structure of the cell wall so formed by cross-linking of the components with the cellulose makes it nearly inaccessible to the hydrolytic enzymes, secreted by the microorganisms, for its conversion to glucose (Khare et al. 2015). Therefore, pretreatment followed by saccharification are necessary steps in the conversion of a biomass to value-added products, as shown in the Figure 3.1.

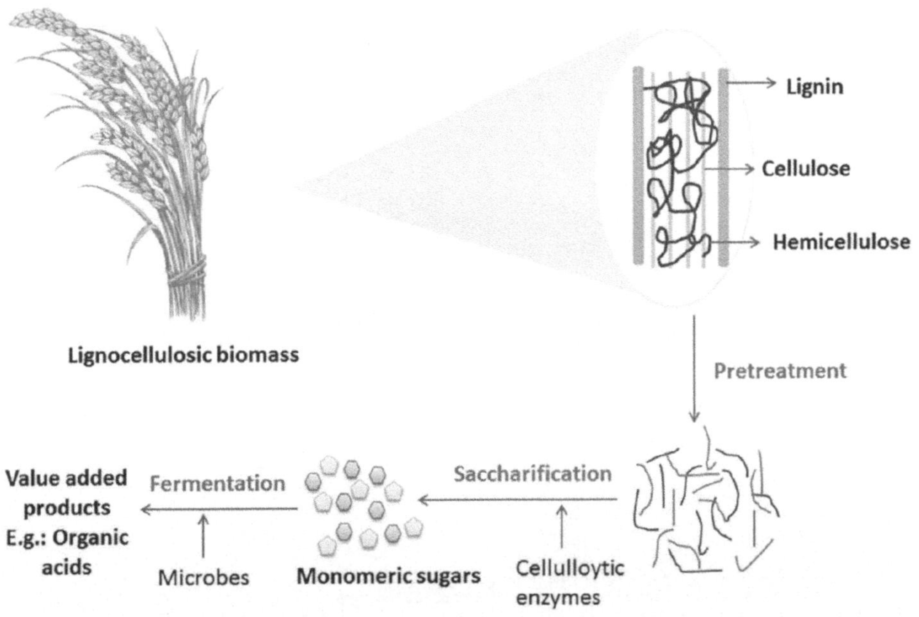

FIGURE 3.1 Basic steps in the conversion of lignocellulosic biomass to value-added products.

3.2.1 Pretreatment

Pretreatment (delignification) is defined as the removal of lignin, which essentially loosens the crystal structure and makes it accessible to the enzymes. An effective pretreatment process facilitates the formation of sugars directly by hydrolysis, avoids loss or degradation of sugars formed, limits the production of inhibitory products, and reduces energy demands and production costs (Kumari and Singh 2018). Several different pretreatment methods have been used by researchers, which can be classified in categories, as shown in Figure 3.2.

1. Physical pretreatment: It involves the use of mechanical energy to break down the biomass to reduce the particle size that ultimately increases the surface for the enzymes in the next hydrolysis step. However, this method is time-consuming and costly (Sharma et al. 2019).
2. Chemical pretreatment: Chemicals such as acids, alkalis, ozone, or peroxide are also used in pretreatment; this method is faster but may produce harmful substances and ultimately increases the expense in the recovery (Fang et al. 2017; Kumari and Singh 2018).
3. Physico-chemical pretreatment: Often, a combination of the mechanical and chemical methods have been reported as cost-effective (Akhtar et al. 2016).
4. Biological pretreatment: This method involves the use of microorganisms such as bacteria and fungi, but it is underexplored due to the low industrial significance and slow technological progress. However, it is a cost-effective technique, requires low energy input, does not use chemicals, and ia ecofriendly (Kumari and Singh 2018; Sharma et al. 2019).

Each pretreatment method has some advantages and disadvantages, and based on discussion on various aspects, steam explosion, acid, and alkaline pretreatment methods are the most commonly used for lignocellulosic biomass pretreatment (Singh et al. 2015).

3.2.2 Saccharification

For the past several decades, it is evident from various studies that simple monosaccharides can be used as raw material for the production of a wide range of compounds by fermentation. The process

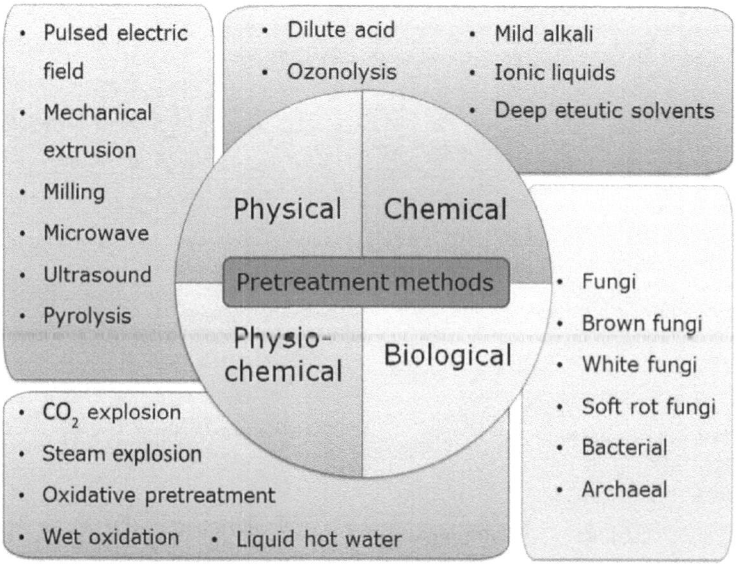

FIGURE 3.2 Methods of pretreatment.

of degradation of complex carbohydrates present in the lignocellulosic biomass into reducing sugars is known as saccharification. Popular methods of saccharification use acidic or enzymatic hydrolysis of biomass, among which enzymatic hydrolysis is the most commonly used (Moe et al. 2012). It is an effective process which is preferred as it releases a significant amount of reducing sugars from the lignocellulosic biomass (Khare et al. 2015; Manisha and Yadav 2017). Enzymes that are generally used for hydrolysis of the lignocellulosic biomass are categorized into three groups, namely a) cellulases; b) hemicellulases; and c) accessory enzymes such as enzymes for debranching of hemicellulose, phenolic acid esterase, and lignin-degrading enzymes. The optimum temperature and pH of the cellulases and hemicellulases is about 50 °C and pH 4.5 to ~5.5. Currently, these enzymes are commercially produced by Novozyme (e.g., CellicCTec/HTec and Celluclast) and Genencor (e.g., Spezyme and Accelarase) (Zeng et al. 2012; Fang et al. 2017). However, there are few limitations of the enzyme hydrolysis while using it at large-scale enzymatic saccharification such as uncooperativeness of the cellulose and hemicellulose, high cost of hydrolytic enzymes, and low enzyme activity for practical use (Chen and Fu 2016; Fang et al. 2017; Gupta et al. 2020). An effective hydrolysis step is dependent on pretreatment of the lignocellulosic biomass; as stated earlier, the sugars in cellulose and hemicellulose are tightly bound by lignin, which makes the hydrolytic enzymes inaccessible to the biomass. A successful and effective saccharification method that can produce monosaccharides in high yields is still a challenge.

3.2.3 Fermentation

The reducing sugars generated by enzymatic hydrolysis are further subjected to specific microbes for fermentation to produce value-added products such as biofuels and platform chemicals. Different microbes and different feedstocks are used for the production of various platform chemicals.

There are different types of approaches used by biorefineries these days, such as simultaneous saccharification and fermentation (SSF) (also known as consolidated bioprocessing [CBP] where saccharification and fermentation are performed together); on the other hand, there is separate hydrolysis and fermentation (SHF), where hydrolysis is followed by fermentation (Xu et al. 2016).

3.2.4 Downstream Processing

Downstream processing includes all the operations that are performed after fermentation to purify the desired end product. It is also known as purification, which is an essential step and contributes significantly to the overall cost of the process (Clarke 2013). For purification of products produced in a small amount, large volumes are required to be processed for recovery, e.g., vitamin B_{12} (produced in a few mg/L). This significantly contributes to the overall cost of production (Clarke 2013). Downstream processing includes several steps depending upon the type of product to be recovered. Before all the steps, the first step is the separation of the microbial cells from the fermentation broth. The most common methods used for separation of cells from the broth are filtration and centrifugation, which depend on the properties of cells that include density, particle diameter, and mechanical strength and the fluid properties such as viscosity and density (López-Garzón and Straathof 2014). Further, steps in the processing are different depending on whether the solid or the liquid stream enters the process. For instance, if the product is intracellular, the cells are to be ruptured and the cell extract has to be separated from the cell fragments. Cells are generally ruptured using either mechanical, nonmechanical methods, or a combination of both. If the product is secreted outside the cell and solubilized in the fermentation broth, the product is concentrated and purified. Depending on molecular size, solubility, and ionic charge of the products, one or more of the techniques such as precipitation, membrane separation, crystallization, adsorption (Cheng et al. 2012a), filtration, electrodialysis (Xu and Huang 2008), reactive extraction, or liquid extraction can be used for purification of the soluble product, either directly from the fermentation broth or from an extract of the cellular contents (López-Garzón and Straathof 2014). Among these methods, although adsorption is the cheapest method, it is not often used, as it requires an excessive amount of water. The use of the crystallization method is limited because of high energy input, and the process includes a number of steps. Electrodialysis, on the other hand, needs lower energy input but is costly. It can thus be concluded that the development of an efficient purification process is still a challenge (Teleky and Vodnar 2019).

3.3 PLATFORM CHEMICALS

Platform chemicals are molecules with multiple functional groups that enable them to be transformed into new families of useful molecules. Some examples of platform chemicals are organic acids (succinic acid, fumaric acid, itaconic acid, malic acid, etc.), diamines (ethylenediamines, cadaverine, putrescine, etc.), and alcohols (glycol, xylitol, butanol, propanediols, butanediols, etc.). These platform chemicals can be used as raw materials for biobased polymer synthesis, which has significantly grown the interest in polymers from the lignocellulosic biomass over the past decade. The US Department of Energy listed the top 12 platform chemicals associated with high-value applications, emerging markets for fine chemicals, medicines, and materials (Werpy and Petersen 2004; Jang et al. 2012; Becker et al. 2015). These platform chemicals have been produced by chemical methods for the last several decades, but with rapid depletion of petrochemical resources and increasing concerns on environmental pollution, the biological method, i.e., biotechnological fermentation, has gained significant attention. Among the platform chemicals, organic acids are used for synthesizing various polymers, plastic materials, and other products. As these organic acids are in great demand, environmentally friendly, more efficient, and cost-effective bioprocessing methods for their production are being developed. Some of the most important organic acids as listed by the Department of Energy are discussed in the following sections. Four carbon 1,4-diacids such as succinic, fumaric, and malic are listed as the topmost platform chemicals and are discussed in the following sections, and some recent studies are listed in Table 3.1.

3.3.1 Succinic Acid

Succinic acid ($C_4H_6O_4$), commonly known as amber acid, is naturally found in food such as broccoli, rhubarb, sugar beets, fresh meat extracts, various types of cheese, and sauerkraut. Its International Union of Pure and Applied Chemistry (IUPAC) name is 1,4-butanedioic acid, and it consists of four carbons and two carboxylic groups. Both prokaryotic and eukaryotic cells synthesize succinic acid as an intermediary in the citric acid cycle.

Succinic acid

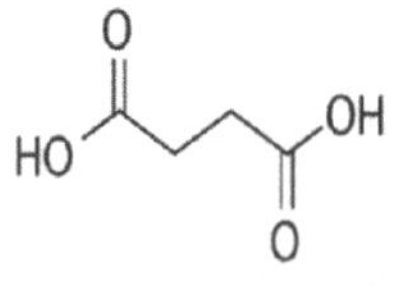

3.3.1.1 Applications

Succinic acid is an important platform chemical, as it is used to derive a number of other compounds such as polybutylene succinate (PBS), 1-4 butanediol (1,4-BDO), gamma- butyrolactone, tetrahydrofuran, n-methyl pyrrolidone, adipic acid, etc. These compounds are applied in several industries such as pharmaceutical, food and agriculture, polymer, cosmetics, etc. Succinic acid is used as an additive and as a sequestrant, buffer, and neutralizing agent in food and agriculture. Furthermore, it is also used to produce synthetic resins in the polymer industry, as a surfactant in detergents, as an ion chelator, etc. (Saxena et al. 2017; Fatima et al. 2022Z). Succinic acid and its derivatives are

TABLE 3.1
Utilization of the Biomass for the Production of Organic Acids

Organic Acid	Feedstocks Used	Microbe	Yield (g/g)/ Productivity (g/L)	References
Succinic acid	Sweet sorghum bagasse (SSB)	*Actinobacillus succinogenes*	17.8 g/L	(Lo et al. 2020)
	Biomass of *Miscanthus*	*A. succinogenes* 130Z (DSM 22257)	0.75 g/g of SA (without nutrient addition), 0.82 g/g of SA with nutrient addition	(Dąbkowska et al. 2019)
	Household kitchen waste	*B. succiniciproducens*	0.46 g/g	(Babaei et al. 2019)
	Mixed food waste (MFW) hydrolysate	*Yarowia lipolytica* PGC202	71.6 g/L	(Li et al. 2019)
	Oil palm fronds (OPF)	*A. succinogenes* DSM 22257	0.71 g/g	(Tan et al. 2018)
	Carob pods	*A. succinogenes* 130Z (ATCC 55618)	0.94 g/g	(Carvalho et al. 2016)
	Oil palm frond (OPF) bagasse	*A. succinogenes* 130Z	0.71 g/g	(Indera et al. 2016)
	Laminaria japonica hydrolysate	*Escherichia coli* BS002	17.44 g/L	(Bai et al. 2015)
Fumaric acid	Cane bagasse	*Rhizopus oryzae* ATCC® 20344	34.20 g/L	(Aita and Deng 2020)
	Soybean cake hydrolysate and glucose	*Rhizopus arrhizus* NRRL 2582	0.7 g/g of total consumed sugars	(Papadaki et al. 2018)
	Lignocellulosic syrup	*R. oryzae* ATCC 20344	34.20 g/L	(Deng and Aita 2018)
	Liquid food waste	*R. arrhizus* RH7-13	32.68 g/L	(Liu et al. 2016)
	Corn cob hydrolysate	*Aspergillus terreus*	1.9% total glucose	(Jiménez-Quero et al. 2016)
	Apple juice extraction waste	*R. oryzae*	25.2 g/L	(Das et al. 2015)
	Apple pomace	*R. oryzae*	138 g/kg dry weight	(Das et al. 2015)
	Brewery wastewater	*R. oryzae*	31.3 g/L	(Das and Brar 2014)
Malic acid	Crude glycerol	*Aspergillus niger* PJR1	92.64 g/L	(Iyyappan et al. 2019)
	Sugarcane molasses	*A. pullulans FJ*-PYC	94.2 g/L	(Feng et al. 2018)
	Miscanthus cellulose hydrolysate	*A. oryzae* DSM 1863	30.8 g/L	(Dörsam et al. 2017)
	Corn starch and glucose	*A. oryzae*	82.3 g/L	(Liu et al. 2017b)
	Jerusalem artichoke tuber hydrolysate	*A. pullulans* HA-4D	131.5 g/L	(Xia et al. 2017)
	Soy molasses	*A. pullulans* ZX-10	71.9 g/L	(Cheng et al. 2017)
	Milled corn stover	*Thermobifida fusca*	21.47	(Deng et al. 2016)
	Sugarcane molasses	*A. pullulans* HA-4D	109.7 g/L	(Xia et al. 2016)
	Corn straw hydrolysate	*R. delemar*	0.96 g/g	(Li et al. 2014)
Itaconic acid	Superfluous algal biomass hydrolysate	*A. niveus*	31.55 g/L	(Gnanasekaran et al. 2019)
	Potato starch	*A. terreus* (C1, isolated)	29.69 g/L	(Bafana et al. 2019)
	Rice wine waste	*Corynebacterium glutamicum* (eng.)	345 mg/L	(Joo et al. 2017)

(Continued)

TABLE 3.1 (Continued)
Utilization of the Biomass for the Production of Organic Acids

Organic Acid	Feedstocks Used	Microbe	Yield (g/g)/ Productivity (g/L)	References
	Potato starch	*Ustilago maydis*	34.52 g/L	(Bafana et al. 2017)
	Rice husk	*A. terreus* ATCC 10020	1.9 g/L	(Pedroso et al. 2017)
	Potato starch	*A. terreus* (C1, isolated)	30.8 g/L	(Bafana et al. 2017)
	Corn cob hydrolysate	*A. oryzae*	0.14% of total glucose	(Jiménez-Quero et al. 2016)
	Corn starch	*A. terreus* CICC 40205	77.1 g/L	(Huang et al. 2014)
	Jatropha seed cake	*A. terreus*	24.46 g/L	(Dowlathabad et al. 2007)
2, 5-Furan dicarboxylic acid (FDCA)	1 g/L (8 mM) HMF	*A. flavus* APLS-1	0.83 g/L	(Rajesh et al. 2019)
	Algal hydrolysate	*Methylobacterium radiotolerans* G-2	459.7 mg/L	(Yang and Huang 2018)
	Algal hydrolysate of *Chaetomorpha linum* and HMF (5-hydroxy methylfurfuraldehyde)	*Burkholderia cepacia*	0.99 g/L	(Yang and Huang 2016)
	Macroalgae *C. linum*	*Burkholderia cepacia* H-2	1276 mg/L FDCA	(Yang and Huang 2016)
3-Hydroxypropionic acid	Glucose	*C. glutamicum* MH15	0.51 g/g	(Chen et al. 2017)
	Glycerol	*Klebsiella pneumonia Δ ldh1Δ ldh2Δ pta(pTAC-puuC*	0.54 g/g	(Li et al. 2016)
	1,3-Propanediol	*Acetobacter* sp. CGMCC 8142	0.96 g/g	(Li et al. 2016)
	Glucose	*S. cerevisiae SCE-R2-200 BcBAPAT EcYDFG TcPAND ScAAT2 ScPYC1/PYC2*	0.55 g/g	(Borodina et al. 2015)
	Glycerol	*K. pneumoniae* WM3 *pUC18kan¬aldHec*	0.41 g/g	(Liu et al. 2014)
	Glycerol	*K. pneumoniae* J2B *ΔdhaT, kgsadH*	0.64 g/g	(Ko et al. 2012)
	Glycerol	*K. pneumoniae* J2B *kgsadH*	0.40 g/g	(Kumar et al. 2012)

also used in the form of dietary supplements in sports foods and sports drinks to strengthen stamina during competitions and in preparation for the recovery process after physical exercises. Therefore, the acid may be called an "elixir of youth" (Voitenko et al. 2018).

3.3.1.2 Chemical Synthesis

Succinic acid was first synthesized in 1546 from amber distillation by Georgius Agricola (Karthik and Rathinamoorthy 2017). Currently, it is primarily obtained by a process known as chemical hydrogenation. In this process, n-butane or benzene is oxidized to produce maleic anhydride, which is further subjected to hydration and produces maleic acid. Maleic acid is then subjected to chemical processes, such as hydrogenation, to produce succinic acid. Despite getting a good yield and high purity and selectivity of succinic acid, the chemical method is regarded to be a complex and expensive method along with environmental concerns (Ge et al. 2018). A more encouraging approach to produce succinic acid is the use of a renewable biomass as feedstock for the fermentation process. Today, various companies, such as Myriant (United States), BioAmber (Canada), Succinity (Spain), and Reverdia (Italy), produce succinic acid from glucose at commercial scale (Ahn et al. 2016; Jiang et al. 2017).

3.3.1.3 Biological Synthesis

Succinic acid is also produced in various microorganisms as an intermediate in one of the following pathways, as discussed here and shown in Figure 3.3:

1. The reductive branch of the citric acid (TCA) cycle, which occurs under anaerobic conditions; here, succinic acid is derived from phosphoenolpyruvate via a series of intermediate compounds, including oxaloacetate, malate, and fumarate (Beauprez et al. 2010; Cheng et al. 2012a).
2. Succinic acid is generated from acetyl-CoA in the oxidative branch of the TCA cycle, which occurs under aerobic conditions, but is afterwards converted to fumarate by succinate dehydrogenase. The enzyme is encoded by the gene *sdhA*, and hence its deletion is necessary to accumulate succinic acid under aerobic conditions (Cheng 2012a; Jiang et al. 2017).
3. Glyoxylate pathway: Two essential enzymes, isocitrate lyase and malate synthase, connect the TCA cycle to the glyoxylate pathway under aerobic conditions. Isocitrate lyase converts isocitrate to succinate and glyoxylate in the TCA. Succinate remains in the cycle, whereas glyoxylate is combined with another acetyl-CoA via malate synthase to produce oxaloacetate, which then enters the regular TCA cycle (Cheng et al. 2012b).

Among lignocellulosic biomass, crop wastes such as corn straw, cotton straw, rice straw, and wheat straw are the most commonly used for biobased production of succinic acid (Tan et al. 2014). These are considered the most promising choices as feedstock because of their abundance and cheap cost. Moreover, these have proved to give a high yield of succinic acid, and third-generation biomasses such as the biomass of *Miscanthus*, *Laminaria japonica* hydrolysate, etc., have also been used for succinic acid production (Bai et al. 2015; Dąbkowska et al. 2019).

The most common microbial cultures used for succinic acid production are *Actinobacillus succinogenes*, *Anaerobiospirillum succiniciproducens*, *Mannheimia succiniciproducens*, *Basfia succiniciproducens*, *Bacteroides fragilis*, *Corynebacterium crenatum*, *Escherichia coli*, *Corynebacterium glutamicum*, and *Yarrowia lipolytica* (Song and Lee 2006; Dessie et al. 2018). Apart from the natural producers, other microbes such as *Saccharomyces cerevisiae*, *C. glutamicum*, and *E. coli* have been engineered by researchers for enhanced production of succinic acid (Jiang et al. 2017). To enhance succinic acid production, other methodologies for metabolic engineering of strains include modulating the expression of certain genes or incorporating additional genes using recombinant DNA technology (Jiang et al. 2017). For example, Gao et al. deleted one of the highly conserved subunits of succinic dehydrogenase, i.e., sdh5 in *Y. lipolytica*. The engineered strain *Y. lipolytica* PGC01003 produced 43 g/L of succinic acid from crude glycerol (Gao et al. 2016). Mao et al. (2018) engineered *C. glutamicum* by incorporating xylose-utilizing genes (*xylA*, xylose isomerase, and *xylB*, xylokinase) from X*anthomonas camestries* 8004 in the plasmid (pX-xcbAB). The group also increased the expression of two endogenous genes, *tkt* and *tal*. This engineered strain of *C. glutamicum* CG55 could produce 98.6 g/L of succinic acid from the biomass of corn stalk (Mao et al. 2018). However, biobased succinic acid production is still in its early stages, and there is a lot of room for expansion at the industrial level.

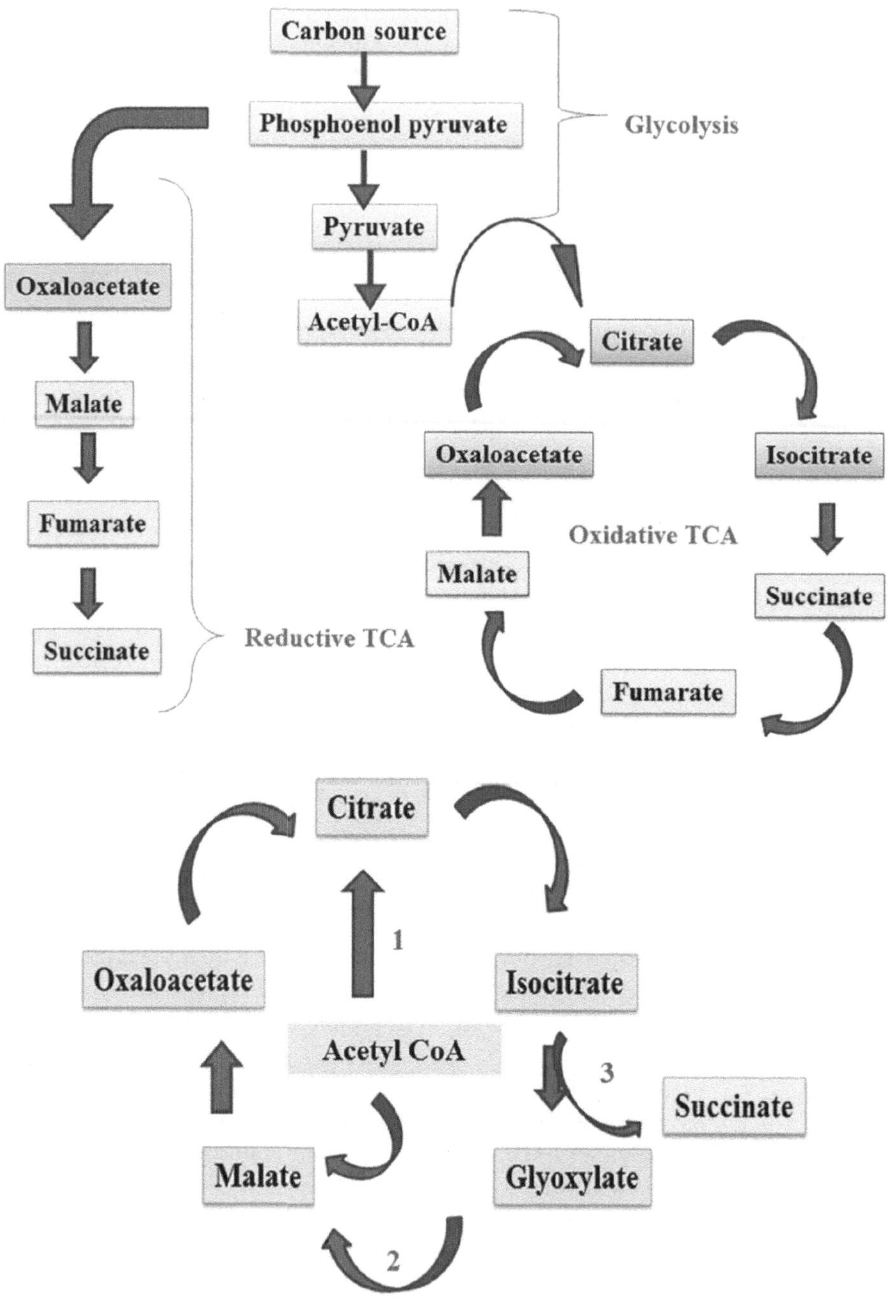

FIGURE 3.3 (top) Reductive and oxidative branch of the TCA cycle. (bottom) Glyoxylate cycle, 1-citrate synthase, 2-malate synthase, and 3-isolcitrate lyase.

3.3.2 Fumaric Acid

Fumaric acid was isolated for the first time from *Fumaria officinalis* (family: Papaveraceae) (Roa et al. 2008). Allomaleic acid, boletic acid, lichenic acid, and tumaric acid are some of the other names for fumaric acid. The IUPAC name of fumaric acid is trans-1,2-ethylene dicarboxylic acid. It is a four-carbon dicarboxylic acid with a carbon-carbon double bond and two carboxylic acid groups, which makes it a multifunctional structure that can be explored for various reactions such as polymerization and esterification (Roa et al. 2008).

Fumaric Acid

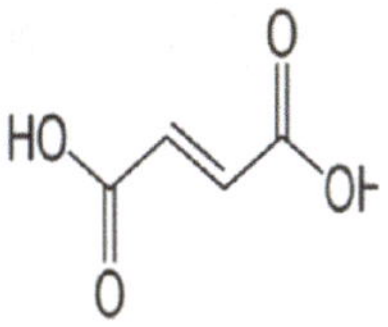

3.3.2.1 Applications

Fumaric acid is employed in a wide range of sectors, including food, chemicals, and pharmaceuticals. The food industry accounts for about 33% of the total fumaric acid consumption, where it is used as an antibacterial agent, a nutritional additive, and an acidulant in various forms in the food (Das et al. 2016). In the dairy industry, fumaric acid is used as an additive in animal feed. The normal digestive process of domesticated livestock and storage of manure of these animals produce large amounts of methane (CH_4) gas. Methane (CH_4) constitutes about 9% of all the greenhouse gases globally. The addition of fumaric acid to the diets of these animals has been shown to lower the emission of methane (Carro and Ranilla 2003). As stated earlier, the structure of fumaric acid enables it to participate in polymerization and esterification reactions, and hence it is used in the manufacturing of polyester resins (by using a condensation reaction between fumaric acid and polyhydric alcohol) at the commercial scale. Application of fumaric acid is also extended to medical sciences, where its ester derivatives (known as fumaric acid esters, FAEs) are used in neurology, immunology, dermatology, bionanotechnology for drug delivery, and tissue engineering (Das et al. 2016).

3.3.2.2 Chemical Synthesis

Fumaric acid is mainly produced by the petrochemical route that includes the isomerization of maleic acid, which is catalyzed by mineral acids, peroxy compounds, or thiourea. This maleic acid is produced from the hydrolysis of maleic anhydride, and maleic anhydride is obtained from the oxidation of benzene. Due to the high production yield of fumaric acid, the chemical method of synthesis is widely used (Felthouse et al. 2001). However, there are certain limitations with the chemical method, i.e., the requirement of a high temperature for the isomerization reaction, the formation of by-products, high expense of the raw material (maleic anhydride, 60–70% of the total production cost), and huge energy consumption (Lohbeck et al. 2000). Moreover, the development of technology for the utilization of low-cost carbon options for the production of value-added products and the concept of sustainable, biobased production of fumaric acid is highly explored these days.

3.3.2.3 Biological Synthesis

To overcome the limitations posed by chemical synthesis methods, the microbial fermentation process has gained attention due to its sustainability and environmentally friendly properties.

In 1911 Felix Ehrlich isolated a filamentous fungal species, *Rhizopus nigricans*, that could produce fumaric acid. Later Foster et al. nominated four genera, *Rhizopus*, *Mucor*, *Cunninghamella*, and *Circinella*, as the best fumaric acid producers (Foster and Waksman 1939). The most common species used for fermentative production of fumaric acid are *Rhizopus nigricans*, *Rhizopus formosa*, *Rhizopus arrhizus*, and *Rhizopus oryzae*, among which *R. oryzae* is the top producer of fumaric acid (Xu et al. 2012).

The lignocellulosic biomasses explored for the production of fumaric acid are cane bagasse soybean cake hydrolysate, wood chips, lignocellulosic syrup, food waste, corn cob hydrolysate, etc. (Liu et al. 2016; Papadaki et al. 2018; Aita and Deng 2020). Among these, corn straw and crude glycerols are known to give a high yield of fumaric acid (Zhou et al. 2014; Das et al. 2016). Low-cost feedstocks contribute towards the biorefinery economy, but commercial production of fumaric acid is still a challenge due to problems in downstream processing such as the increase in the viscosity of the fermented broth, mass, heat, and low oxygen supply due to filamentous growth of fungi (Das et al. 2016). Therefore, the current need of the hour in fumaric acid research is the development of efficient purification methods.

3.3.2.4 Metabolic Pathways

There are three metabolic pathways in which fumaric acid is produced, i.e., the reductive branch of the TCA cycle (major contributor for fumaric acid accumulation) and the oxidative branch of the TCA cycle. The third potential pathway in which fumaric acid is synthesized is the glyoxylate route; here, the isocitric acid (intermediate of the TCA cycle) is broken down by isocitrate lyase into succinic acid and glyoxylic acid. Eventually, glyoxylic acid combines with acetyl-CoA and produces L-malic acid by malate synthetase. Malic acid is further converted to fumaric acid.

Metabolic pathways of several strains such as *Saccharomyces cerevisiae*, *R. oryzae*, and *E. coli* have also been engineered to increase the fumaric acid yield. For example, *S. cerevisiae* was engineered by introducing the reductive TCA cycle by adding two genes encoding *mdh* and *fum* from *R. oryzae* and overexpressing the endogenous gene *pyc*. The recombinant strain was able to produce a higher titer of fumaric acid as compared to the natural strain. However, due to an energy imbalance, productivity was low at 3.18 g/L (Liu et al. 2016). Likewise, the noncyclic glyoxylate pathway has also been engineered to produce fumaric acid in *E. coli*. For instance, Song and the group in 2013 knocked out the gene *iclR* (isocitrate lyase repressor) to enhance the carbon flow towards the glyoxylate route. Simultaneously, genes *fumA*, *fumB*, and *fumC* were also deleted to accumulate fumaric acid in the cell. Moreover, the endogenous gene *ppc* (encoding *pep*C) was overexpressed to increase the reductive TCA cycle flux. The final engineered strain could produce 28.2 g/L of fumaric acid, with a productivity of 0.448 g/L per hour (Song et al. 2013). However, biobased fumaric acid production is still in its early stages, and stated yields must be improved to compete with similar chemical synthesis from petrochemical feedstock.

3.3.3 Malic Acid

Malic acid ($C_4H_6O_5$) was first isolated from apples and was later also found in other fruits such as bananas, litchis, and plums. It contributes to the sour taste of fruits (Wyman and Palmer 1964). It is a dicarboxylic acid with IUPAC name 2-hydroxybutanedioic acid, also known as hydroxybutanedioic acid, 2-hydroxysuccinic acid, (*S*)-hydroxybutanedioic acid, and (*R*)-hydroxybutanedioic acid. There are two stereoisomeric forms, namely L- and D-enantiomers, but only the L-isomer exists naturally.

Malic acid

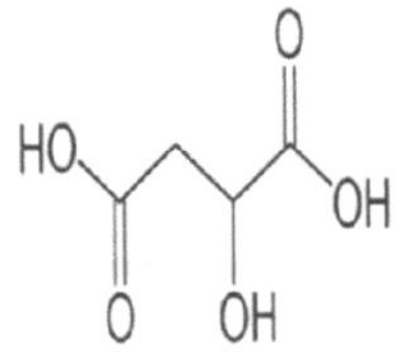

3.3.3.1 Applications

Malic acid has applications in food, beverage, pharmaceuticals, and the cosmetic industry. It is used as an acidulant to increase the flavor of candy and foods. Some of the commercial applications of malic acid include metal cleaning, paints, finishing of textile, electroless plating, etc. (Goldberg et al. 2006). In pharmaceuticals, it is used to produce infusions of amino acid to treat hyperammonemia and liver dysfunction (Chi et al. 2016). Malic acid is also used as a precursor for many useful compounds during chemical synthesis of special polymers, biodegradable polymers (polymalic acid [PMA]), optically pure 4-bromo-2- hydroxybutanoic acid esters, and structure mimetics (Chi et al. 2016). In agriculture, malic acid is used to solubilize aluminum phosphate in the soil.

Malic acid is produced by three methods, i.e., chemical, enzymatic, and biotechnological. Among these methods, the chemical and enzymatic methods are well developed, while biotechnological production processes are still in the early stages (Iyyappan et al. 2019). Currently, malic acid is produced by chemical methods commercially.

3.3.3.2 Chemical Synthesis

The chemical synthesis of malic acid includes the reaction that converts petrochemically derived compounds such as maleic anhydride, maleic acid, and fumaric acid to malic acid. For example, hydration of fumaric acid and maleic acid produces malic acid under high-pressure and -temperature conditions (Goldberg et al. 2006) (Wang et al. 2016). Malic acid is also produced from the oxidation of benzene or butane (West 2017).

3.3.3.3 Biological Synthesis

1. Enzymatic synthesis: Enzymatically, malic acid production includes the use of fumarase, an enzyme that converts fumaric acid to malic acid through hydration (Naude and Nicol 2018). The enzyme can be used in different ways, such as a) substrate fumaric acid can be directly subjected to purified fumarase enzyme and b) permeabilized or lyophilized cells or even whole cells can be added to the substrate fumaric acid (Giorno et al. 2001; Presečki et al. 2007; Pandurić et al. 2017).
2. The biological synthesis: Biological production of malic acid offers two main advantages over chemical or enzymatic methods. First, a pure L-form of malic acid is produced by microbes. Second, the versatility of this method lies in the fact that a combination of different microbes with suitable renewable feedstock can be used to make the fermentation process efficient and environmentally friendly. Fungal strains such as *A. oryzae* and *A. niger* are generally used for malic acid production. Similar to the previously mentioned acids such as succinic acid and fumaric acid, even malic acid is produced via the TCA cycle.

Renewable, inexpensive, and abundant waste biomass would be an ideal feedstock for malic acid production. Several lignocellulosic wastes and algal biomasses have been used to produce malic acid such as sugarcane molasses, corn starch, soy molasses, milled corn, crude glycerol, Miscanthus cellulose hydrolysate, Jerusalem artichoke, etc. (Xia et al. 2017; Feng et al. 2018; Iyyappan et al. 2019). Among the substrates listed, molasses and crude glycerol are easily available and cheap. They also show an advantage over other substrates, as they can be used directly without pretreatment, which further cuts the cost of pretreatment. It has been reported that for producing 1 ton of malic acid, crude glycerol worth US$169 and molasses worth US$52 are required (Kövilein et al. 2020).

The recent research on malic acid production mainly focuses on the development of efficient downstream processing methods. The most commonly used method is precipitation with calcium ions ($CaCO_3$ is added during fermentation), which is associated with the accumulation of gypsum in large amounts along with malic acid (Kövilein et al. 2020).

Four metabolic pathways have malic acids as an intermediate, namely the reductive TCA pathway, the oxidative TCA cycle, the glyoxylate cycle, and the noncyclic glyoxylate shunt (Yin et al. 2015). As discussed earlier also, metabolic pathways have been engineered by scientists and researchers all over the world to enhance the production and accumulation of the desired products. Similarly, for obtaining a high titer of malic acid, *E. coli* KJ060 was engineered by deleting two genes, i.e., *poxB* and *mgsA*, to generate strain KJ071, which could produce malic acid at a titer of 62.9 g/L, with a yield of 1.42 mol/mol. However, the strain also produced succinic acid as a by-product in large amounts (Jantama et al. 2008). Zhang et al. (2011) also developed an *E. coli* KJ060 strain for malic acid production. The group altered the carbon flow by deleting fumB, fumAC (fumarase isoenzymes), and specifically fumarate reductase (frdBC) and redirecting it to malic acid. Knocking off eight more genes abolished the production of by-products, including lactate, acetate, and ethanol. The final mutant strain, XZ658, produced malic acid with a yield of 1.42 mol/mol glucose (Zhang et al. 2011).

3.3.4 Itaconic Acid

Itaconic acid is a naturally occurring 5-C dicarboxylic acid, which is a derivative of citric acid. The IUPAC name of itaconic acid is 2-methylenesuccinic acid. It is also known by other names such as methylenesuccinic acid, methylenebutanedioic acid, and 1-propene-2-3-dicarboxylic acid. Itaconic acid has a trifunctional structure with two carboxylic acid functional groups and an α,β-unsaturated double bond (Geilen et al. 2010). The unsaturation in the structure enables itaconic acid to be used for polymerization, esterification with alcohols, production of anhydride, salt formation with metals, and esterification reactions (Larsen and Eimhjellen 1955). Moreover, the structure of itaconic acid and its esters such as dimethyl itaconate and dibutyl itaconate resemble acrylic acid and methacrylic acid.

Itaconic acid

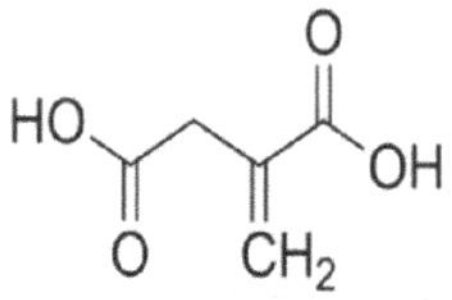

3.3.4.1 Applications

Itaconic acid, polyitaconic acid, and esters of itaconic acid are applied in a large number of industries such as the food industry, polymer industry, pharmaceuticals, and medical applications. In polymer industries, itaconic acid can be used as a substitute for existing polymers such as acrylic acid, methacrylic acid, fumaric acid, and maleic anhydride, as it is biodegradable and nontoxic (Geilen et al. 2010) (Delidovich et al. 2016). Itaconic acid is widely used in synthesizing styrene-butadiene rubber (SBR) latexes, synthetic latex, superabsorbent polymers, chelant dispersant agents, unsaturated polyester resins, and phosphate-free detergents. Esters of itaconic acid are used as plastics, adhesives, elastomers, or coatings (De Carvalho et al. 2018). In food industries, itaconic acid is used for active packaging. Since itaconic acid has antimicrobial properties, it is also used as an antibacterial and antifungal agent (Teleky and Vodnar 2019). Due to these potential applications of itaconic acid, its demand in the market is expected to increase by 5.5% every year to 2023.

3.3.4.2 Chemical Synthesis

Itaconic acid was first produced by the thermal decomposition of citric acid (Baup 1836). The most common chemical method used to synthesize itaconic acid was the dry distillation and treatment of anhydride with water. Other chemical methods were also studied, including isoprene oxidation, mesityl oxide oxidation, isomerization of the produced citric acid, etc. (Baup 1836; Teleky and Vodnar 2019). However, none of the chemical processes could take on the biological methods of synthesis, which is based on fermentation by fungi owing to its high productive rate (Baup 1836; Tate 1967; Ahmed et al. 2022Z).

3.3.4.3 Biological Synthesis

In 1932, a fungal strain, *Aspergillus itaconicus*, producing itaconic acid was isolated (Kinoshita 1932). Later Calam et al. in 1939 discovered a better itaconic acid–producing strain of *Aspergillus*, namely *A. terreus* (Calam et al. 1939). Since then, *A. terreus* is considered a top itaconic acid–producing microorganism. Other natural producers of itaconic acid are *Pseudozyma antarctica* and *Ustilago maydis*. Currently, itaconic acid is produced commercially on an industrial scale via fermentation with *A. terreus* in aerobic conditions using glucose as feedstock (Teleky and Vodnar 2019).

Due to the concept of sustainable development and bioeconomy, lignocellulosic hydrolysates and starchy wastes—for example, wheat bran, beech wood, rice husk, potato starch, and molasses—are being explored to replace glucose as the feedstock for the production of itaconic acid (Pedroso et al. 2017; Gnanasekaran et al. 2019; Bafana et al. 2019). However, the yields from crude substrates are generally lower than those from sugars. Several other limitations associated with itaconic acid production are as follows., filamentous growth of fungus blocks the supply of oxygen during fermentation and expensive purification requirements; production of by-products such as malic acid, acetic acid, oxalic acid; etc. (Kim et al. 2017; Moon et al. 2018).

3.3.4.4 Metabolic Pathway

Itaconic acid is synthesized by *A. terreus* through the TCA cycle, where pyruvate is converted to acetyl-CoA, which further enters the TCA cycle and gets metabolized to oxaloacetate, as shown in Figure 3.4. Three key enzymes that are involved in itaconic acid production in *A. terreus* are cadA, mttA, and mfsA. Cis-aconitate is an intermediate in the TCA cycle that takes place in mitochondria. The mitochondrial tricarboxylate transporter (encoded by the *mttA* gene) transports cis-aconitate from mitochondria to the cytosol, where cis-aconitate is decarboxylated to produce itaconic acid by cis-aconitate decarboxylase (encoded by the *cadA* gene). Itaconic acid is transported outside the cell by the facilitator superfamily protein (encoded by the *mfsA* gene).

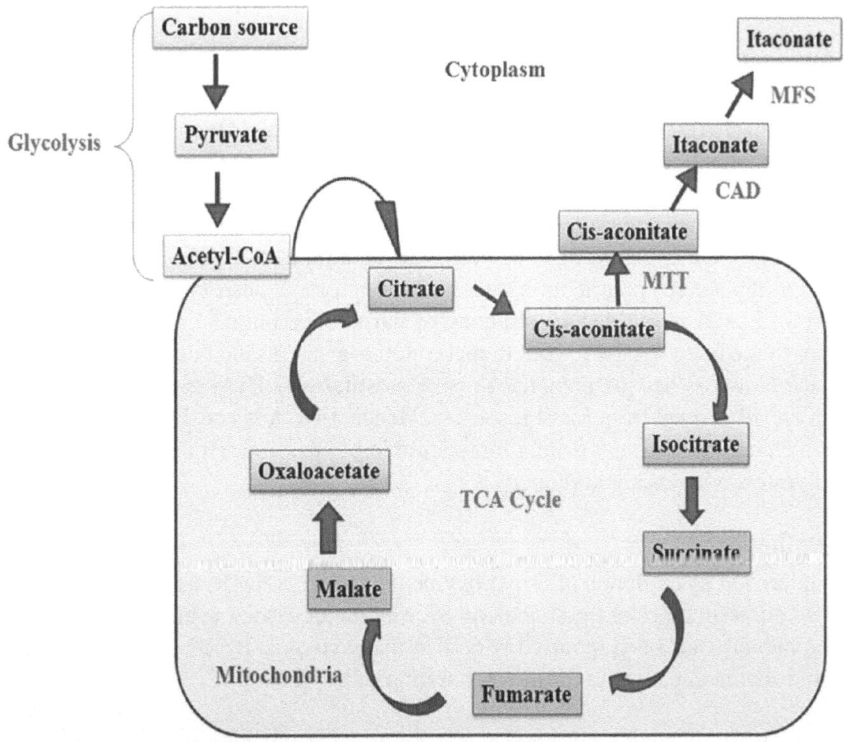

FIGURE 3.4 Pathway of itaconic acid synthesis. MTT (mitochondrial tricarboxylate transporter), CAD (cis-aconitate decarboxylase), MFF (mitochondrial facilitator superfamily).

Several strains have been engineered to enhance itaconic acid production such as *E. coli*, *C. glutamicum*, and *A. niger*. For instance, Hossain and colleagues used an engineered strain of *A. niger* AB1.13 which had the essential genes for itaconic acid production, i.e., *cad*, *mfs*, and *mtt*. The group further overexpressed the cytosolic citrate synthase citB gene that resulted in enhanced itaconic acid production, i.e., up to 26.2 g/L IA (Hossain et al. 2016). Furthermore, the same group overexpressed ATP-citrate lyase acl12 to improve glycolytic flux that resulted in a final titer up to 42.7 g/L (Hossain et al. 2019). Another study also reported enhanced itaconic acid production by developing recombinant *E. coli* as a whole-cell biocatalyst. They expressed aconitase and the cis-aconitate decarboxylase gene in *E. coli* and used citrate as the substrate and reported 500 mM of citrate conversion to 319.8 mM itaconate (Kim et al. 2017).

3.3.5 2,5-Furan Dicarboxylic Acid

2,5-Furandicarboxylic acid (FDCA), also known as furan-2,5-dicarboxylic acid or dehydromucic acid, is an organic acid with two carboxylic acid groups attached to a central furan ring. The structure of FDCA enables it to react and get converted to various other important chemical compounds such as succinic acid, 2,5-furandicarbaldehyde, and 2,5-dihydroxymethylfuran (Werpy and Petersen 2004).

2,5 Furan DicarboxylicAcid

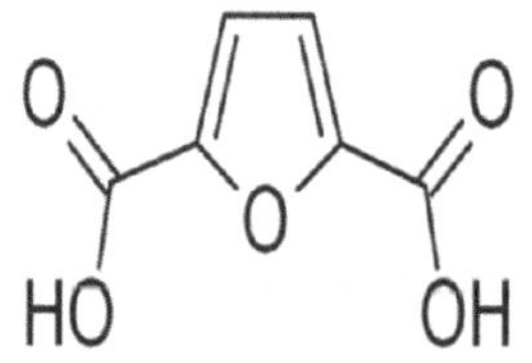

3.3.5.1 Applications

FDCA is used as an aromatic monomer to make unsaturated polyesters (UPs) which show exceptional mechanical, thermal, and corrosion-resistant performances. Owing to such properties UPs have a wide range of applications in transportation, construction, aerospace, marine, and auto industries. Currently, the commercial production of UPs is based on petroleum-based monomers such as fumaric acid, glycols, maleic anhydrides, and phthalic acids (Biswas et al. 2018). Due to the depletion of fossil fuels and the increasing concern for the environment, biobased dicarboxylic acids are being studied to replace the petroleum-based monomers for synthesis of UPs (Farmer et al. 2015). The structure of FDCA resembles the structure of purified terephthalic acid (PTA), which is used as a precursor to the polyester PET, used to make clothing and plastic bottles. FDCA (derived from lignocellulosic biomass) has the potential to be a substitute for PTA, isophthalic acid, and adipic acids, which are all derived from fossil resources. Hence, FDCA is considered a potential platform chemical, which can be obtained from a biorenewable biomass. It can be used in the synthesis of biobased polyesters with excellent properties.

3.3.5.2 Synthesis of FDCA

FDCA is synthesized by oxidation of 5-hydroxymethylfurfural (HMF) using a variety of reactions, pathways, and different types of the stoichiometric amount of oxidizing agents. Production of HMF derived from biomass has been reported widely in many studies. Based on the raw material used, there are three chemical pathways for the preparation of FDCA:

1. FDCA preparation from diglycolic acid: Here, diglycolic acid (raw material) is first esterified with methanol in the presence of thionyl chloride to produce dimethyl diglycolate. This compound is further reacted with 1,2-dicarbonyl compounds in the presence of potassium

hydroxide to produce FDCA. Following this, downstream techniques such as acidification and extraction are used to purify the FDCA (Zhang et al. 2015).

2. FDCA preparation from monosaccharides derived from lignocellulose: The monosaccharide is first dehydrated to produce 2-furoic acid, which is subsequently transformed to FDCA by esterification, chloromethylation, hydrolysis, and oxidation or disproportionation (Jöbstl et al. 2010).
3. FDCA preparation from renewable biomass resources: The biomass is hydrolyzed to produce polysaccharides, which on further hydrolysis produce sugars such as glucose and fructose. The monosaccharides obtained are first dehydrated to HMF followed by oxidation of the hydroxymethyl group into the aldehyde group, which is further oxidized to carboxyl to generate the FDCA (Jöbstl et al. 2010).

3.3.6 3-Hydroxypropionic Acid

Hydroxypropionic acid ($C_3H_6O_3$) is also known as hydracrylic acid or 3-hydroxypropanoic acid. It can also be called ethylene lactic acid, as 3-HPA is a structural isomer of lactic acid (2-hydroxypropionic acid). It is a three-carbon, nonchiral carboxylic acid with two functional groups, i.e., a carboxyl group and a hydroxyl group (at the β position), which makes it a highly reactive molecule. In nature, 3-HPA is found in thermophilic archaea and bacteria, where it is involved in autotrophic carbon fixation cycles (Liu et al. 2017a).

3-HPA

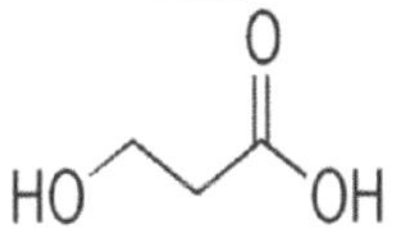

3.3.6.1 Applications

Due to the high reactivity of 3-HPA, it can be easily converted into many valuable compounds such as acrylic acid, 1,3-propanediol, acrylamide, ethyl 3-HP, methyl acrylate, malonic acid, propiolactone, and acrylonitrile (Rathnasingh et al. 2009). The bifunctionality of 3-HPA allows its polymerization to produce poly-3-hydroxypropionate. It is a novel biocompatible and biodegradable polymer with higher mechanical strength and elongation capability than polylactic acid (2-HPA) (Kumar et al. 2013; Liu et al. 2017a). The polymer of 3-HPA, i.e., poly-3-hydroxypropionate, is also used in the medical field for the manufacturing of surgical products and drug-releasing materials (Jöbstl et al. 2010; Zhang et al. 2015). In chemical industries, 3-HPA is used as a cross-linking agent for polymer coatings, metal lubricants, and antistatic agents for textiles (Della et al. 2011; Liu et al. 2017a). Due to the wide range of applications, 3-HPA is gaining attention from researchers in recent years.

3.3.6.2 Chemical Synthesis

To produce 3-HPA, several chemical synthesis routes have been described, including the use of expensive starting material such as ketene, acrylic acid, β-hydroxypropionitrile, 3-HPA, vinyl acetate, and allyl alcohol (Valdehuesa et al. 2013). For instance, ketene, when reacted with formaldehyde in the presence of Lewis acids ($AlCl_3$, $ZnCl_2$, BF_3), produces beta-propiolactone, which when further subjected to the alkali and acid-catalyzed hydrolysis produces 3-HPA (Della et al. 2011). It can also be produced from used acrylic acid by a hydration reaction in the presence of alkali or by acid catalysis (Della et al. 2011).

The overall production cost of chemical synthesis is too high due to the high cost of the starting materials and the high operation costs. High production cost, low yield, and environmental incompatibility of the chemical methods have limited the application of 3-HPA as a bulk chemical.

3.3.6.3 Biological Synthesis

Biobased synthesis of 3-HPA has come up as a way to overcome the limitations of traditional chemical synthesis. Currently, the biosynthesis routes for the synthesis of 3-HPA is through the utilization of glucose or glycerol as starting materials; the pathways are discussed next and shown in Figure 3.5.

1. Glucose based pathway: In this route glucose is first converted to acetyl-CoA, which is further converted to propionyl-CoA by the enzyme acetyl-CoA carboxylase, followed by conversion to 3-HPA by malonyl-CoA reductase (Li and Tian 2015). Another pathway has also been used in which glucose is first converted to glycerol, which is then catalyzed to create 3-HP, as indicated in the second point that follows (Meynial et al. 2007).
2. Glycerol-based pathway: Here, glycerol dehydratase converts glycerol to 3-HPA, which is then transformed to 3-HP by aldehyde dehydrogenase (AldH) and finally to 1,3-propanediol by 1,3-propanediol oxidoreductase (Li and Tian 2015).

Natural producers of 3-HPA have been reported for species includinkg *Bacillus*, *Citrobacter*, *Clostridium*, *Enterobacter*, *Klebsiella*, and *Lactobacillus* (Lindlbauer et al. 2017). Among these, *Klebsiella pneumonia* is the most important producer due to its biochemical characteristics, i.e., the extraordinary capacity to metabolize glycerol, to synthesize vitamin B_{12} (cofactor of glycerol dehydratase), and active cell proliferation (Forage and Foster 1982; Lindlbauer et al. 2017). Some of the microorganisms such as *Byssochlamys* sp., *Alcaligenes faecalis*, *Rhodococcus erythropolis*, *Geotrichum* sp., and *Trichoderma* sp. can reconvert acrylic acid to 3-HP (Vidra and Németh 2017).

Biological methods of synthesis from renewable resources have certain merits over conventional chemical methods such as reduction of the dependency on petroleum resources, the very low

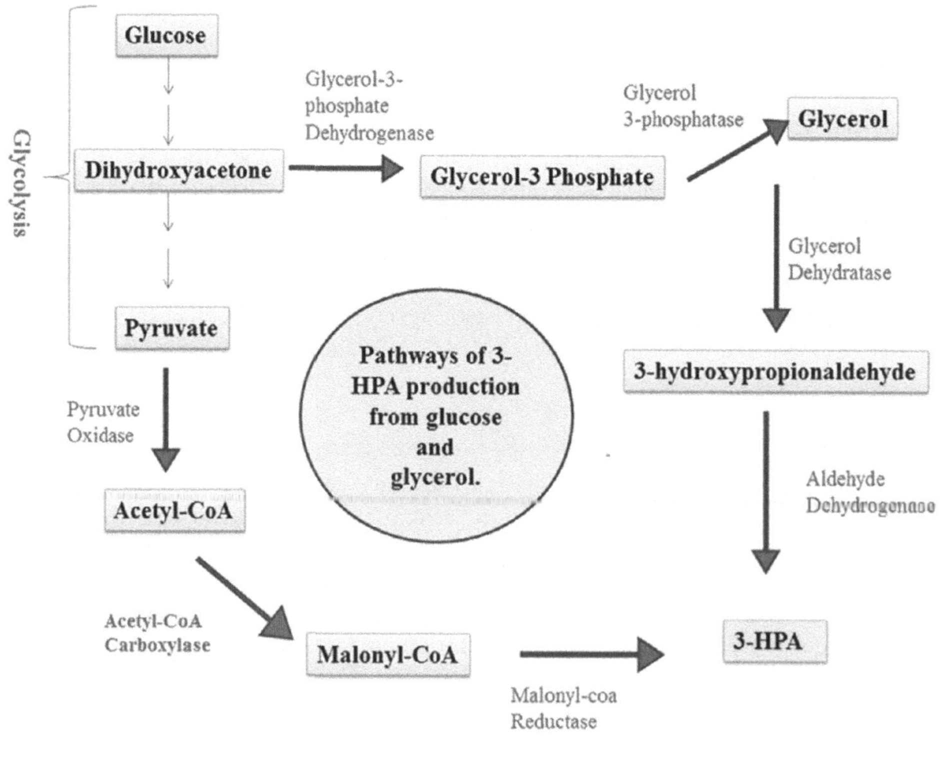

FIGURE 3.5 Pathways of 3-hydroxypropionic acid production.

contribution of carbon dioxide into the atmosphere, and hence less environmental pollution (Vidra and Németh 2017).

3.4 CONCLUSIONS AND FUTURE PERSPECTIVE

The depletion of fossil fuels and related environmental problems have emerged as major concerns. In this regard, meeting the energy demands of the future generation is a big challenge, and therefore development of alternative sources of energy is a goal of current research in the area. Lignocellulosic biomass has been identified as the most abundant and cheap (cheaper than crude oil) renewable feedstock available, and therefore it appears to be the most promising option for the production of energy and platform chemicals. The functioning of any biorefinery is the production of fuel as the main product and platform chemicals as value-added products. To develop an economically sustainable biorefinery, it is important to produce value-added products; hence, the development of biomass-derived platform chemicals is gaining the attention of the scientific community. Organic acids are among the top 12 platform chemicals listed by the Department of Energy in the United States due to a large number of their derivatives and wild range of applications making them economically important. In the last several years various methods have been evolved for biobased production of organic acids from lignocellulosic biomass. However, these biobased methods are still cost-ineffective in comparison to the conventional chemical methods. Therefore, it can be concluded that the efficient conversion of lignocellulosic biomass to organic acids is essential to make the biorefinery economically sustainable. However, the development of cost-effective and efficient technologies for the conversion of lignocellulosic biomass to important organic acids is still a challenge, and there is a lot of scope in the development of better processes. Increasing interest in developing various methods and the use of methodologies such as metabolic engineering can pave a way for making the biobased synthesis of various platform chemicals cost-effective. Moreover, it can also tackle the ever-increasing problem of pollution by utilizing the agricultural waste lignocellulosic biomass as the raw materials, thereby making the process more sustainable.

ACKNOWLEDGMENTS

The authors gratefully acknowledge the financial grant provided by the National Agricultural Science Fund (NASF), ICAR, Government of India, for carrying out this study. HF is grateful for the financial support provided by the Ministry of Science and Technology; Department of Science and Technology (DST) INSPIRE Fellowship (Government of India) [no. IF190004]. Author SK is grateful to the Council of Scientific and Industrial Research (CSIR, Government of India) for Senior Research Associateship (Scientists' Pool Scheme).

REFERENCES

Ahmed, S. F., M. Mofijur, S. N. Chowdhury, M. Nahrin, N. Rafa, Anika Tasnim Chowdhury, Samiha Nuzhat and Hwai Chyuan Ong. 2022. Pathways of Lignocellulosic Biomass Deconstruction for Biofuel and Value-Added Products Production. *Fuel* 318 (June): 123618. https://doi.org/10.1016/j.fuel.2022Z.123618

Ahn, J. H., Y. S. Jang and S. Y. Lee. 2016. Production of Succinic Acid by Metabolically Engineered Microorganisms. *Current Opinion in Biotechnology* 42 (December): 54–66. https://doi.org/10.1016/j.copbio.2016.02.034.

Aita, G. M. and F. Deng. 2020. Bagasse as a Source for Fumaric Acid Production. *International Sugar Journal* 122: 34–39.

Akhtar, N. K. Gupta, D. Goyal and A. Goyal. 2016. Recent Advances in Pretreatment Technologies for Efficient Hydrolysis of Lignocellulosic Biomass. *Environmental Progress & Sustainable Energy* 35 (2): 489–511. https://doi.org/10.1002/ep.12257.

Babaei, M., Panagiotis T., M. Alvarado-Morales, M. Hosseini, S. Ebrahimi, A. Niaei and I. Angelidaki. 2019. Valorization of Organic Waste with Simultaneous Biogas Upgrading for the Production of Succinic Acid. *Biochemical Engineering Journal* 147 (July): 136–145. https://doi.org/10.1016/j.bej.2019.04.012.

Bafana, R., S. Sivanesan and R. A. Pandey. 2017. Itaconic Acid Production by Filamentous Fungi in Starch-Rich Industrial Residues. *Indian Journal of Microbiology* 57 (3): 322–328. https://doi.org/10.1007/s12088-017-0661-5.

Bafana, R., S. Sivanesan and R. A. Pandey. 2019. Optimization and Scale up of Itaconic Acid Production from Potato Starch Waste in Stirred Tank Bioreactor. *Biotechnology Progress* 35 (3): e2774. https://doi.org/10.1002/btpr.2774.

Bai, B., J. Zhou, M. Yang, Y. Liu, X. Xu and J. Xing. 2015. Efficient Production of Succinic Acid from Macroalgae Hydrolysate by Metabolically Engineered *Escherichia Coli*. *Bioresource Technology* 185 (June): 56–61. https://doi.org/10.1016/j.biortech.2015.02.081.

Balan, V., D. Chiaramonti and Sandeep Kumar. 2013. Review of US and EU Initiatives toward Development, Demonstration, and Commercialization of Lignocellulosic Biofuels. *Biofuels, Bioproducts and Biorefining* 7 (6): 732–759. https://doi.org/10.1002/bbb.1436.

Baup, S. 1836. Ueber eine neue Pyrogen-Citronensäure, und über Benennung der Pyrogen-Säuren überhaupt. *Annalen der Pharmacie* 19 (1): 29–38. https://doi.org/10.1002/jlac.18360190107.

Beauprez, J. J., M. D. Mey and W. K. Soetaert. 2010. Microbial Succinic Acid Production: Natural Versus Metabolic Engineered Producers. *Process Biochemistry* 45 (7): 1103–1114.

Becker, J., A. Lange, J. Fabarius and C. Wittmann. 2015. Top Value Platform Chemicals: Bio-Based Production of Organic Acids. *Current Opinion in Biotechnology* 36 (December): 168–175. https://doi.org/10.1016/j.copbio.2015.08.022.

Bhuvaneshwari, S., H. Hettiarachchi and J. Meegoda. 2019. Crop Residue Burning in India: Policy Challenges and Potential Solutions. *International Journal of Environmental Research and Public Health* 16 (5): 832. https://doi.org/10.3390/ijerph16050832.

Biswas, B., S. Chabri, B. C. Mitra, N. R. Bandyopadhyay and A. Sinha. 2018. Mechanical Behaviour of Aluminium Dispersed Unsaturated Polyester/Jute Composites for Structural Applications. *Journal of The Institution of Engineers (India): Series C* 99 (5): 525–530. https://doi.org/10.1007/s40032-016-0329-7.

Borodina, I., R. K. Kildegaard, N. B. Jensen, T. H. Blicher, J. Maury, S. Sherstyk and K. Schneider. 2015. Establishing a Synthetic Pathway for High-Level Production of 3-Hydroxypropionic Acid in Saccharomyces Cerevisiae via β-Alanine. *Metabolic Engineering* 27 (January): 57–64. https://doi.org/10.1016/j.ymben.2014.10.003.

Calam, C. T., A. E. Oxford, and H. Raistrick. 1939. Studies in the Biochemistry of Micro-Organisms: Itaconic Acid, a Metabolic Product of a Strain of *Aspergillus Terreus* Thom. *The Biochemical Journal* 33 (9): 1488–1495. https://doi.org/10.1042/bj0331488.

Carro, M. D., and M. J. Ranilla. 2003. Influence of Different Concentrations of Disodium Fumarate on Methane Production and Fermentation of Concentrate Feeds by Rumen Micro-Organisms in *Vitro. British Journal of Nutrition* 90 (3): 617–623. https://doi.org/10.1079/BJN2003935.

Carvalho, M., C. Roca and A.M Maria. 2016. Improving Succinic Acid Production by Actinobacillus Succinogenes from Raw Industrial Carob Pods. *Bioresource Technology* 218 (October): 491–497. https://doi.org/10.1016/j.biortech.2016.06.140.

Chaturvedi, S., A. Bhattacharya, and S. K. Khare. 2018a. Trends in Oil Production from Oleaginous Yeast using Biomass: Biotechnological Potential and Constraints. *Applied Biochemistry and Microbiology* 54 (4): 361–369. https://doi.org/10.1134/S000368381804004X.

Chaturvedi, S., A. Kumari, A. Bhatacharya, A. Sharma, L. Nain and S. K. Khare. 2018b. Banana Peel Waste Management for Single-Cell Oil Production. *Energy, Ecology and Environment* 3 (5): 296–303. https://doi.org/10.1007/s40974-018-0101-3.

Chen, H. and Fu, X. 2016. Industrial Technologies for Bioethanol Production from Lignocellulosic Biomass. *Renewable and Sustainable Energy Reviews* 57: 468–478.

Chen, Z., J. Huang, Y. Wu, W. Wu, Y. Zhang and D. Liu. 2017. Metabolic Engineering of Corynebacterium Glutamicum for the Production of 3-Hydroxypropionic Acid from Glucose and Xylose. *Metabolic Engineering* 39 (January): 151–158. https://doi.org/10.1016/j.ymben.2016.11.009.

Cheng, C., Y. Zhou, M. Lin, P.Wei and S. T. Yang. 2017. Polymalic Acid Fermentation by Aureobasidium Pullulans for Malic Acid Production from Soybean Hull and Soy Molasses: Fermentation Kinetics and Economic Analysis. *Bioresource Technology* 223 (January): 166–174. https://doi.org/10.1016/j.biortech.2016.10.042.

Cheng, K. K., X. B. Zhao, J. Zeng, R. C. Wu, Y. Z. Xu, D. H. Liu and J. A. Zhang. 2012a. Downstream Processing of Biotechnological Produced Succinic Acid. *Applied Microbiology and Biotechnology* 95 (4): 841–850. https://doi.org/10.1007/s00253-012-4214-x.

Cheng, K. K., X. B. Zhao, J. Zeng and J. A Zhang. 2012b. Biotechnological Production of Succinic Acid: Current State and Perspectives. *Biofuels, Bioproducts and Biorefining* 6 (3): 302–318. https://doi.org/10.1002/bbb.1327.

Chi, Z., Z. P. Wang, G. Y. Wang, I. Khan and Z. M. Chi. 2016. Microbial Biosynthesis and Secretion of L -Malic Acid and Its Applications. *Critical Reviews in Biotechnology* 36 (1): 99–107. https://doi.org/10.3109/07388551.2014.924474.

Clarke, K. G. 2013. *Bioprocess Engineering: An Introductory Engineering and Life Science Approach*. Elsevier.

Dąbkowska, K., M. A. Morales, M. Kuglarz and I. Angelidaki. 2019. Miscanthus Straw as Substrate for Biosuccinic Acid Production: Focusing on Pretreatment and Downstream Processing. *Bioresource Technology* 278 (April): 82–91. https://doi.org/10.1016/j.biortech.2019.01.051.

Das, R. K., and S. K. Brar. 2014. Enhanced Fumaric Acid Production from Brewery Wastewater and Insight into the Morphology of Rhizopus Oryzae 1526. *Applied Biochemistry and Biotechnology* 172 (6): 2974–2988. https://doi.org/10.1007/s12010-014-0739-z.

Das, R. K., S. K. Brar and M. Verma. 2015. A Fermentative Approach towards Optimizing Directed Biosynthesis of Fumaric Acid by Rhizopus Oryzae 1526 Utilizing Apple Industry Waste Biomass. *Fungal Biology* 119 (12): 1279–1290. https://doi.org/10.1016/j.funbio.2015.10.001.

Das, R. K., S. K. Brar and M. Verma. 2016. Fumaric Acid. In *Platform Chemical Biorefinery*, 133–157. Elsevier. https://doi.org/10.1016/B978-0-12-802980-0.00008-0.

De Carvalho, J. C., A. I. Magalhães and C. R. Soccol. 2018. Biobased Itaconic Acid Market and Research Trends—Is It Really a Promising Chemical. *Chimica Oggi-Chemical Today* 36: 56–58.

Delidovich, I. P., J. C. Hausoul, L. Deng, R. Pfützenreuter, M. Rose and R. Palkovits. 2016. Alternative Monomers Based on Lignocellulose and Their Use for Polymer Production. *Chemical Reviews* 116 (3): 1540–1599. https://doi.org/10.1021/acs.chemrev.5b00354.

Della P. C. Cristina, E. Falletta and M. Rossi. 2011. A Green Approach to Chemical Building Blocks. The Case of 3-Hydroxypropanoic Acid. *Green Chemistry* 13 (7): 1624. https://doi.org/10.1039/c1gc15052a.

Deng, F., and G. M. Aita. 2018. Fumaric Acid Production by Rhizopus Oryzae ATCC® 20344™ from Lignocellulosic Syrup. *BioEnergy Research* 11 (2): 330–340. https://doi.org/10.1007/s12155-018-9899-y.

Deng, Y., Y. Mao and X. Zhang. 2016. Metabolic Engineering of a Laboratory-Evolved *Thermobifida Fusca* MuC Strain for Malic Acid Production on Cellulose and Minimal Treated Lignocellulosic Biomass. *Biotechnology Progress* 32 (1): 14–20. https://doi.org/10.1002/btpr.2180.

Dessie, W., F. Xin, W. Zhang, Y. Jiang, H. Wu, J. Ma and M. Jiang. 2018. Opportunities, Challenges, and Future Perspectives of Succinic Acid Production by Actinobacillus Succinogenes. *Applied Microbiology and Biotechnology* 102 (23): 9893–9910. https://doi.org/10.1007/s00253-018-9379-5.

Dörsam, S., J. Fesseler, O. Gorte, T. Hahn, S Zibek, C. Syldatk and K. Ochsenreither. 2017. Sustainable Carbon Sources for Microbial Organic Acid Production with Filamentous Fungi. *Biotechnology for Biofuels* 10 (1): 242. https://doi.org/10.1186/s13068-017-0930-x.

Dowlathabad, M. R., M. D. J Hussain, S. P. V. Rangadu, Subramanyam K, S. G. Krishna and V. N. Swamy. 2007. Fermentatative Production of Itaconic Acid by Aspergillus Terreus Using Jatropha Seed Cake. *African Journal of Biotechnology* 6 (18): 2140–2142. https://doi.org/10.5897/AJB2007.000-2333.

Fang, Z., R. L. Smith and X. Qi, eds. 2017. Production of Platform Chemicals from Sustainable Resources. In *Biofuels and Biorefineries*. Singapore: Springer Singapore. https://doi.org/10.1007/978-981-10-4172-3.

Farmer, T., R. Castle, J. Clark and D. Macquarrie. 2015. Synthesis of Unsaturated Polyester Resins from Various Bio-Derived Platform Molecules. *International Journal of Molecular Sciences* 16 (12): 14912–14932. https://doi.org/10.3390/ijms160714912.

Fatima, H., S. Chaturvedi, J. Grewal and S. K. Khare. 2022. Sustainable Production of Succinic Acid by Utilization of Agricultural Wastes. In *Biomass, Biofuels, Biochemicals*, 463–480. Elsevier. https://doi.org/10.1016/B978-0-323-89855-3.00004-2.

Felthouse, T. R., C. J. Burnett, B. Horrell, M. J. Mummey and Y. J. Kuo. 2001. Maleic Anhydride, Maleic Acid, and Fumaric Acid. In *Kirk-Othmer Encyclopedia of Chemical Technology*. John Wiley & Sons, Inc., https://doi.org/10.1002/0471238961.1301120506051220.a01.pub2.

Feng, J., J. Yang, W. Yang, J. Chen, M. Jiang and Xiang Zou. 2018. Metabolome- and Genome-Scale Model Analyses for Engineering of Aureobasidium Pullulans to Enhance Polymalic Acid and Malic Acid Production from Sugarcane Molasses. *Biotechnology for Biofuels* 11 (1): 94. https://doi.org/10.1186/s13068-018-1099-7.

Forage, R G, and M A Foster. 1982. Glycerol Fermentation in Klebsiella Pneumoniae: Functions of the Coenzyme B12-Dependent Glycerol and Diol Dehydratases. *Journal of Bacteriology* 149 (2): 413–419. https://doi.org/10.1128/jb.149.2.413-419.1982.

Foster, J. W., and S. A. Waksman. 1939. The Production of Fumaric Acid by Molds Belonging to the Genus *Rhizopus*. *Journal of the American Chemical Society* 61 (1): 127–135. https://doi.org/10.1021/ja01870a043.

Gao, C., X. Yang, H. Wang, C. P. Rivero, C. Li, Z. Cui, Q. Qi and C. S. K. Lin. 2016. Robust Succinic Acid Production from Crude Glycerol using Engineered *Yarrowia Lipolytica*. *Biotechnology for Biofuels* 9 (1): 179. https://doi.org/10.1186/s13068-016-0597-8.

Ge, X., C. Chang, L. Zhang, S. Cui, X. Luo, S. Hu, Y. Qin and Yebo Li. 2018. Conversion of Lignocellulosic Biomass into Platform Chemicals for Biobased Polyurethane Application. In *Advances in Bioenergy*, 3: 161–213. Elsevier. https://doi.org/10.1016/bs.aibe.2018.03.002.

Geilen, F. M. A., B. Engendahl, A. Harwardt, W. Marquardt, J. Klankermayer and Walter Leitner. 2010. Selective and Flexible Transformation of Biomass-Derived Platform Chemicals by a Multifunctional Catalytic System. *Angewandte Chemie International Edition* 49 (32): 5510–5514. https://doi.org/10.1002/anie.201002060.

Ginni G., S. Kavitha, Y. Kannah, S. K. Bhatia, A. Kumar, M. Rajkumar, G. Kumar, A. Pugazhendhi, N. T. L. Chi and R. Banu. 2021. Valorization of Agricultural Residues: Different Biorefinery Routes. *Journal of Environmental Chemical Engineering* 9 (4): 105435. https://doi.org/10.1016/j.jece.2021.105435.

Giorno, L., E. Drioli, G. Carvoli, A. Cassano and L. Donato. 2001. Study of an Enzyme Membrane Reactor with Immobilized Fumarase for Production of L-Malic Acid. *Biotechnology & Bioengineering* 72 (1): 77–84. https://doi.org/10.1002/1097-0290(20010105)72:1<77::AID-BIT11>3.0.CO;2-L.

Gnanasekaran, R., B. Dhandapani and J. Iyyappan. 2019. Improved Itaconic Acid Production by Aspergillus Niveus Using Blended Algal Biomass Hydrolysate and Glycerol as Substrates. *Bioresource Technology* 283 (July): 297–302. https://doi.org/10.1016/j.biortech.2019.03.107.

Goldberg, I., J. S. Rokem and O. Pines. 2006. Organic Acids: Old Metabolites, New Themes. *Journal of Chemical Technology & Biotechnology* 81 (10): 1601–1611. https://doi.org/10.1002/jctb.1590.

Gupta, M. N., M. Perwez and M. Sardar. 2020. Protein Crosslinking: Uses in Chemistry, Biology and Biotechnology. *Biocatalysis and Biotransformation* 38 (3) 178–201.

Hossain, A. H., R. V. Gerven, K. M. Overkamp, P. S. Lübeck, H. Taşpınar, M. Türker and P. J. Punt. 2019. Metabolic Engineering with ATP-Citrate Lyase and Nitrogen Source Supplementation Improves Itaconic Acid Production in Aspergillus Niger. *Biotechnology for Biofuels* 12 (1): 233. https://doi.org/10.1186/s13068-019-1577-6.

Hossain, A. H., A. Li, A. Brickwedde, L. Wilms, M. Caspers, K. Overkamp and Peter J. Punt. 2016. Rewiring a Secondary Metabolite Pathway towards Itaconic Acid Production in Aspergillus Niger. *Microbial Cell Factories* 15 (1): 130. https://doi.org/10.1186/s12934-016-0527-2.

Huang, X., M. Chen, X. Lu, Y. Li, X. Li and Jian-Jun Li. 2014. Direct Production of Itaconic Acid from Liquefied Corn Starch by Genetically Engineered Aspergillus Terreus. *Microbial Cell Factories* 13 (1): 108. https://doi.org/10.1186/s12934-014-0108-1.

Indera L., A. Amru, J. M. Jahim, S. Harun, J. P. Tan and A. W. Mohammad. 2016. Biorefinery Approach towards Greener Succinic Acid Production from Oil Palm Frond Bagasse. *Process Biochemistry* 51 (10): 1527–1537. https://doi.org/10.1016/j.procbio.2016.08.011.

Ingle, A. P., A. K. Chandel and S. S. Silva. 2020. Biorefining of Lignocellulose into Valuable Products. In *Lignocellulosic Biorefining Technologies*, edited by Avinash P. Ingle, Anuj Kumar Chandel and Silvio Silvério Silva, 1st ed., 1–5. Wiley. https://doi.org/10.1002/9781119568858.ch1.

Iyyappan, J., G. Baskar, E. Gnansounou, A. Pandey, J. K. Raaman, B. Bharathiraja and R. Praveenkumar. 2019. Recent Advances in Microbial Production of Malic Acid from Renewable Byproducts. *Reviews in Environmental Science and Bio/Technology* 18 (3): 579–595. https://doi.org/10.1007/s11157-019-09503-2.

Iyyappan, J., B. Bharathiraja, G. Baskar, and E. Kamalanaban. 2019. Process Optimization and Kinetic Analysis of Malic Acid Production from Crude Glycerol Using Aspergillus Niger. *Bioresource Technology* 281 (June): 18–25. https://doi.org/10.1016/j.biortech.2019.02.067.

Jang, Y. S., B. Kim, J. H. Shin, Y. J. Choi, S. Choi, C. W. Song, J. Lee, H. G.Park and S. Y. Lee. 2012. Bio-Based Production of C2-C6 Platform Chemicals. *Biotechnology and Bioengineering* 109 (10): 2437–2459. https://doi.org/10.1002/bit.24599.

Jantama, K., M. J. Haupt, S. A. Svoronos, X. Zhang, J. C. Moore, K. T. Shanmugam, and L. O. Ingram. 2008. Combining Metabolic Engineering and Metabolic Evolution to Develop Nonrecombinant Strains of Escherichia Coli C that Produce Succinate and Malate. *Biotechnology and Bioengineering* 99 (5): 1140–1153. https://doi.org/10.1002/bit.21694.

Jiang, M, J. Ma, M. Wu, R. Liu, L. Liang, F. Xin, W. Zhang, H. Jia and W. Dong. 2017. Progress of Succinic Acid Production from Renewable Resources: Metabolic and Fermentative Strategies. *Bioresource Technology* 245 (December): 1710–1717. https://doi.org/10.1016/j.biortech.2017.05.209.

Jiménez Q., Amparo, E. Pollet, M. Zhao, E. Marchioni, L. Avérous and V. Phalip. 2016. Itaconic and Fumaric Acid Production from Biomass Hydrolysates by Aspergillus Strains. *Journal of Microbiology and Biotechnology* 26: 1557–1565.

Jöbstl, D., T. Husøy, J. Alexander, T. Bjellaas, E. Leitner and M. Murkovic. 2010. Analysis of 5-Hydroxymethyl-2-Furoic Acid (HMFA) the Main Metabolite of Alimentary 5-Hydroxymethyl-2-Furfural (HMF) with HPLC and GC in Urine. *Food Chemistry* 123 (3): 814–818. https://doi.org/10.1016/j.foodchem.2010.05.015.

Joo, Y. C., S. K. You, S. K. Shin, Y. J. Ko, K. H. Jung, S. A. Sim and S. O. Han. 2017. Bio-Based Production of Dimethyl Itaconate from Rice Wine Waste-Derived Itaconic Acid. *Biotechnology Journal* 12 (11): 1700114. https://doi.org/10.1002/biot.201700114.

Kamm, B., P. R. Gruber and M. Kamm, eds. 2006. *Biorefineries-Industrial Processes and Products*, Vol. 2. Wiley-VCH.

Karthik, T., and R. Rathinamoorthy. 2017. Sustainable Synthetic Fibre Production. In *Sustainable Fibres and Textiles*, 191–240. Elsevier. https://doi.org/10.1016/B978-0-08-102041-8.00008-1.

Khare, S. K., A. Pandey and C. Larroche. 2015. Current Perspectives in Enzymatic Saccharification of Lignocellulosic Biomass. *Biochemical Engineering Journal* 102 (October): 38–44. https://doi.org/10.1016/j.bej.2015.02.033.

Kim, J., H. M. Seo, S. K. Bhatia, H. S. Song, J. H. Kim, J. M. Jeon and K. Y. Choi. 2017. Production of Itaconate by Whole-Cell Bioconversion of Citrate Mediated by Expression of Multiple Cis-Aconitate Decarboxylase (CadA) Genes in Escherichia Coli. *Scientific Reports* 7 (1): 39768. https://doi.org/10.1038/srep39768.

Kinoshita, K. 1932. Über die Produktion von Itaconsäure und Mannit durch einen neuen Schimmelpilz Aspergillus itaconicus. *Acta Phytochim* 5: 271–287.

Ko, Y., S. Ashok, S. Zhou, V. Kumar and S. Park. 2012. Aldehyde Dehydrogenase Activity Is Important to the Production of 3-Hydroxypropionic Acid from Glycerol by Recombinant Klebsiella Pneumoniae. *Process Biochemistry* 47 (7): 1135–1143. https://doi.org/10.1016/j.procbio.2012.04.007.

Komesu, A., J. Oliveira, D. K. T. Moreira, A. H. Khalid, J. M. Neto and L. H. S.2022Z. Martins. Biorefinery Approach for Production of Some High-Value Chemicals. In *Advanced Biofuel Technologies*, 409–429. Elsevier.

Kövilein, A., C. Kubisch, L. Cai and K Ochsenreither. 2020. Malic Acid Production from Renewables: A Review. *Journal of Chemical Technology & Biotechnology* 95 (3): 513–526. https://doi.org/10.1002/jctb.6269.

Kumar, V., S. Ashok and S. Park. 2013. Recent Advances in Biological Production of 3-Hydroxypropionic Acid. *Biotechnology Advances* 31 (6): 945–961. https://doi.org/10.1016/j.biotechadv.2013.02.008.

Kumar, V., M. Sankaranarayanan, K. Jae, M. Durgapal, S. Ashok, Y. Ko, R. Sarkar and S. Park. 2012. Co-Production of 3-Hydroxypropionic Acid and 1,3-Propanediol from Glycerol Using Resting Cells of Recombinant Klebsiella Pneumoniae J2B Strain Overexpressing Aldehyde Dehydrogenase. *Applied Microbiology and Biotechnology* 96 (2): 373–383. https://doi.org/10.1007/s00253-012-4187-9.

Kumari, D., and R. Singh. 2018. Pretreatment of Lignocellulosic Wastes for Biofuel Production: A Critical Review. *Renewable and Sustainable Energy Reviews* 90 (July): 877–891. https://doi.org/10.1016/j.rser.2018.03.111.

Larsen, H., and K. E. Eimhjellen. 1955. The Mechanism of Itaconic Acid Formation by Aspergillus Terreus. 1. The Effect of Acidity. *The Biochemical Journal* 60 (1): 135–139. https://doi.org/10.1042/bj0600135.

Lee, R. A., and J. M. Lavoie. 2013. From First- to Third-Generation Biofuels: Challenges of Producing a Commodity from a Biomass of Increasing Complexity. *Animal Frontiers* 3 (2): 6–11. https://doi.org/10.2527/af.2013-0010.

Li, C., K. L. Ong, X. Yang and C. S. K. Lin. 2019. Bio-Refinery of Waste Streams for Green and Efficient Succinic Acid Production by Engineered Yarrowia Lipolytica without PH Control. *Chemical Engineering Journal* 371 (September): 804–812. https://doi.org/10.1016/j.cej.2019.04.092.

Li, J., H. Zong, B. Zhuge, X. Lu, H. Fang and Jin Sun. 2016. Immobilization of Acetobacter Sp. CGMCC 8142 for Efficient Biocatalysis of 1, 3-Propanediol to 3-Hydroxypropionic Acid. *Biotechnology and Bioprocess Engineering* 21 (4): 523–530. https://doi.org/10.1007/s12257-016-0022-y.

Li, X., Y. Liu, Y. Y. H. Zhang, H. W., Y. Wu and M. Zhangl. 2014. High Levels of Malic Acid Production by the Bioconversion of Corn Straw Hydrolyte Using an Isolated Rhizopus Delemar Strain. *Biotechnology and Bioprocess Engineering* 19 (3): 478–492. https://doi.org/10.1007/s12257-014-0047-z.

Li, Y., and P. Tian. 2015. Contemplating 3-Hydroxypropionic Acid Biosynthesis in Klebsiella Pneumoniae. *Indian Journal of Microbiology* 55 (2): 131–139. https://doi.org/10.1007/s12088-015-0513-0.

Li, Y., X. Wang, X. Ge and P. Tian. 2016. High Production of 3-Hydroxypropionic Acid in Klebsiella Pneumoniae by Systematic Optimization of Glycerol Metabolism. *Scientific Reports* 6 (1): 26932. https://doi.org/10.1038/srep26932.

Lindlbauer, K. A., H. Marx and M. Sauer. 2017. 3-Hydroxypropionaldehyde Production from Crude Glycerol by Lactobacillus Diolivorans with Enhanced Glycerol Uptake. *Biotechnology for Biofuels* 10 (1): 295. https://doi.org/10.1186/s13068-017-0982-y.

Liu, C., Y. Ding, M. Xian, M. Liu, H. Liu, Q. Ma and Guang Zhao. 2017a. Malonyl-CoA Pathway: A Promising Route for 3-Hydroxypropionate Biosynthesis. *Critical Reviews in Biotechnology* 37 (7): 933–941. https://doi.org/10.1080/07388551.2016.1272093.

Liu, C., W. Shi, H. Li, Z. Lei, L. He and Z. Zhang. 2014. Improvement of Methane Production from Waste Activated Sludge by On-Site Photocatalytic Pretreatment in a Photocatalytic Anaerobic Fermenter. *Bioresource Technology* 155 (March): 198–203. https://doi.org/10.1016/j.biortech.2013.12.041.

Liu, H., J. Ma, M. Wang, W. Wang, L. Deng, K. Nie, X.Yue, F.Wangand T. Tan. 2016. Food Waste Fermentation to Fumaric Acid by Rhizopus Arrhizus RH7-13. *Applied Biochemistry and Biotechnology* 180 (8): 1524–1533. https://doi.org/10.1007/s12010-016-2184-7.

Liu, J., Z.g Xie, H. D. Shin, J. Li, G. Du, J. Chen and L. Liu. 2017b. Rewiring the Reductive Tricarboxylic Acid Pathway and L-Malate Transport Pathway of Aspergillus Oryzae for Overproduction of L-Malate. *Journal of Biotechnology* 253 (July): 1–9. https://doi.org/10.1016/j.jbiotec.2017.05.011.

Lo, E., L. B. Catala, I. Dogaris, E. M. Ammar and G. P. Philippidis. 2020. Biochemical Conversion of Sweet Sorghum Bagasse to Succinic Acid. *Journal of Bioscience and Bioengineering* 129 (1): 104–109. https://doi.org/10.1016/j.jbiosc.2019.07.003.

Lohbeck, K., H. Haferkorn, W. Fuhrmann and N. Fedtke. 2000. Maleic and Fumaric Acids. In *Ullmann's Encyclopedia of Industrial Chemistry*. Wiley-VCH Verlag GmbH & Co. KGaA. https://doi.org/10.1002/14356007.a16_053.

López-Garzón, C. S., and J.J A. Straathof. 2014. Recovery of Carboxylic Acids Produced by Fermentation. *Biotechnology Advances* 32 (5): 873–904. https://doi.org/10.1016/j.biotechadv.2014.04.002.

Manisha and S. K. Yadav. 2017. "Technological Advances and Applications of Hydrolytic Enzymes for Valorization of Lignocellulosic Biomass." Bioresource Technology 245 (December): 1727–1739. https://doi.org/10.1016/j.biortech.2017.05.066.

Mao, Y., G. Li, Z. Chang, R.Tao, Z. Cui, Z. Wang, Y. J. Tang, T. Chen and X. Zhao. 2018. Metabolic Engineering of Corynebacterium Glutamicum for Efficient Production of Succinate from Lignocellulosic Hydrolysate. *Biotechnology for Biofuels* 11 (1): 95. https://doi.org/10.1186/s13068-018-1094-z.

Meynial S., I., N. Forchhammer, C. Croux, L. Girbal and P. Soucaille. 2007. Evolution of a Saccharomyces Cerevisiae Metabolic Pathway in Escherichia Coli. *Metabolic Engineering* 9 (2): 152–159. https://doi.org/10.1016/j.ymben.2006.09.002.

Moe, S. T., K. K. Janga, T. Hertzberg, M. B. Hägg, K. Øyaas and Nils Dyrset. 2012. Saccharification of Lignocellulosic Biomass for Biofuel and Biorefinery Applications – A Renaissance for the Concentrated Acid Hydrolysis? *Energy Procedia* 20: 50–58. https://doi.org/10.1016/j.egypro.2012.03.007.

Moon, Y. M., R. Gurav, J. Kim, Y. G. Hong, S. K. Bhatia, H. R. Jung, and J. W. Hong. 2018. Whole-Cell Immobilization of Engineered Escherichia Coli JY001 with Barium-Alginate for Itaconic Acid Production. *Biotechnology and Bioprocess Engineering* 23 (4): 442–447. https://doi.org/10.1007/s12257-018-0170-3.

Naude, A., and W. Nicol. 2018. Malic Acid Production through the Whole-Cell Hydration of Fumaric Acid with Immobilised Rhizopus Oryzae.*Biochemical Engineering Journal* 137 (September): 152–161. https://doi.org/10.1016/j.bej.2018.05.022.

Özdenkçi, K., C. D. Blasio, H. R. Muddassar, K. Melin, P. Oinas, J. Koskinen, G. Sarwar and Mika Järvinen. 2017. A Novel Biorefinery Integration Concept for Lignocellulosic Biomass. *Energy Conversion and Management* 149 (October): 974–987. https://doi.org/10.1016/j.enconman.2017.04.034.

Pandurić, N., A. Šalić and Bruno Zelić. 2017. Fully Integrated Biotransformation of Fumaric Acid by Permeabilized Baker's Yeast Cells with in Situ Separation of L-Malic Acid Using Ultrafiltration, Acidification and Electrodialysis. *Biochemical Engineering Journal* 125 (September): 221–229. https://doi.org/10.1016/j.bej.2017.06.005.

Papadaki, A., H. Papapostolou, M. Alexandri, N. Kopsahelis, S. Papanikolaou, A. M. de Castro, D. M. G. Freire and A. A. Koutinas. 2018. Fumaric Acid Production Using Renewable Resources from Biodiesel and Cane Sugar Production Processes. *Environmental Science and Pollution Research* 25 (36): 35960–35970. https://doi.org/10.1007/s11356-018-1791-y.

Pedroso, G. B., S. Montipó, D. A. N. Mario, S. H. Alves and A. F. Martins. 2017. Building Block Itaconic Acid from Left-over Biomass. *Biomass Conversion and Biorefinery* 7 (1): 23–35. https://doi.org/10.1007/s13399-016-0210-1.

Presečki, A. V., B. Zelić, and Đ. V. Rački. 2007. Comparison of the L-Malic Acid Production by Isolated Fumarase and Fumarase in Permeabilized Baker's Yeast Cells. *Enzyme and Microbial Technology* 41 (5): 605–612. https://doi.org/10.1016/j.enzmictec.2007.05.007.

Rajesh, R. O., T. K. Godan, A. K. Rai, D. Sahoo, A. Pandey and P. Binod. 2019. Biosynthesis of 2,5-Furan Dicarboxylic Acid by Aspergillus Flavus APLS-1: Process Optimization and Intermediate Product Analysis. *Bioresource Technology* 284 (July): 155–160. https://doi.org/10.1016/j.biortech.2019.03.105.

Rathnasingh, C., S. M. Raj, J. E. Jo and Sunghoon Park. 2009. Development and Evaluation of Efficient Recombinant *Escherichia Coli* Strains for the Production of 3-Hydroxypropionic Acid from Glycerol. *Biotechnology and Bioengineering*, n/a-n/a. https://doi.org/10.1002/bit.22429.

Roa E., A. Carol, J. J. A. Straathof, T. W. Zijlmans, W. M. Gulik and L. A. M. van der Wielen. 2008. Fumaric Acid Production by Fermentation. *Applied Microbiology and Biotechnology* 78 (3): 379–389. https://doi.org/10.1007/s00253-007-1341-x.

Saxena, R. K., S. Saran, J. Isar and R. Kaushik. 2017. Production and Applications of Succinic Acid. In *Current Developments in Biotechnology and Bioengineering*, 601–630. Elsevier. https://doi.org/10.1016/B978-0-444-63662-1.00027-0.

Sharma, H. K., C. Xu and W. Qin. 2019. Biological Pretreatment of Lignocellulosic Biomass for Biofuels and Bioproducts: An Overview. *Waste and Biomass Valorization* 10 (2): 235–251. https://doi.org/10.1007/s12649-017-0059-y.

Singh, J., M. Suhag and A. Dhaka. 2015. Augmented Digestion of Lignocellulose by Steam Explosion, Acid and Alkaline Pretreatment Methods: A Review. *Carbohydrate Polymers* 117 (March): 624–631. https://doi.org/10.1016/j.carbpol.2014.10.012.

Song, C., W. D. Kim, S. Choi, J. W. Jang and Sang Yup Lee. 2013. Metabolic Engineering of *Escherichia Coli* for the Production of Fumaric Acid. *Biotechnology and Bioengineering* 110 (7): 2025–2034. https://doi.org/10.1002/bit.24868.

Song, H. and S. Y. Lee. 2006. Production of Succinic Acid by Bacterial Fermentation. *Enzyme and Microbial Technology* 39 (3): 352–361.

Takkellapati, S., T. Li and M. A. Gonzalez. 2018. An Overview of Biorefinery-Derived Platform Chemicals from a Cellulose and Hemicellulose Biorefinery. *Clean Technologies and Environmental Policy* 20 (7): 1615–1630. https://doi.org/10.1007/s10098-018-1568-5.

Tan, J. P., J. M. Jahim, T. Y. Wu, S. Harun, B. H. Kim and A. W. Mohammad. 2014. Insight into Biomass as a Renewable Carbon Source for the Production of Succinic Acid and the Factors Affecting the Metabolic Flux toward Higher Succinate Yield. *Industrial & Engineering Chemistry Research* 53 (42): 16123–16134. https://doi.org/10.1021/ie502178j.

Tan, J. P., A. A. I. Luthfi, S. F. A. Manaf, T. Wu and J. M. Jahim. 2018. Incorporation of CO2 during the Production of Succinic Acid from Sustainable Oil Palm Frond Juice. *Journal of CO2 Utilization* 26 (July): 595–601. https://doi.org/10.1016/j.jcou.2018.06.006.

Tate, B. E. 1967. Polymerization of itaconic acid and derivatives. In *Fortschritte der Hochpolymeren-Forschung* 5/2: 214–232. Advances in Polymer Science. Springer-Verlag. https://doi.org/10.1007/BFb0051282.

Teleky, B. and D. Vodnar. 2019. Biomass-Derived Production of Itaconic Acid as a Building Block in Specialty Polymers. *Polymers* 11 (6): 1035. https://doi.org/10.3390/polym11061035

Valdehuesa, K., N. G., H. Liu, G. M. Nisola, W. Chung, S. H. Lee and Si Jae Park. 2013. Recent Advances in the Metabolic Engineering of Microorganisms for the Production of 3-Hydroxypropionic Acid as C3 Platform Chemical. *Applied Microbiology and Biotechnology* 97 (8): 3309–3321. https://doi.org/10.1007/s00253-013-4802-4.

Vidra, A., and Á. Németh. 2017. Bio-Based 3-Hydroxypropionic Acid: A Review. *Periodica Polytechnica Chemical Engineering* 62 (2): 156. https://doi.org/10.3311/PPch.10861.

Voitenko, V. L., L. M. Gunina, V. G. Oleshko, and E. V. Nosach. 2018. Assessing Action Mechanisms of Pharmacological Agent Based on Succinic Acid Derivative under Physical Loads of Submaximal Intensity. *World of Medicine and Biology* 14 (65): 028. https://doi.org/10.26724/2079-8334-2018-3-65-28-32.

Wang, J., M. Lin, M. Xu, and S. T. Yang. 2016. Anaerobic Fermentation for Production of Carboxylic Acids as Bulk Chemicals from Renewable Biomass. In *Anaerobes in Biotechnology*, edited by Rajni Hatti-Kaul, Gashaw Mamo, and Bo Mattiasson, 156: 323–361. Advances in Biochemical Engineering/ Biotechnology. Springer International Publishing. https://doi.org/10.1007/10_2015_5009.

Werpy, T. and G Petersen. 2004. Top Value-Added Chemicals from Biomass: Volume I -- Results of Screening for Potential Candidates from Sugars and Synthesis Gas. *DOE/GO-102004-1992*, 15008859. https://doi.org/10.2172/15008859.

West, T. 2017. Microbial Production of Malic Acid from Biofuel-Related Coproducts and Biomass. *Fermentation* 3 (2): 14. https://doi.org/10.3390/fermentation3020014.

Wyman, H. and J. K. Palmer. 1964. Organic Acids in the Ripening Banana Fruit. *Plant Physiology* 39 (4): 630–633. https://doi.org/10.1104/pp.39.4.630.

Xia, J., J. Xu, L. Hu and X. Liu. 2016. Enhanced Poly(L-Malic Acid) Production from Pretreated Cane Molasses by *Aureobasidium Pullulans* in Fed-Batch Fermentation. *Preparative Biochemistry & Biotechnology* 46 (8): 798–802. https://doi.org/10.1080/10826068.2015.1135464.

Xia, J., J. Xu, X. Liu, J. Xu, X. Wang and X. Li. 2017. Economic Co-Production of Poly(Malic Acid) and Pullulan from Jerusalem Artichoke Tuber by Aureobasidium Pullulans HA-4D. *BMC Biotechnology* 17 (1): 20. https://doi.org/10.1186/s12896-017-0340-y.

Xu, H., B. Li and Xindong Mu. 2016. Review of Alkali-Based Pretreatment To Enhance Enzymatic Saccharification for Lignocellulosic Biomass Conversion. *Industrial & Engineering Chemistry Research* 55 (32): 8691–8705. https://doi.org/10.1021/acs.iecr.6b01907.

Xu, Q., S. Li, H. Huang and J. Wen. 2012. Key Technologies for the Industrial Production of Fumaric Acid by Fermentation. *Biotechnology Advances* 30 (6): 1685–1696. https://doi.org/10.1016/j.biotechadv.2012.08.007.

Xu, T. and C. Huang. 2008. Electrodialysis-Based Separation Technologies: A Critical Review. *AIChE Journal* 54 (12): 3147–3159. https://doi.org/10.1002/aic.11643.

Yang, C. and C. Huang. 2016. Biotransformation of 5-Hydroxy-Methylfurfural into 2,5-Furan-Dicarboxylic Acid by Bacterial Isolate Using Thermal Acid Algal Hydrolysate. *Bioresource Technology* 214 (August): 311–318. https://doi.org/10.1016/j.biortech.2016.04.122.

Yang, C. F. and C. R. Huang. 2018. 2018. Isolation of 5-Hydroxymethylfurfural Biotransforming Bacteria to Produce 2,5-Furan Dicarboxylic Acid in Algal Acid Hydrolysate. *Journal of Bioscience and Bioengineering* 125 (4): 407–412. https://doi.org/10.1016/j.jbiosc.2017.11.005.

Yin, X., J. Li, H. Shin, G. Du, L. Liu and J. Chen. 2015. Metabolic Engineering in the Biotechnological Production of Organic Acids in the Tricarboxylic Acid Cycle of Microorganisms: Advances and Prospects. *Biotechnology Advances* 33 (6): 830–841. https://doi.org/10.1016/j.biotechadv.2015.04.006.

Zeng, M., E. Ximenes, M. R. Ladisch, N. S. Mosier, W. Vermerris, C. P. Huang and D. M. Sherman. 2012. Tissue-Specific Biomass Recalcitrance in Corn Stover Pretreated with Liquid Hot-Water: Enzymatic Hydrolysis (Part 1). *Biotechnology and Bioengineering* 109 (2): 390–397.

Zhang, J., J. Li, Y. Tang, L. Lin and M. Long. 2015. Advances in Catalytic Production of Bio-Based Polyester Monomer 2,5-Furandicarboxylic Acid Derived from Lignocellulosic Biomass. *Carbohydrate Polymers* 130 (October): 420–428. https://doi.org/10.1016/j.carbpol.2015.05.028.

Zhang, X., X. Wang, K. T. Shanmugam and L. O. Ingram. 2011. L-Malate Production by Metabolically Engineered *Escherichia Coli. Applied and Environmental Microbiology* 77 (2): 427–434. https://doi.org/10.1128/AEM.01971-10.

Zhou, Y., K. Nie, X. Zhang, S. Liu, M. Wang, L. Deng, F. Wang and T. Tan. 2014. Production of Fumaric Acid from Biodiesel-Derived Crude Glycerol by Rhizopus Arrhizus. *Bioresource Technology* 163: 48–53.

4 Poultry Waste Biorefinery

Opportunities for Sustainable Management

*Arijit Jana, Diptarka Dasgupta,
Thallada Bhaskar, and Debashish Ghosh*

CONTENTS

4.1 INTRODUCTION

The Food and Agriculture Organization (FAO) of the United Nations estimates over 1.3 billion tons of foods are wasted every year (FAO, 2011). Meanwhile, the fast world population growth and the increasing average income over the last 50 years have intensified food demand. This growth will require a 60%-70% increase in food production by 2050 (Moraes et al., 2014). Crop production alone cannot resolve the food crisis and poverty of the fast-growing population (Dal ólio et al., 2017). Poultry production has been the appropriate answer for the fulfillment of requirements of a balanced diet (Singh et al., 2018). It provides a key source of protein and fulfills the basic needs to improve the standard of living, including poor people (Brandelli et al., 2015). The requirement for animal-sourced protein for human consumption is increasing globally at an unprecedented rate (Van Boeckel et al., 2015; Molaey et al., 2018). Poultry products, including eggs, chicken, and turkey meat, represent an important protein source in the diets of most people. Increased production of poultry for human consumption nonetheless results in the generation of huge amounts of waste materials, including offal (feathers, entrails, and organs of slaughtered birds), processing wastewater, and biosolids. Any kind of waste in excess concentration can be hazardous for humans, animals, and vegetation (Seadi and Holm-Nielsen, 2004). The irresponsible disposal of wastes has an acute negative impact on all environmental health. With the Science and Technology (S&T) interventions and systematic approach, most of these wastes can be recycled, reused, and channeled towards the production of value-added products, regardless of flock size.

The management of agricultural wastes is a top priority today in order to achieve sustainable development (Martin-Luengo et al., 2011). Nowadays, we have to focus on a sustainable bioeconomy,

DOI: 10.1201/9781003335740-4

where no waste is generated at the end of the chain. The biorefinery concept is a technical application of this principle in which exploitation of waste biomass is applied to get more economical and environmental benefits. The biorefinery approach is considered sustainable because the processes produce minimal waste to the environment, in fact decreasing the environmental burden and the pressure on ecosystem services. Day-to-day centralization of poultry processing has intensified the problem of disposal of poultry waste, the volume of which is sufficient to attract any processing companies. This chapter will provide the state of the art of the biorefinery concept in poultry waste implementation in the bioeconomy and conceptualize a baseline for potential biorefinery concept integration from a commercial viewpoint.

4.2 POULTRY INDUSTRY: STRUCTURE AND SIZE

The term "poultry" indicates all domestic avian species such as chicken, duck, emu, geese, guinea fowl, ostrich, pigeon, quail, and turkey, which are generally reared for eggs (called "layers") and meat (called "broilers"). Since the 1980s, broiler production has been greatly increased for the nutritional benefits of chicken meat in comparison to other meats (Brandelli et al., 2015). As a cost-effective and healthy protein resource, regular consumption of poultry and its by-products increases annually through various sections of society encompassing diverse cultures, traditions, and religions. The global human population is estimated to reach 9.6 billion in 2050 (Mottet and Tempio, 2017), and poultry emerges as the most efficient natural resource of protein to supply the global growing demand. This increase has been also accredited to the poultry industry, which supplies processed products that are easier for the customer to prepare. Livestock operations are now a multibillion-dollar trade industry in the global marketplace, and the growth of the global livestock sector is expected to continue. This industry is constantly growing, and the largest producers worldwide—namely the United States, Brazil, and China—account for more than 40 million tons (MT) per year in terms of production (USDA, 2014). The European Union annual production is approximately 11 MT (AVEC, 2014). Alexandratos and Bruinsma (2012) projected that the demand for animal-derived food could grow by 70% between 2005 and 2050. While beef and pork demand could increase by 66% and 43%, respectively, poultry meat is expected to have the highest growth, with 121%, and demand for eggs could increase by 65%. The world has over 23 billion poultry—about three birds per person on the planet and about five times more than 50 years ago. Based on the Global Livestock Environmental Assessment Model (GLEAM 2.0, 2016), backyard systems contribute 8% of global egg production and 2% of global meat, whereas specialized broiler systems contribute 92% of global poultry meat production. Backyard systems make a significant contribution to eggs and poultry meat production in Eastern Europe, South Asia, sub-Saharan Africa, and, to a lesser extent, East Asia and Latin America and the Caribbean. Poultry is particularly important for small holders and poor rural and urban communities in most of the developing countries and contributes about 98% of poultry products consumed in villages. At the global level, three main types of poultry production systems can be considered: broilers, layers, and backyard. There are also other interlinked segments, and many of them owned by the same company. This type of ownership is called vertical integration (Figure 4.1).

Until the 1950s, poultry keeping was largely a backyard venture, including native birds such as Aseel, Kadaknath, and other nondescriptive breeds in India (Sasidhar and Suvedi, 2016). Hybrid layer strains were introduced in 1955 and broiler strains in 1961 in India. Commercial broiler production is only 59 years old. At present, the poultry industry in India has become a vibrant and self-reliant industry that is booming and emerging as the world's second-largest market (Sharma, 2020). However, poultry production in India has taken a quantum leap in the last four decades, emerging from an entirely unorganized and unscientific farming practice to a full-fledged commercial enterprise/production system with state-of-the-art technological interventions. Now, poultry is the most systemized sector in animal husbandry in India (Bashir et al., 2020). The poultry industry is growing and providing direct or indirect employment to 6.5 million people in India. Agricultural

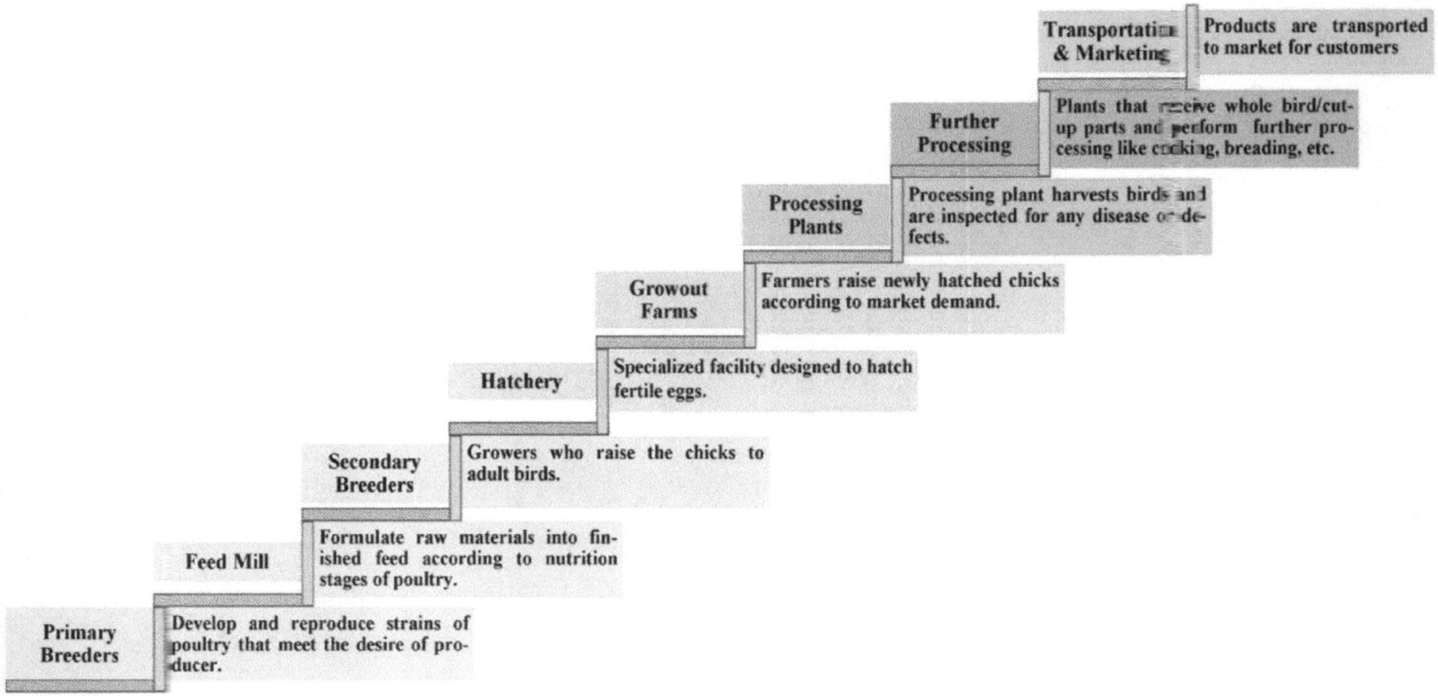

FIGURE 4.1 Poultry integration organization setup.

production has been expanding at a rate around 1.5%-2% per annum, whereas poultry production is increasing at the rate of about 8%-10% per annum, with an annual turnover of US$7500 million (Mehta et al., 2002). It provides about 1% of India's gross domestic product (GDP) and 10% of the total GDP attributed to livestock products. In India, per capita poultry meat consumption is around 3.6 kg per year (Jayathilakan et al., 2012). The Indian poultry market, encompassing broilers and eggs, was worth INR 1750 billion in 2018 and was worth INR 2049 billion in 2019 and is further expected to hit INR 4340 billion by 2024, growing at a compound annual growth rate (CAGR) of 16.2% during 2019–2024 (Sharma, 2020). A number of factors have triggered such progressive growth like religious restrictions, high per capita income growth, changing food habits and preferences (due to health consciousness), limited availability of fish, developmental support from state and central government, research and development support from research institutes, international collaborations, private-sector participation, and relatively low price (Mehta et al., 2002). Increasing household production and growing integration have pushed poultry meat prices downward. The future outlook for Indian poultry also appears to be very favorable. As of now, India has become the third-largest egg producer (after China and the United States), and the 19th largest broiler producer in the world (Mehta et al., 2002).

The structure of India's poultry farming and marketing practices in India varies from region to region. Such industries are generally divided in two sectors: an organized poultry sector that fulfills nearly 70% of the total market demand, and the nonorganized sector which delivers the rest 30% (Figure 4.2). Among total poultry production, approximately 66.7% is supplied as meat by the broiler sectors and the rest is used for egg production. India has exported 3,50,817.80 MT of poultry products to the world at a value of Rs. 574.61 Crores/80.34 USD million during the period 2019–2020. The live poultry markets in India mainly export their production to the South Asian Association for Regional Cooperation (SAARC) countries. Oman, Kuwait, the United Arab

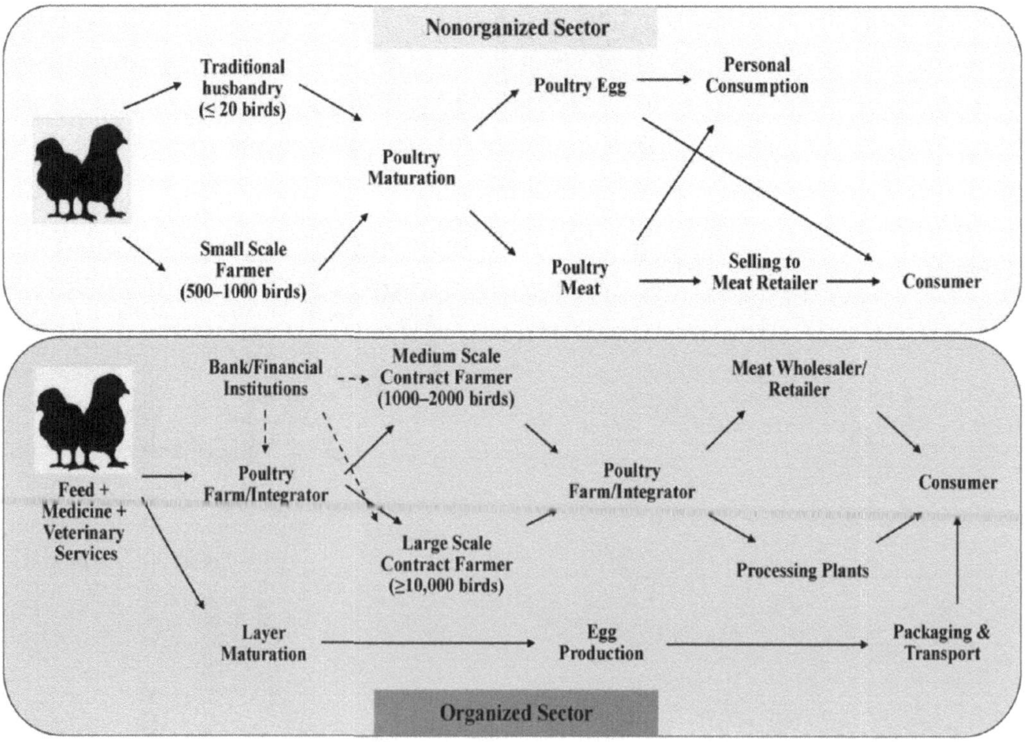

FIGURE 4.2 Structure of the Indian poultry industry.

Emirates, Saudi Arabia, and Yemen are the most important importers of India's table and hatching eggs. Germany, Austria, Japan, the Netherlands, and the Republic of Korea mainly import India's egg powder. The leading states in India for poultry production are Andhra Pradesh, Tamil Nadu, Maharashtra, Haryana, Karnataka, Maharashtra, Gujarat, Madhya Pradesh, Orissa, West Bengal, and North Eastern States. Poultry industries such as Venkateshwara Hatcheries, Suguna, Godrej, Shanti, Taffa, Arambagh, and Skylark play leading roles in distributing the latest techniques of poultry keeping and animal health care in India.

4.3 POULTRY WASTE: THE INDIAN SCENARIO

Waste is defined as anything that is no longer useful and needs to be disposed of. All farming processes generate different kinds of wastes. A large amount of wastes is also produced from the poultry industries, including solid waste and wastewater streams. The nature and quantity of the waste vary in different regions. Globally, over 90% of poultry waste is spread on land close to the poultry farms (Moore et al., 1995). Assuming an average chicken meat yield rate of 60%, the amount of live chicken consumption implies the generation of around 48 MT of processing by-products globally each year. Pelletier (2017) reported 4.6 kg of mortalities and 96.4 kg of spent hens per tons of eggs produced in Canada, as well as in the United States (Pelletier et al., 2013). In Maryland and Delaware, 523 million chickens are produced each year, which generates over 42 million cubic feet of chicken waste, enough to fill the dome of the US Capitol about 50 times. In Iran, the waste materials from poultry slaughterhouses are about 12.6 thousand tons per year (Geshlog-Olyayee et al., 2010). About 23.54 million metric tons (MMt) of poultry manure, 9486 Mt of hatchery waste, and 1.15 MMt of slaughterhouse waste are produced in India every year (Tensingh et al., 2016). These figures are bound to increase in the years to come owing to the fast growth rate witnessed by the Indian poultry industry.

4.4 PUBLIC HEALTH AND ENVIRONMENTAL IMPACT ASSESSMENT

Presently, the poultry industries are facing difficulties caused by the large amount of waste that is creating different environmental, social, and economic problems. Traditional disposal strategies of poultry waste are very expensive and problematic. The waste is often burned in incineration plants, buried in landfills, or recycled into low-quality animal feeds. These disposal methods generate greenhouse gases (GHG) and are dangerous to the human health and the, soil, water, and air environments. However, if it is assumed that poultry waste is contributing 33.7 MMt of CO_2 eq./yr, or 0.0337 giga tons (Gt) CO_2 eq./yr, improper disposal of poultry wastes causes following environmental impacts:

1. Emission of ammonia, hydrogen sulfide, volatile organic compounds (VOCs), and dust particles.
2. GHG emissions and health effects associated with nuisance odorants.
3. Wastes from animal production are directly applied into the soil, resulting in nitrate (NO_3) and phosphate (PO_4) contamination of groundwater.
4. NO_3 contamination in drinking water resulting in blue baby syndrome, cancer, and respiratory illness in humans and fetal abortions in livestock.
5. Nutrients in animal waste that enter streams also may lower oxygen levels and kill fish and other wildlife.
6. Bacteria in animal waste can contaminate drinking water and may cause potentially serious illnesses.
7. Land disposal of poultry waste causes undesirable nutrient imbalances followed by reduced crop production.

Poultry waste is also a potential reservoir of bacterial, viral, prion, and parasitic pathogens, capable of infecting both animals and humans (Franke-Whittle and Insam, 2013). Due to high market demand, poultry accumulate various metals (As, Pb, Zn, Ni, Cd, Cu, Mn), drugs, and other chemicals added in their feed as xanthophylls, antiprotozoals, antioxidants, probiotics, polychlorinated phenols, tetrachlorodibenzopdioxin, and hormones for higher growth yield (Haapapuro et al., 1997; Jackson et al., 2003). Apart from that, a huge amount of antibiotics and other drugs are used to treat poultry diseases (Marshall and Levy, 2011). The average annual ingestion of antimicrobials per kilogram of animal was estimated at about 148 mg/kg for chickens (Van Boeckel et al., 2015). Haphazard use of antibiotics, along with poor biosecurity and waste management systems in livestock as well as poultry production, create a reservoir of resistant bacteria with onward transmission to humans through consumption of contaminated food, contaminated environment (i.e., through water, soil, air), and direct contact with livestock (Price et al., 2005; Graham et al., 2009; Smith et al., 2013; Verraes et al., 2013; Van Boeckel et al., 2015). Farm and slaughterhouse workers and veterinarians are directly at risk of exposure to resistant bacteria (Marshall and Levy, 2011). The most common microorganisms associated with antimicrobial resistance in poultry operations are *Salmonella* spp., *Campylobacter* spp., *Listeria* spp., and *Cryptosporidium* spp. (USEPA, 2013). They have shown resistance mostly against tetracycline (46%), trimethoprim-sulfamethoxazole (27%), nalidixic acid (26%), ampicillin (26%), and streptomycin (21%) (Hasan et al., 2011). *Campylobacter* infections affect over 800,000 people annually, while *Salmonella* infections are responsible for an estimated 1 million foodborne illnesses a year. Potential bacterial pathogens, including *Salmonella* spp. and *Campylobacter jejuni*, are highly prevalent in US poultry (Hutchison et al., 2004; Petersen et al., 2001; USEPA, 2013). In Mexico, avian chlamydiosis is included in the list of notifiable diseases of animals. Growing global concerns of antibiotic resistance in livestock and subsequent transport to the environment have prompted the need for additional research.

4.5 CLASSIFICATION AND CHARACTERIZATION OF POULTRY WASTE

The consequences of poultry production are usually empty shells, infertile eggs, dead embryos, viscous liquid from eggs, decaying tissue, manure, bedding materials, hair, feather, food waste, blood, offal, bone, fat, on-farm mortalities, and considerable quantities of wastewater. According to the structure and farming practice of the poultry industry, waste is mainly generated from four different sectors in various forms with distinct characteristics (Table 4.1 and Figure 4.3).

4.6 WASTE PROCESSING

A summary table is provided to focus the comparative pros and cons of processes available and that were reported as potentially feasible at present for poultry waste valorization (Table 4.2).

a. ***Litter***

The main waste stream generated from the poultry industry is poultry litter, which consists of manure (excreta), bedding materials (wood shavings, sawdust, wheat straw, peanut hulls, rice hulls, etc.), spilled feed, and feathers (Chen and Jiang, 2014; Huang et al., 2015). Perkins et al. (1964) reported that an average of 1460 kg (dry basis) of litter can be harvested from 1000 birds in their 10-week growing cycle. The composition of poultry litter is largely water and carbon (C) with smaller amounts of nitrogen (N) and phosphorous (P) and trace levels of chlorine (Cl), calcium (Ca), magnesium (Mg), sodium (Na), manganese (Mn), iron (Fe), copper (Cu), zinc (Zn), and arsenic (As). Given the high nutrient content, it has traditionally been used as a soil amendment (Shih, 1993; Simpson, 1991; Billen et al., 2015). However, overapplication of this material can lead to eutrophication of water bodies, pathogens spread, production of phytotoxic substances, air pollution, and emission of GHG. With increasing production of poultry, litter disposal is becoming an urgent problem for the poultry industry (Dal ólio et al., 2017). Alternative, ecofriendly disposal routes, with probable economic benefits, may lie in large-scale biomass-to-energy schemes that can also provide

TABLE 4.1
Characteristics of Wastes Produced in Poultry Farming and Slaughterhouses

	TS	VS	Organic N	Protein	Lipids	Methane Potential	Methane Potential
	%	% of TS				$m^3 = kg\ VS_{added}$	$m^3 = kg$ wet weight
Manure	20–47 (Huang and Shih, 1981; Webb and Hawkes, 1985b)	60–76 (Huang and Shih, 1981; Webb and Hawkes, 1985b)	4.6–6.7 (Huang and Shih, 1981; Webb and Hawkes, 1985b)	NA	1.5–2.1 (Mackie et al., 1991)	0.2–0.3 (Huang and Shih, 1981; Safley et al., 1987)	0.04–0.06 (Huang and Shih, 1981; Safley et al., 1987)
Litter	52–81 (Webb and Hawkes, 1985a)	61–65 (Webb and Hawkes, 1985a)	3.2–5.7 (Webb and Hawkes, 1985a)	NA	NA	0.14–0.22 (Webb and Hawkes, 1985a)	0.10–0.15 (Webb and Hawkes, 1985a)
Feather	24.3 (Salminen et al., 2003)	96.7 (Salminen et al., 2003)	15 (Salminen et al., 2003)	91 (Salminen et al., 2003)	1–10 (Bourne, 1993)	0.2 (Salminen et al., 2003)	0.05 (Salminen et al., 2003)
Blood	22 (Salminen et al., 2003)	91 (Salminen et al., 2003)	7.6 (Salminen et al., 2003)	48 (Salminen et al., 2003)	2 (Salminen et al., 2003)	0.5 (Salminen et al., 2003)	0.10 (Salminen et al., 2003)
Carcass	37 (Chen and Shyu, 1998)	NA	NA	NA	NA	NA	0.20–0.25 (Chen, 1999; Chen and Shyu, 1998)
Offal, feet, and head	39 (Salminen et al., 2003)	95 (Salminen et al., 2003)	5.3 (Salminen et al., 2003)	32 (Salminen et al., 2003)	54 (Salminen et al., 2003)	0.7–0.9 (Salminen et al., 2003)	0.3 (Salminen et al., 2003)
Trimmings and bone	22.4 (Salminen et al., 2003)	68 (Salminen et al., 2003)	68.6 (Salminen et al., 2003)	51 (Salminen et al., 2003)	22 (Salminen et al., 2003)	0.6–0.7 (Salminen et al., 2003)	0.15–0.17 (Salminen et al., 2003)

TS: Total solid; VS: Volatile solid; NA: Not available.
Source: Adapted from (Salminen and Rintala 2002)

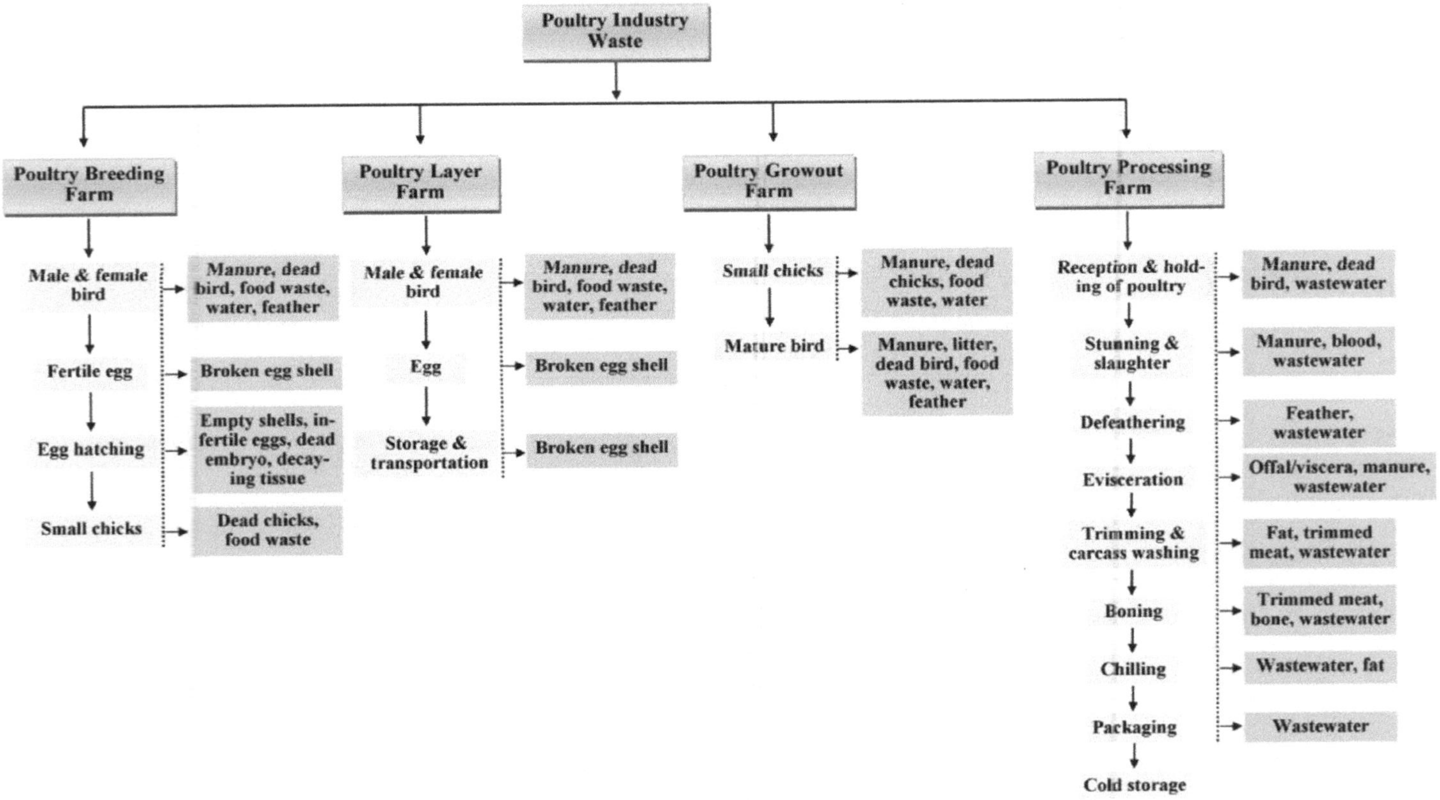

FIGURE 4.3 Organic solic and liquid materials produced in poultry farming and slaughtering.

TABLE 4.2
The Pros and Cons of Promising Poultry Waste Processing Methods

Sl. No.	Method	Applied For	Advantage	Disadvantage	Reference
1	Natural disposal	Feather, soft waste of slaughterhouse, poultry litter	1. Easy and inexpensive	1. Illegal in several areas 2. Air, water, and soil contamination 3. Probable chance of community disease outbreak 4. Increased predation of animals	(Gwyther et al., 2011)
2	Incineration: High-temperature combustion of organic substances contained in waste into ash, flue gas, and heat	In-farm mortalities	2. Easy and economic 3. High volume of waste reduction 4. Most effective methods for destroying potentially infectious agents 5. Eliminating the threat of disease 6. The residue is mostly harmless and does not attract rodents or insects	1. Atmospheric pollution 2. The air emission, process conditions, and the disposal of solid and liquid residues need to be strictly controlled 3. It requires proper sources of fuel and supervision; otherwise, smoke and odor can create nuisance complaints 4. Expensive equipment required	(Ritter and Chinside, 1995; Blake et al., 2008; Cai et al., 1994; Bhari et al., 2018)
3	Burning: High-temperature burning using fuels such as wood, tires, diesel, etc.	In-farm mortalities	1. Easy and economic 2. Eliminates the threat of disease outbreak	1. Leads to atmospheric pollution	(Anon, 2005; Brandelli et al., 2015)
4	Burial: Pit filled with waste, followed by covering with lime and soil	In-farm mortalities, feather, soft waste of slaughterhouse, poultry litter	1. Lasting containment for disease outbreaks 2. Simplest and economical method 3. Safe method	1. Poor site selection poses a pollution threat to groundwater, surface water, soil, and air 2. Tracking of sites required	(Blake, 2004; Thyagarajan et al., 2013)
5	Rendering: Different heating with high temperature and pressure to remove fat	Feather, soft waste of slaughterhouse, poultry litter	1. Rendering products can be used as feed and fertilizer 2. Fat product can be used for soaps, washing powders, cosmetics, and fuel 3. Best for on-farm disease control	1. Emission of gas and odor leads environmental concern 2. Not readily available everywhere 3. Demolition of essential amino acids 4. Poor digestibility and low nutritional value of animal feed	(Swan, 1992; Kirkpinar et al., 2004; Gwyther et al., 2011; Tiwary and Gupta, 2012)

(Continued)

TABLE 4.2 (Continued)
The Pros and Cons of Promising Poultry Waste Processing Methods

Sl. No.	Method	Applied For	Advantage	Disadvantage	Reference
6	Gasification: Thermochemical process converts carbon-rich materials into syngas (CO, H_2, CO_2 and CH_4) using a catalyst	Poultry manure	1. Easy temperature control 2. Can handle a variety of substrates 3. High (90–97%) conversion efficiency 4. Decreased formation of NOx 5. Suitable for low-value fuels 6. Lower environmental impacts 7. Zero wastewater discharge	1. High operational temperature (700–900 °C) 2. Higher tar/biochar production 3. Complex design 4. Pretreatment (drying) required for moist (>30%) substrate	(Dai et al., 2015; Taupe et al., 2016)
7	Composting: Aerobic microbial process to decompose organic material	Poultry slaughterhouse wastes, manure, litter, feathers	1. Resultant products are much more environmentally acceptable 2. Economic method 3. Kills pathogens 4. Controls disease outbreaks 5. Reduces the risk of nitrogen and phosphorus entering the water systems 6. Reduces the volume of feed and stabilizes trace minerals 7. Less odor 8. Can be done any time of the year 9. Less costly equipment required, hence it is economical	1. Loss of nutrients like nitrogen 2. Land area required for the compositing 3. Emission of greenhouse gases such as methane and nitrous oxide	(Turnell et al., 2007; Moreki and Chiripasi, 2011; Glatz et al., 2011)
8	Anaerobic digestion: In the presence of microorganisms, degradation, and stabilization of organic material under anaerobic conditions lead to formation of CH_4 and CO_2	Poultry litter, feathers, soft waste of slaughterhouse	1. Renewable energy production 2. Pathogen reduction 3. Organic fertilizer production	1. Expensive infrastructure 2. Technical 3. Economies of scale	(Lin et al., 1986; Williams, 1999; Glatz et al., 2011; Braber, 1995; Shih, 1993)
9	Chemical hydrolysis: Acid or alkali treatment with high temperature	Feathers, soft waste of slaughterhouse	1. Fewer gas emissions 2. Pathogen reduction 3. Beneficial by-product	1. Relatively expensive 2. High-volume acidic or alkaline effluent discharge 3. Limited capacity for large volumes	(Gwyther et al., 2011; Coward-Kelly, 2006; Kim and Patterson, 2000)

Sl. No.	Method	Applied For	Advantage	Disadvantage	Reference
10	Microbial enzyme: Microbial fermentation of protein through enzymes (i.e., proteases, keratinases) into peptides and amino acids	Feathers, slaughterhouse	1. Simple and ecofriendly methods 2. Beneficial varieties of by-product 3. High conversion efficiency	1. Limited capacity for large volumes 2. High operational costs 3. Trained manpower required	(Kim et al., 2001; Paul et al., 2016; Bhari et al., 2018; Brandelli et al., 2015)
11	Pyrolysis: Thermal degradation of organic material in the partial or complete absence of an oxidizing agent	Poultry litter	1. Convert biowaste into bio-oil/tar, char, and syngas 2. Works well with high-moisture contents 3. Simple, flexible, and easy to scale up 4. Fast pyrolysis produces larger volume of gas and reduces production of char in sort period	1. High operational costs 2. Low (30–40%) conversion efficiency	(Kim et al., 2009; Basu, 2010; Chan et al., 2008; Hussein et al., 2017)
12	Calcination: Conversion of $CaCO_3$ to CaO under high temperature (1000 °C) and open air conditions	Eggshell	1. Produce two types of nanocatalysts, CaOCeH-D and CaDG 2. Yields CaO with high purity (99.06% w/w)	1. Requires high temperature 2. Relative expensive equipment 3. High operational costs	(Tangboriboon et al., 2012; Pandit and Fulekar, 2017; Gupta and Rathod, 2018; Kanani et al., 2020)

an easier-to-handle fertilizer as a by-product. Currently, poultry litter waste is converted to valuable by-products through composting, anaerobic digestion (AD), and pyrolysis.

b. ***Feathers***

A second waste stream, unique to the poultry industry, is discarded feathers, since they account for approximately 8% of the adult chicken weight (Onifade et al., 1998). Globally, 8.5 billion tons of feather waste are produced annually, among which India alone constitutes about 350 million tons of feathers from the slaughterhouses and poultry industries (Kowalczyk et al., 2018). This huge pile of feathers and their localized accumulation create a serious disposal problem leading to environmental threats. Landfilling and burning are the most common practices adopted for feather disposal in India, as they require less effort. But these kinds of strategies are restricted or banned in many countries, as the consequence leads to soil infertility, emission of toxic gases (NH_3, H_2S, N_2O, etc.), and groundwater pollution (Adeoye et al., 2014). Improper feather disposal also causes various human ailments, including chlorosis, mycoplasmosis, fowl cholera, Newcastle disease, air sacculitis, keratoconjunctivitis, and deadly avian influenza (Saber et al., 2010; Sarkar et al., 2012; Bagust, 2013; Singh et al., 2018). Apart from that, feathers are a bioresource with a high protein content (above 750g/kg crude protein) (Dalev, 1994). Keratin is the most abundant recalcitrance protein found in feathers in the form of α-keratin, β-keratin, and amorphous keratins (Poopathi and Abidha, 2008). Owing to the high protein content, researchers have shown their interest in converting these wastes into many value-added products such as bedding material, decorative products, microbial enzymes, digestible proteins, amino acids, bioactive peptides, animal feed, fertilizers, biofilm, biofuel and biogas, etc. (Jayathilakan et al., 2012; Thyagarajan et al., 2013; Mezes et al., 2015; Brandelli et al., 2015; Campos et al., 2017; Ozdemir and Yetilmezsoy, 2020). Nowadays, feathers are converted widely into value-added products by physical (steam pressure cooking, incineration), chemical (strong alkali and acid hydrolysis), mechanical (milling, grinding), and biological (microbes) treatments. However, current mainstream strategies are not up to the mark. Hence, there is a dire need of a promising alternative approach that might overcome the drawbacks of the existing methods of waste feather processing.

c. ***Soft waste***

Poultry soft waste includes mainly viscera, blood, intestines, heads, feet, gizzard contents, bones, skin, trimmed meat, and fat, which are not suitable for human consumption (Owens et al., 2001; Ferreira et al., 2018). It is the main by-product from poultry slaughterhouses. It is consider that 22–23% of poultry waste is considered soft waste (Sari et al., 2016). These categories of by-products and wastes are mainly composed of proteins and fats (Table 4.1). Dry poultry blood also contains about 95% protein with high nutritional quality (Márquez et al., 2005). Blood is generally dried to become a concentrated protein source known as blood meal, which contains about 95% protein with a high content of basic (Lys and Arg) and sulfur (Cys and Met) amino acids (Márquez et al., 2005). To process a value-added new product, proper management practices based on regulations are vital for poultry soft wastes.

d. ***Eggshell***

A third waste stream unique to the poultry industry is discarded eggs and eggshells. Apart from the poultry industry, eggs are indispensable part of our food, whose production results in several daily tons of eggshell waste. About 250,000 tons of eggshell waste are produced annually worldwide (Verma et al., 2012). Most of the eggshell wastes are either accumulated on-site without any pretreatment or sent to landfills, resulting in methane pollution and nitrogen leaching into the air and soil, respectively (Stadelman, 2000; Tsai et al., 2008a). Other traditional disposal options include composting, rendering, and incineration (Das et al., 2002). However, the propogation of unpleasant odors, disease-causing microbes, and scarcity of proper disposal sites are the major drawbacks of such waste management systems (Tsai et al., 2008b). The composition of the eggshell has been reported chemically (by weight) as follows: $CaCO_3$ (94%), $MgCO_3$ (1%), $Ca_3(PO_4)_2$ (1%), and organic matter (4%) (Stadelman, 2000). Valorization of the waste produced by eggshells can be used as a biological template for catalysis, antibacterial agents, food additives, soil conditioners, cosmetics base material, pure calcium carbonate, and biomaterials (Dudusola, 2010; King'ori, 2011; Oliveira et al., 2013; Abdulrahman et al., 2014; Ajala et al., 2018). Separation of the eggshell membrane also increases the value of the resulting products. The eggshell membrane contains about 10% collagen, which has a market price of US$1000/gram (Glatz et al., 2011).

c. ***Wastewater***

Apart from the solid wastes, a substantial amount of wastewater is also generated from poultry farming and processing. Mainly wastewater is generated from the poultry processing during stunning, slaughtering, scalding, feathering, evisceration, meat trimming, and further processing. According to the literature, up to 26 L of water per bird are used during poultry processing (Kiepper et al., 2008; Avula et al., 2009; Malmali et al., 2018). Even for transport of inedible poultry by-products for further processing, water is required (Dalev, 1994). Poultry processing wastewater (PPW) contains high organic matter, chemical oxygen demand (COD), biochemical oxygen demand (BOD), total suspended solids (TSS), total organic carbon (TOC), fat, oil, grease, nitrogen, phosphorus, cleaning detergents, and different species of microorganisms (Ozdemir and Yetilmezsoy, 2020). The PPW is composed of manures, blood, fat, feathers, and undigested feed substances (Debik and Coskun, 2009; Bayar et al., 2011). These PPW require proper treatment before their disposal into the environment (Shih and Kozink, 1980; Sayed and de Zeeuw, 1988; Rusten et al., 1998). After a preliminary treatment to eliminate settled solid materials and floating materials, the wastewater proceeds to the coagulation-floatation unit to remove grease and finally the activated sludge system for the final treatment as per industrial wastewater treatment regulations (Yetilmezsoy et al., 2011; Ozdemir and Yetilmezsoy, 2020). The methods for wastewater disposal include sending it to a landfill, using it for irrigation, disposing it directly into wastewater lagoons, and AD (Glatz et al., 2011).

4.7 SYSTEM AND PROCESS SUSTAINABILITY

The term sustainability implies the conscious application of techniques and tools that positively impact society's three pillars, economy, social responsibility, and the environment, without any adverse effect (zero harm goal). The processes used to achieve sustainability aim to reduce waste through resource conservation and material reuse with a strong impetus on value creation. At the same time, these processes should be reliable and easily replicated with measurable outcomes. Thermochemical and biochemical technologies have the potential to valorize poultry wastes to generate energy and nonenergy products with reduced environmental impacts. The hydrothermal conversion route has gained significant attention due to its ability to process the wet biomass feedstock without intensive drying (Libra et al., 2011). Kantarli et al. (2016) reported hydrothermal liquefaction and pyrolysis of poultry litter to transform it into bio-oil. The bio-oil generated could be further upgraded into liquid petroleum fractions through secondary conversion processes. Hydrothermal carbonization of these waste materials can generate solid biochar with high carbon content and hydrophobic behavior and could be used for carbon sequestration when applied to the soil (Sermyagina et al., 2015). With its multiple advantages in terms of simplicity and flexibility, pyrolysis can be a viable option in places that do not have an established infrastructure and skillset for more sophisticated technologies. Alternatively, AD has been viewed as a plausible technology for poultry waste, although it has not been extensively applied in the poultry industry. Due to the higher nitrogen content in the poultry wastes, it is technically a challenging feedstock for AD (Beausang et al., 2020). However, with the advent of new processes such as ammonia stripping, multistage digestion, bioaugmentation, and pretreatment techniques, AD coupled with these methods shows significant promise for the conversion of the wastes into useful products (Hassanein et al., 2019). According to the reports of Hijazi et al. (2016), who evaluated life cycle analysis (LCA) studies on biogas production from different waste management practices, AD of manure waste from the egg industry reduced the overall methane and nitrous oxide emissions overall and lessened the need for additional fertilizer when the AD material was applied on land (Hijazi et al., 2016). Gasification, a nearly two-century-old technique for biomass valorization, has also been suggested as a suitable technology for valorizing poultry wastes. An upward draft gasifier can be the preferred reactor choice for feather waste, while fluidized bed gasifiers are more appropriate for poultry manure as the feedstock (Lappas et al., 2016; Dal ólio et al., 2017). Billen et al. (2015) concluded that ~655 kg of CO_2 released into the environment could be prevented by gasifying manure rather than conventional coal-based technology.

New technologies such as microbial degradation of poultry wastes have also been investigated that benefit from lower energy requirements and mild process conditions (Perlack et al., 2005). Cell-free extracts containing keratinases offer controlled hydrolysis of chicken feathers to obtain useful products such as compost material, nitrogen-rich fertilizer, and also a source of bioactive peptides. The keratinase is now being extensively used in agriculture and industry. In a different approach, scientists from the University of Nevada have established an environmentally friendly process for biodiesel production from poultry feather meal (Kondamudi et al., 2009). The biodiesel demonstrated good quality compared to the petro-derived diesel as the reference.

The critical approach to sustainability is maximizing profit by minimizing production cost and substantially reducing the negative environmental impacts. A detailed techno-economic evaluation of the process combined with "cradle to grave" LCA is an effective approach to measure sustainability. Bora et al., 2020 carried out a detailed techno-economic analysis (TEA) and LCA of the different thermochemical conversion technologies for treating poultry litter and reported that pyrolysis technologies contributed to lower GHG emissions. They concluded that a centralized fast pyrolysis plant in New York State would be more economical than other processes, considering the overall process variables (Bora et al., 2020). Likewise, Zhao et al. (2020) developed a life cycle optimization (LCO) model with multiobjective functions (for pyrolysis of poultry litter) that take into account the processes' economic and environmental performances. The model incorporates the techno-economics and LCA data from the process into its model as a seed. Through optimization, it provides a set of optimal solutions through Pareto chart analysis. Based on the analysis, a trade-off between the two indicators was considered for the sustainable design of an end-to-end process based on a local poultry litter supply chain (Zhao et al., 2020). Although the literature reviews indicate significant environmental benefits from a life cycle perspective for valorization of poultry wastes through different technologies, still there is some doubt on industrial-scale adoption of these technologies due to lack of large-scale datasets. The reason is that the parameters optimized in the lab perform poorly when scaled up. Thus, it is essential to carry out detailed life cycle costing of the processes with an analysis of the technology readiness levels at the pilot scale to comment on the operational "sustainability".

4.8 CONCEPTUALIZATION OF ZLD-BASED BIOREFINERY TOWARDS THE CIRCULAR ECONOMY

In current years, waste valorization has gained importance as an unconventional choice as opposed to waste disposal and landfill. Since all the waste has a specific property of having an integral net positive energy, this energy can be reclaimed and recycled by the biochemical and thermochemical reactions in a closed loop. Waste valorization is central to the concept of a circular economy (CE), which is currently gaining considerable significance within the European Union, United Kingdom, China, Japan, and Canada as well as in the business world (Bugge et al., 2016; Korhonen et al., 2018). The policy- and business-focused CE approach stresses product, material reuse, remanufacturing, repair, and upgrading (Rashid et al., 2013). CE is also relevant to achieving several Sustainable Development Goals (SDGs), including affordable and clean energy, industry, innovation and infrastructure, and responsible consumption and production, among others (Heimann, 2019). Biorefineries are the linking hub which permits the industry to shift from a high-carbon-utilizing economy toward a circular and low-carbon bioeconomy (Mohan et al., 2016). It is considered an infrastructure facility wherein various conversion technologies such as physical, thermochemical, biochemical, and biological platforms are integrated to efficiently produce biofuels, biochemicals, bioenergy, and other high-valued by-products (Cherubini, 2010; Ferreira, 2017). The biorefinery concept has been recently designed and utilized to route various biomass feedstocks that include a green biorefinery, forestry biorefinery, agricultural biorefinery, oilseed biorefinery, lingocellulosic biorefinery, and industrial waste biorefinery (Monte et al., 2009; Mongkhonsiri et al., 2018; Ubando et al., 2020). The approach in the biorefinery must focus on the process simplification rather than installing multiple complex process units. And the by-products of one unit can be used as feed in other units, or the intermediate products can be sent as a resource for other industries (Banerjee et al., 2020). The poultry waste can be treated through the zero-liquid discharge (ZLD)–based biorefinery route to utilize every component of the waste. However, implementation of biorefineries can be done through two routes. The first one is the construction of an integrated biorefinery at the same premises where waste is generated. And the second one is the presence of a biorefinery near to waste generation sites; through this, waste of more than one source can be transported and reused (Sailwal et al., 2020). According to the structure of the global poultry industry, long-term planning is requiring before conceptualizing and installing a biorefinery. Two main criteria always have to be remembered before installing a poultry waste biorefinery. One is the continuous supply of feed (poultry waste) throughout the year, and the other is good economic returns. Therefore, without understanding the geographical location, raw material flow, energy loss and recovery, and product type, biorefinery installation and its successful running are difficult.

Up to now, substantial technological progress has been achieved, along with numerous associated research publications (Zhao et al., 2009). There is comparatively limited literature available in the area of LCA of waste valorization technologies that focus specifically on the poultry sector. Comprehensive reviews of LCA studies of technologies for poultry waste valorization options, including nonlitter waste streams, are scanty in the literature. Furthermore, it is unclear which waste valorization technologies are better for poultry waste streams and hence merit deeper scrutiny using comparative LCA. According to our study, installation of an integrated biorefinery or treatment plant in the same or nearby premises of the poultry industry would be more significant to address the increasing pollution due to poultry waste. As poultry waste contains large amounts of organic material, valorization of waste through thermochemical and biochemical routes will add more value and good economic return for an integrated biorefinery or treatment plant. According to the existing literature and our current study, the design of a poultry waste biorefinery must be produced by an interdisciplinary approach with the consideration of the principles of CE, green chemistry, and industrial ecology (Figure 4.4).

The biorefinery and CE concept is only possible when each and every person knows about the consequences of mismanagement and nonscientific treatment and has knowledge about waste processing and awareness on making extra money from the waste. Broiler slaughterhouse waste has turned into a major source of pollution in all states of India. According to a report, it was estimated that about 5.3 lakh/day broilers are slaughtered in Kerala, which leads to the production of 350 tons of waste per day (Agrahari and Wadhwa, 2010; Ali, 2013; Abraham et al., 2015). According to a report, the Kannur district of Kerala, India, has become the first poultry-waste-free district in the country (Mohanan, 2019). According to the Kerala model (Suchitwa Mission, Government of Kerala), broiler meat retailers are now selling the slaughter waste at 7.00Rs/kg to a collection agent on a daily basis. This is further processed and utilized as fertilizer (70%), biogas (33.33%), fish feed (23.33%), pig feed (10%), and animal feed (3.33%) (Bashir et al., 2020). Installation of a biorefinery in this region helped in the socioeconomic development with waste valorization in the CE model (Figure 4.5).

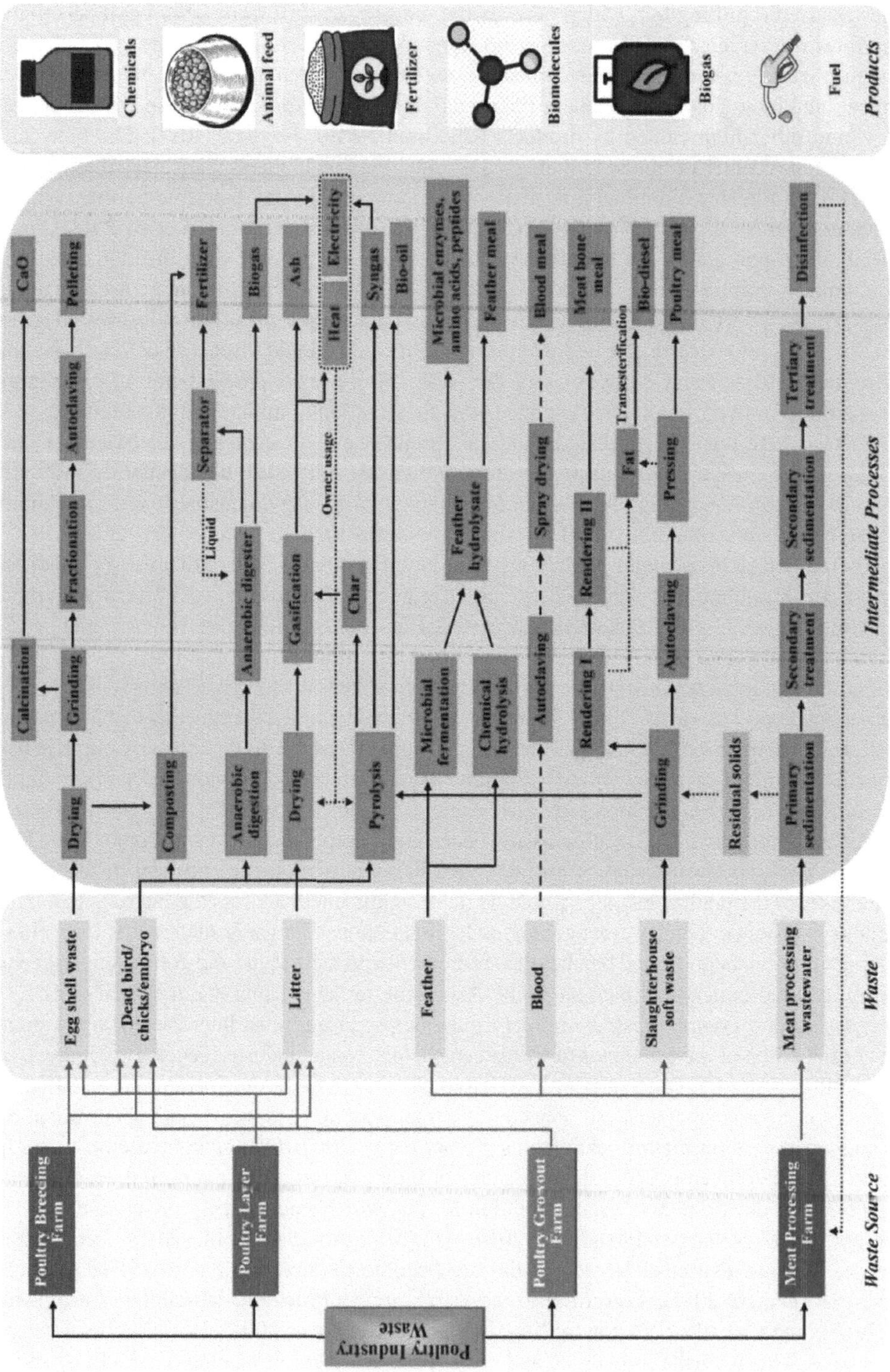

FIGURE 4.4 Biorefinery concept for the poultry industry.

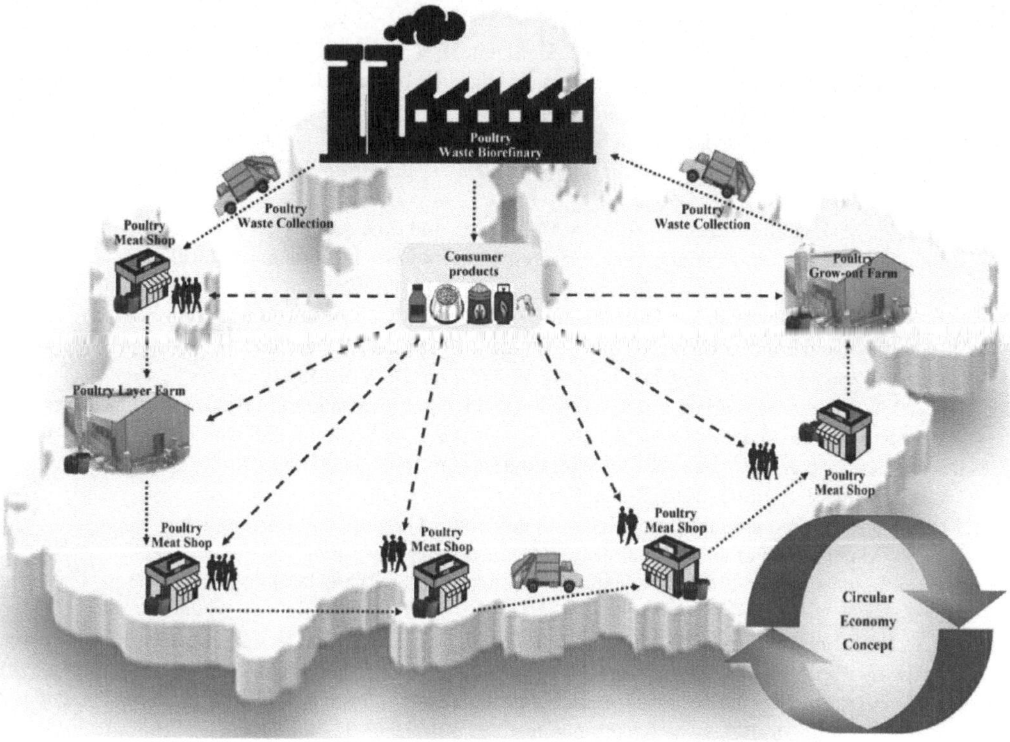

FIGURE 4.5 Conceptualized model of poultry waste-based biorefinery and circular economy

4.9 CONCLUSION

Poultry waste is a major pollutant and needs proper disposal in an ecofriendly manner with little or no environmental impact. With the increase in the human population (particularly in the last decade), there has been an upsurge in poultry processing resulting in huge volumes of solid and liquid waste streams. New and improved technologies employing different conversion routes (biological, chemical, and thermochemical) have been developed especially to mitigate the issue and simultaneously generate wealth out of the waste. However, to be deemed sustainable for large-scale deployment, the technology must include a detailed life cycle costing of the entire process with ZLD.

ACKNOWLEDGMENT

The authors thankfully acknowledge Dr. Anjan Ray, director CSIR-Indian Institute of Petroleum, for his constant encouragement and support.

FUNDING SOURCES

CSIR AN&B FBR Grant MLP-1167 & MLP-1168

CONFLICT OF INTEREST

The authors declare that they have no competing interests

REFERENCES

Abdulrahman, I., H.I. Tijani, D.A. Mohammed, H. Saidu, H. Yusuf, M.N. Jibrin, and S. Mohammed. 2014. "From garbage to biomaterials: an overview on egg shell based hydroxyapatite." *Journal of Materials* 802467: 6. http://doi.org/10.1155/2014/802467

Abraham, J., R.S. Kumar, F. Xavier, and D. Mathew. 2015. "Biodiesel production from broiler chicken waste." *International Journal of Agricultural and Biosystems Engineering* 9(12): 1190. http://doi.org/10.5281/zenodo.1110157

Adeoye, P.A., C.M. Hasfalina, M.S.M. Amin, A.M. Thamer, and C.O. Akinbile. 2014. "Environmental implication of poultry waste generation and management techniques in Minna, semi-arid region of Nigeria." *Annual Research and Review in Biology* 4(10): 1669–1681. http://doi.org/10.9734/ARRB/2014/8525

Agrahari, S., and N. Wadhwa. 2010. "Degradation of chicken feather a poultry waste product by keratinolytic bacteria isolated from dumping site at Ghazipur poultry processing plant." *International Journal of Poultry Science* 9(5): 482–489. http://doi.org/10.3923/ijps.2010.482.489

Ajala, E.O., O.A.A. Eletta, M.A. Ajala, and S.K. Oyeniyi. 2018. "Characterization and evaluation of chicken eggshell for use as a bio-resource." *Arid Zone Journal of Engineering, Technology and Environment* 14(1): 26–40.

Alexandratos, N., and J. Bruinsma. 2012. "World agriculture towards 2030/2050: the 2012 revision." ESA Working paper No. 12-03, 2012, Rome, FAO.

Ali, N. 2013. "Personal communication." Kerala State Poultry Development Corporation.

Anon. 2005. "Management of animal mortality in Georgia." January 2005.

AVEC. 2014. "Annual report 2014." Association of Poultry Processors and Poultry Trade in the EU Countries, Brussels.

Avula, R.Y., H.M. Nelson, and R.K. Singh. 2009. "Recycling of poultry process wastewater by ultrafiltration." *Innovative Food Science and Emerging Technologies* 10: 1–8. http://doi.org/10.1016/j.ifset.2008.08.005

Bagust, T.J. 2013. "Poultry health and disease control in developing countries." In *Poultry Development Review*. FAO, pp 96–100. www.fao.org/publications, ISBN 978-92-5-108067-2 (PDF).

Banerjee, A., T. Bhaskar, and D. Ghosh. 2020. "A biorefinery approach for sewage sludge." In T. Bhaskar, A. Pandey, E.R. Rene, and D.C.W. Tsang (Eds.), *Waste Biorefinery*. Elsevier, pp. 393–421, ISBN 9780128182284. http://doi.org/10.1016/B978-0-12-818228-4.00015-0

Bashir, B., N.K. Lubna, and J. Krishna. 2020. "A study on poultry slaughter house waste management in Malappuram and Kozhikode districts of Kerala." *Indian Journal of Social Research* 60: 757–764.

Basu, P. 2010. *Biomass Gasification and Pyrolysis: Practical Design and Theory*. Academic Press.

Bayar, S., Y.Ş. Yildiz, A.E. Yılmaz, and Ş. İrdemez. 2011. "The effect of stirring speed and current density on removal efficiency of poultry slaughterhouse wastewater by electrocoagulation method." *Desalination* 280(1–3): 103–107. http://doi.org/10.1016/j.desal.2011.06.061

Beausang, C., K. McDonnell, and F. Murphy. 2020. "Anaerobic digestion of poultry litter: a consequential life cycle assessment." *Science of the Total Environment* 2020;139494. http://doi.org/10.1016/j.scitotenv.2020.139494

Bhari, R., M. Kaur, R.S. Singh, A. Pandey, and C. Larroche. 2018. "Bioconversion of chicken feathers by *Bacillus aerius* NSMk2: a potential approach in poultry waste management." *Bioresource Technology Reports* 3: 224–230. http://doi.org/10.1016/j.biteb.2018.07.015

Billen, P., J. Costa, L. Van der Aa, J. Van Caneghaem, and C. Vandecasteele. 2015. "Electricity from poultry manure: a cleaner alternative to direct land application." *Journal of Cleaner Production* 96: 467–475. http://doi.org/10.1016/j.jclepro.2014.04.016

Blake, J.P. 2004. "Methods and technologies for handling mortality losses." *World's Poultry Science Journal* 60(1): 489–499. http://doi.org/10.1079/WPS200432

Blake, J.P., J.B. Carey, A.K.M. Haque, G.W. Malone, P.H. Patterson, N.L. Tablante, and N.G. Zimmermann. 2008. "Poultry carcass disposal options for routine and catastrophic mortality." Council for Agricultural Science and Technology (CAST) Issue Paper No. 40, October 2008.

Bora, R.R., Y. Tao, J. Lehmann, J.W. Tester, R.E. Richardson, and F. You. 2020. "Techno-economic feasibility and spatial analysis of thermochemical conversion pathways for regional poultry waste valorization." *ACS Sustainable Chemistry & Engineering* 8(14): 5763–5775. http://doi.org/10.1021/acssuschemeng.0c01229

Bourne, T.F. 1993. "Biodegradation of keratins and phenolic compounds." Ph.D. Thesis, Georgia Institute of Technology.

Braber, K. 1995. "Anaerobic digestion of municipal solid waste: a modern waste disposal option on the verge of breakthrough." *Biomass Bioenergy* 9: 365–376. http://doi.org/10.1016/0961-9534(95)00103-4

Brandelli, A., L. Sala, and S.J. Kalil. 2015. Microbial enzymes for bioconversion of poultry waste into added-value products. *Food Research International* 73: 3–12. http://doi.org/10.1016/j.foodres.2015.01.015

Bugge, M.M., T. Hansen, and A. Klitkou. 2016. "What is the bioeconomy? A review of the literature." *Sustainability* 8(7): 691. http://doi.org/10.3390/su8070691

Cai, T., O.C. Pancorbo, W.C. Merka, J.E. Sander, and H.M. Barnhart. 1994. "Stabilization of poultry processing by-products and waste and poultry carcasses through lactic acid fermentation." *Journal of Applied Poultry Research* 3(1): 17–25. http://doi.org/10.1093/japr/3.1.17

Campos, I., Matos, E., Marques, A., and L.M. Valente. 2017. "Hydrolyzed feather meal as a partial fishmeal replacement in diets for European seabass (*Dicentrarchus labrax*) juveniles." *Aquaculture* 476: 152–159. http://doi.org/10.1016/j.aquaculture.2017.04.024

Chan, Y.K., L. Van Zwieten, I. Meszaros, A. Downie, and S. Joseph. 2008. "Using poultry litter biochars as soil amendments." *Australian Journal of Soil Research* 46: 437–444. http://doi.org/10.1071/SR08036

Chen, T.H. 1999. "Anaerobic treatment of poultry mortalities in a modified LB-UASB system." *Journal of Environmental Science and Health, Part B* 34(3): 527–547. http://doi.org/10.1080/03601239909373212

Chen, T.H., and W.H. Shyu. 1998. "Chemical characterization of anaerobic digestion treatment of poultry mortalities." *Bioresource Technology* 63: 37–48. http://doi.org/10.1016/S0960-8524(97)00093-X

Chen, Z., and X. Jiang. 2014. "Microbiological safety of chicken litter or chicken litter-based organic fertilizers: a review." *Agriculture* 4(1): 1–29. http://doi.org/10.3390/agriculture4010001

Cherubini, F. 2010. "The biorefinery concept: using biomass instead of oil for producing energy and chemicals." *Energy Conversion and Management* 51(7): 1412–1421. http://doi.org/10.1016/j.enconman.2010.01.015

Coward-Kelly, G., V.S. Chang, F.K. Agbogbo, and M.T. Holtzapple. 2006. "Lime treatment of keratinous materials for the generation of highly digestible animal feed: 1. Chicken feathers." *Bioresource Technology* 97(11): 1337–1343. http://doi.org/10.1016/j.biortech.2005.05.021

Dai, J., J. Saayman, J.R. Grace, and N. Ellis. 2015. "Gasification of woody biomass." *Annual Review of Chemical and Biomolecular Engineering* 6: 77–99. http://doi.org/10.1146/annurev-chembioeng-061114-123310

Dal olio, F.S., J.N. da Silva, A.C.C. de Oliveira, F.F. Tinôco I de, R.C. Barbosa, M. de Oliveira Resende, L.F.T. Albino, and S.T. Coelho. 2017. "Poultry litter as biomass energy: a review and future perspectives." *Renewable and Sustainable Energy Reviews* 76: 941–949. http://doi.org/10.1016/j.rser.2017.03.104

Dalev, P.G. 1994. "Utilisation of waste feathers from poultry slaughter for production of a protein concentrate." *Bioresource Technology* 48: 265–267. http://doi.org/10.1016/0960-8524(94)90156-2

Das, K.C., M.Y. Minkara, N.D. Melear, and E.W. Tollner. 2002. "Effect of poultry litter amendment on hatchery waste compositing." *Journal of Applied Poultry Research* 11: 282–290. http://doi.org/10.1093/japr/11.3.282

Debik, E., and T. Coskun. 2009. "Use of the static granular bed reactor (SGBR) with anaerobic sludge to treat poultry slaughterhouse wastewater and kinetic modeling." *Bioresource Technology* 100: 2777–2782. http://doi.org/10.1016/j.biortech.2008.12.058

Dudusola, I.O. 2010. "Comparative evaluation of internal and external qualities of eggs from quail and guinea fowl." *International Research Journal of Plant Science* 1: 112–115.

FAO. 2011. "Global food losses and food waste extent, causes and prevention." Food and Agriculture Organization of the United Nations (FAO), Rome. ISBN 978-92-5-107205-9.

Ferreira, A.F. 2017. "Biorefinery concept." In M. Rabaçal, A.F. Ferreira, C.A.M. Silva, and M. Costa (Eds.), *Biorefineries: Targeting Energy, High Value Products and Waste Valorisation*. Springer International Publishing, pp. 1–20.

Ferreira, A., S.S. Kunh, P.A. Cremonez, J. Dieter, J.G. Teleken, S.C. Sampaio, and P.D. Kunh. 2018. "Brazilian poultry activity waste: destinations and energetic potential." *Renewable and Sustainable Energy Reviews* 81: 3081–3089. http://doi.org/10.1016/j.rser.2017.08.078

Franke-Whittle, I.H., and H. Insam. 2013. "Treatment alternatives of slaughterhouse wastes, and their effect on the inactivation of different pathogens: a review." *Critical Reviews in Microbiology* 39(2): 139–151. http://doi.org/10.3109/1040841X.2012.694410

Geshlog-Olyayee, M., H. Janmohammadi, A. Taghizadeh, and S.A. Rafat. 2010. "Effects of poultry by-product meal on performance, egg quality and blood parameters of commercial laying hens in the 42–54 weeks of age." [In Persian] *Iranian Journal of Applied Animal Science* 21: 29–42.

Glatz, P., Z. Miao, and B. Rodda. 2011. "Handling and treatment of poultry hatchery waste: a review." *Sustainability* 3: 216–237. http://doi.org/doi.org/10.3390/su3010216

GLEAM 2.0. 2016. "Global livestock environmental assessment model." FAO, Rome. http://www.fao.org/gleam.

Graham, J.P., S.L. Evans, L.B. Price, and E.K. Silbergeld. 2009. "Fate of antimicrobial-resistant enterococci and staphylococci and resistance determinants in stored poultry litter." *Environmental Research* 109(6): 682–689. http://doi.org/10.1016/j.envres.2009.05.005

Gupta, A.R., and V.K. Rathod. 2018. "Waste cooking oil and waste chicken eggshells derived solid base catalyst for the biodiesel production: optimization and kinetics." *Waste Management* 79: 169–178. http://doi.org/10.1016/j.wasman.2018.07.022

Gwyther, C.L., A.P. Williams, P.N. Golyshin, G. Edwards-Jones, and D.L. Jones. 2011. "The environmental and biosecurity characteristics of livestock carcass disposal method: a review." *Waste Management* 31: 767–778. http://doi.org/10.1016/j.wasman.2010.12.005

Haapapuro, E.R., N.D. Barnard, and M. Simon. 1997. "Review-animal waste used as livestock feed: danger to human health." *Preventive Medicine* 26: 599–602. http://doi.org/10.1006/pmed.1997.0220

Hasan, B., R. Faruque, M. Drobni, J. Waldenstrom, A. Sadique, K.U. Ahmed, Z. Islam, M.B. Hossain Parvez, B. Olsen, and M. Alam. 2011. High prevalence of antibiotic resistance in pathogenic *Escherichia coli* from large- and small-scale poultry farms in Bangladesh. *Avian Diseases* 55(4): 689–692. http://doi.org/10.1637/9686-021411-Reg.1

Hassanein, A., S. Lansing, and R. Tikekar. 2019. "Impact of metal nanoparticles on biogas production from poultry litter." *Bioresource Technology* 275: 200–206. http://doi.org/10.1016/j.biortech.2018.12.048

Heimann, T. 2019. "Bioeconomy and SDGs: does the bioeconomy support the achievement of the SDGs?" *Earth's Future* 7(1): 43–57. http://doi.org/10.1029/2018EF001014

Hijazi, O., S. Munro, B. Zerhusen, and M. Effenberger. 2016. "Review of life cycle assessment for biogas production in Europe." *Renewable and Sustainable Energy Reviews* 54: 1291–1300. http://doi.org/10.1016/j.rser.2015.10.013

Huang, J.J.H., and J.C.H. Shih. 1981. "The potential of biological methane generation from chicken manure." *Biotechnology and Bioengineering* 23: 2307–2314. http://doi.org/10.1002/bit.260231013

Huang, Y., M. Anderson, D. McIlveen-Wright, G.A. Lyons, W.C. McRoberts, Y.D. Wang, A.P. Roskilly, and N.J. Hewitt. 2015. "Biochar and renewable energy generation from poultry litter waste: a technical and economic analysis based on computational simulations." *Applied Energy* 160: 656–663. http://doi.org/10.1016/j.apenergy.2015.01.029

Hussein, M.S., K.G. Burra, R.S. Amano, and A.K. Gupta. 2017. "Temperature and gasifying media effects on chicken manure pyrolysis and gasification." *Fuel* 202: 36–45. http://doi.org/10.1016/j.fuel.2017.04.017

Hutchison, M.L., L.D. Walters, S.M. Avery, B.A. Synge, and A. Moore. 2004. "Levels of zoonotic agents in British livestock manures." *Letters in Applied Microbiology* 39: 207–214. http://doi.org/10.1111/j.1472-765X.2004.01564.x.

Jackson, B.P., P.M. Bertsch, M.L. Cabrera, J.J. Camberato, J.C. Seaman, and C.W. Wood. 2003. "Trace element speciation in poultry litter." *Journal of Environmental Quality* 32: 535–540. http://doi.org/10.2134/jeq2003.5350

Jayathilakan, K., K. Sultana, K. Radhakrishna, and A.S. Bawa. 2012. "Utilization of byproducts and waste materials from meat, poultry and fish processing industries: a review." *Journal of Food Science and Technology* 49: 278–293. http://doi.org/10.1007/s13197-011-0290-7

Kanani, F., M.D. Heidari, B.H. Gilroyed, and N. Pelletier. 2020. "Waste valorization technology options for the egg and broiler industries: a review and recommendations." *Journal of Cleaner Production* 262: 121–129. http://doi.org/10.1016/j.jclepro.2020.121129

Kantarli, I.C., A. Kabadayi, S. Ucar, and J. Yanik. 2016. "Conversion of poultry wastes into energy feedstocks." *Waste Management* 56: 530–539. http://doi.org/10.1016/j.wasman.2016.07.019

Kiepper, B.H., W.C. Merka, and D.L. Fletcher. 2008. "Proximate composition of poultry processing wastewater particulate matter from broiler slaughter plants." *Poultry Science* 87: 1633–1636. http://doi.org/10.3382/ps.2007-00331

Kim, J.M., W.J. Lim, and H. Suh. 2001. Feather-degrading *Bacillus* species from poultry waste. *Process Biochemistry* 37: 287–291. http://doi.org/10.1016/S0032-9592(01)00206-0

Kim, S.S., F.A. Agblevor, and J. Lim. 2009. "Fast pyrolysis of chicken litter and turkey litter in a fluidized bed reactor." *Journal of Industrial and Engineering Chemistry* 15: 247–252. http://doi.org/10.1016/j.jiec.2008.10.004

Kim, W.K., and P.H. Patterson. 2000. "Recycling dead hens by enzyme or sodium hydroxide pretreatment and fermentation." *Poultry Science* 79: 879–885. http://doi.org/10.1093/ps/79.6.879

King'ori, A.M. 2011. "A review of the uses of poultry eggshells and shell membranes." *International Journal of Poultry Science* 10: 908–912. http://doi.org/10.3923/ijps.2011.908.912

Kirkpinar, F., Z. Acikgoz, M. Bozkurt, and V. Ayhan. 2004. "Effects of inclusion of poultry by-product meal and enzyme-prebiotic supplementation in grower diets on performance and feed digestibility of broilers." *British Poultry Science* 45: 273–279. http://doi.org/10.1080/00071660410001715885

Kondamudi, N., J. Strull, M. Misra, and S.K. Mohapatra. 2009. "A green process for producing biodiesel from feather meal." *Journal of Agricultural and Food Chemistry* 57(14): 6163–6136. http://doi.org/10.1021/jf900140e

Korhonen, J., A. Honkasalo, and J. Seppälä. 2018. "Circular economy: the concept and its limitations." *Ecological Economics* 143: 37–46. http://doi.org/10.1016/j.ecolecon.2017.06.041

Kowalczyk, P., S. Mahdi-Oraibi, A. Misiewicz, N. Gabzdyl, A. Miskiewicz, and G. Szparecki. 2018. "Feather-degrading bacteria: their biochemical and genetic characteristics." *Arabian Journal for Science and Engineering* 43: 33–41. http://doi.org/10.1007/s13369-017-2700-2

Lappas, A., and E. Heracleous. 2016. "Production of biofuels via Fischer-Tropsch synthesis: biomass-to-liquids." In *Handbook of Biofuels Production*. Elsevier, pp. 549–593. http://doi.org/10.1016/B978-0-08-100455-5.00018-7

Libra, J.A., K.S. Ro, C. Kammann, A. Funke, N.D. Berge, Y. Neubauer, M.M. Titirici, C. Fühner, O. Bens, J. Kern, and K.H. Emmerich. 2011. "Hydrothermal carbonization of biomass residuals: a comparative review of the chemistry, processes and applications of wet and dry pyrolysis." *Biofuels* 2(1): 71–106. http://doi.org/10.4155/bfs.10.81

Lin, C.Y., K. Sato, T. Noike, and J. Matsumoto. 1986. "Methanogenic digestion using mixed substrate of acetic, propionic and butyric acids." *Water Research* 20: 385–394. http://doi.org/10.1016/0043-1354(86)90088-6

Mackie, R.I., B.A. White, and M.P. Bryant. 1991. "Lipid metabolism in anaerobic ecosystems." *Critical Reviews in Microbiology* 17: 449–478. http://doi.org/10.3109/10408419109115208

Malmali, M., J. Askegaard, K. Sardari, S. Eswaranandam, A. Sengupta, and S.R. Wickramasinghe. 2018. "Evaluation of ultrafiltration membranes for treating poultry processing wastewater." *Journal of Water Process Engineering* 22: 218–226. http://doi.org/10.1016/j.jwpe.2018.02.010

Márquez, E., M. Bracho, A. Archile, L. Rangel, and B. Benítez. 2005. "Proteins, isoleucine, lysine and methionine content of bovine, porcine and poultry blood and their fractions." *Food Chemistry* 93: 503–505. http://doi.org/10.1016/j.foodchem.2004.10.030

Marshall, B.M., and S.B. Levy. 2011. "Food animals and antimicrobials: impacts on human health." *Clinical Microbiology Reviews* 24(4): 718–733. http://doi.org/10.1128/CMR.00002-11

Martin-Luengo, M.A., M. Yates, M. Ramos, J.L. Salgado, R.M. Martin Aranda, F. Plou, J. Lopez Sanz, et al. 2011. "Renewable raw materials for advanced applications." In *Proceedings of the World Congress on Sustainable Technologies (WCST '11)*. IEEE, pp. 19–22. http://doi.org/10.1109/WCST19361.2011.6114229

Mehta, R., R.G. Nambiar, S.K. Singh, S. Subrahmanyam, and C. Ravi. 2002. "Livestock industrialization, trade and social-health-environment issues for the Indian poultry sector." Annex II, Part of IFPRI-FAO Livestock Industrialization Project: Phase I.

Mezes, L., A. Nagy, B. Galya, and J. Tamas. 2015. "Poultry feather wastes recycling possibility as soil nutrient." *Eurasian Journal of Soil Science* 4(4): 244. http://doi.org/10.18393/ejss.2015.4.244-252

Mohan, S.V., S.K. Butti, K. Amulya, S. Dahiya, and J.A. Modestra. 2016. "Waste biorefinery: a new paradigm for a sustainable bioelectro economy." *Trends in Biotechnology* 34(11): 1–3. http://doi.org/10.1016/j.tibtech.2016.06.006

Mohanan, P.V. 2019. "Kannur is becoming the first poultry waste free district in the country." *Clean India Journal* 23 Apr 2019.

Molaey, R., A. Bayrakdar, R.O. Sürmeli, and B. Çalli. 2018. "Anaerobic digestion of chicken manure: mitigating process inhibition at high ammonia concentrations by selenium supplementation." *Biomass Bioenergy* 108: 439–446. http://doi.org/10.1016/j.biombioe.2017.10.050

Mongkhonsiri, G., R. Gani, P. Malakul, and S. Assabumrungrat. "Integration of the biorefinery concept for the development of sustainable processes for pulp and paper industry." *Computers & Chemical Engineering* 119: 70–84. http://doi.org/10.1016/j.compchemeng.2018.07.019

Monte, M.C., E. Fuente, A. Blanco, and C. Negro. 2009. "Waste management from pulp and paper production in the European Union." *Waste Management* 29: 293–308. http://doi.org/10.1016/j.wasman.2008.02.002

Moore, P.A., T.C. Daniel, A.N. Sharpley, and C.W. Wood. 1995. "Poultry manure management: environmentally sound options." *Journal of Soil and Water Conservation* 50(3): 321–327.

Moraes, C.A., I.J. Fernandes, D. Calheiro, A.G. Kieling, F.A. Brehm, M.R. Rigon, J.A. Berwanger Filho, I.A. Schneider, and E. Osorio. 2014. "Review of the rice production cycle: by-products and the main applications focusing on rice husk combustion and ash recycling." *Waste Management & Research* 32(11): 1034–1048. http://doi.org/10.1177/0734242X14557379

Moreki, J.C., and S.C. Chiripasi. 2011. "Poultry waste management in Botswana: a review." *Online Journal of Animal and Feed Research* 1(6): 285–292.

Mottet, A., and G. Tempio. 2017. "Global poultry production: current state and future outlook and challenges." *World's Poultry Science Journal* 73: 245–256. http://doi.org/10.1017/S0043933917000071

Oliveira, D.A., P. Benelli, and E.R. Amante. 2013. "A literature review on adding value to solid residues: egg shells." *Journal of Cleaner Production* 46: 42–47. http://doi.org/10.1016/j.jclepro.2012.09.045

Onifade, A.A., N.A. Al-Sane, A.A. Al-Musallam, and S. Al-Zarban. 1998. A review: potentials for biotechnological applications of keratin-degrading microorganisms and their enzymes for nutritional improvement of feathers and other keratins as livestock feed resource. *Bioresource Technology* 66: 1–11. http://doi.org/10.1016/S0960-8524(98)00033-9

Owens, C.M., C. Alvarado, and A.R. Sams (Eds.) 2001. *Poultry Meat Processing* (1st ed.). CRC Press. http://doi.org/10.1201/9781420042177

Ozdemir, S., and K. Yetilmezsoy. 2020. "A mini literature review on sustainable management of poultry abattoir wastes." *Journal of Material Cycles and Waste Management* 22: 11–21. http://doi.org/10.1007/s10163-019-00934-1

Pandit, P.R., and M.H. Fulekar. 2017. "Egg shell waste as heterogeneous nanocatalyst for biodiesel production: optimized by response surface methodology." *Journal of Environmental Management* 198: 319–329. http://doi.org/10.1016/j.jenvman.2017.04.100

Paul, T., A. Jana, A.K. Mandal, A. Mandal, P.K.D. Mohpatra, and K.C. Mondal. 2016. "Bacterial keratinolytic protease, imminent starter for NextGen leather and detergent industries." *Sustainable Chemistry and Pharmacy* 3: 8–22. http://doi.org/10.1016/j.scp.2016.01.001

Pelletier, N. 2017. "Life cycle assessment of Canadian egg products, with differentiation by hen housing system type." *Journal of Cleaner Production* 152: 167–180. http://doi.org/10.1016/j.jclepro.2017.03.050

Pelletier, N., M. Ibarburu, and H. Xin. 2013. "A carbon footprint analysis of egg production and processing supply chains in the Midwestern United States." *Journal of Cleaner Production* 54: 108–114. http://doi.org/10.1016/j.jclepro.2013.04.041

Perkins, H.F., M.B. Parker, and M.L. Walker. 1964. "Chicken manure: its production, composition, and use as a fertilizer." Bulletin N.S. 193, December 1964. University of Georgia College of Agriculture. https://www.nurserycoop.auburn.edu/PDF%20files/manure.pdf.

Perlack, R.D., L.L. Wright, A.F. Turhollow, and R.L. Graham. 2005. "Biomass as feedstock for a bioenergy and bioproducts industry: the technical feasibility of a billion-ton annual supply." Oak Ridge National Laboratory. http://doi.org/10.2172/885984

Petersen, L., E.M. Nielsen, J. Engberg, S.L. On, and H.H. Dietz. 2001. "Comparison of genotypes and serotypes of *Campylobacter jejuni* isolated from Danish wild mammals and birds and from broiler flocks and humans." *Applied Environmental Microbiology* 67(7): 3115–3121. http://doi.org/10.1128/AEM.67.7.3115-3121.2001.

Poopathi, S., and S. Abidha. 2008. "Biodegradation of poultry waste for the production of mosquitocidal toxins." *International Biodeterioration & Biodegradation* 62: 479–482. http://doi.org/10.1016/j.ibiod.2008.03.005

Price, L.B., E. Johnson, R. Vailes, and E. Silbergeld. 2005. "Fluoroquinolone-resistant Campylobacter isolates from conventional and antibiotic-free chicken products." *Environmental Health Perspectives* 113(5): 557–560. http://doi.org/10.1289/ehp.7647.

Rashid, A., F.M.A. Asif, P. Krajnik, and C.M. Nicolescu. 2013. "Resource conservative manufacturing: an essential change in business and technology paradigm for sustainable manufacturing." *Journal of Cleaner Production* 57: 166–177. http://doi.org/10.1016/j.jclepro.2013.06.012

Ritter, W.F., and A.E.M. Chinside. 1995. "Impact of dead bird disposal pits on groundwater quality on the DelmarvaPeninsula." *BioresourceTechnology*53: 105–111. http://doi.org/10.1016/0960-8524(95)00057-L

Rusten, B., J.G. Siljudalen, A. Wien, D. Eidem. 1998. "Biological pretreatment of poultry processing wastewater." *Water Science and Technology* 38: 19–28. http://doi.org/10.1016/S0273-1223(98)00493-4

Saber, W.I.A., M.M. El-Metwally, and M.S. El-Hersh. 2010. "Keratinase production and biodegradation of some keratinous wastes by *Alternaria tenuissima* and *Aspergillus nidulans*." *Research Journal of Microbiology* 5: 21–35. http://doi.org/10.17311/jm.2010.21.35

Safley, J.L.M., R.L. Vetter, and D. Smith. 1987. "Operating a full-scale poultry manure anaerobic digester." *Biological Wastes* 19: 79–90. http://doi.org/10.1016/0269-7483(87)90102-9

Sailwal, M., A. Banerjee, T. Bhaskar, and D. Ghosh. 2020. "Integrated biorefinery concept for Indian paper and pulp industry." In T. Bhaskar, A. Pandey, E.R. Rene, and D.C.W. Tsang (Eds.), *Waste Biorefinery*. Elsevier, pp. 631–658, ISBN 9780128182284. http://doi.org/10.1016/B978-0-12-818228-4.00023-X

Salminen, E., J. Einola, and J. Rintala. 2003. "The methane production of poultry slaughtering residues and effects of pre-treatments on the methane production of poultry feather." *Environmental Technology* 24(9): 1079–1086. http://doi.org/10.1080/09593330309385648

Salminen, E., and J. Rintala. 2002. "Anaerobic digestion of organic solid poultry slaughterhouse waste-a review." *Bioresource Technology* 83: 13–26. http://doi.org/10.1016/S0960-8524(01)00199-7

Sari, O.F., S. Ozdemir, and A. Celebi. 2016. "Utilization and management of poultry slaughter house wastes with new methods." In *Eurasia 2016 Waste Management Symposium*. 02–04 May 2016, Istanbul, Turkey.

Sarkar, P., E. Dutta, P. Sen, and R. Banerjee. 2012. "Isolation and molecular characterization of extracellular keratinase from *Xanthomonas* sp: a potential approach in feather waste management." *Asian Journal of Applied Science and Technology* 1: 37–46.

Sasidhar, P.V.K. and M. Suvedi. 2016. "Assessment of core competencies of livestock extension professionals in India." *Modernizing Extension and Advisory Services Project*. https://meas.illinois.edu/wp-content/uploads/2016/11/MEAS-EVAL-2016-Core-Competencies-Livestock-Extension-Suvedi-and-SasidharJuly-2015.pdf

Sayed, S., and W. de Zeeuw. 1988. "The performance of a continuously operated flocculent sludge UASB reactor with slaughterhouse wastewater." *Biological Wastes* 24: 199–212. http://doi.org/10.1016/0269-7483(88)90062-6

Seadi, T.A., and J.B. Holm-Nielsen. 2004. "Agricultural wastes." In I. Twardowska and J.L. William (Eds.), *Waste Management Series*. Elsevier. http://doi.org/10.1016/S0713-2743(04)80011-4

Sermyagina, E., J. Saari, J. Kaikko, and E. Vakkilainen. 2015. "Hydrothermal carbonization of coniferous biomass: effect of process parameters on mass and energy yields." *Journal of Analytical and Applied Pyrolysis* 113: 551–556. http://doi.org/10.1016/j.jaap.2015.03.012

Sharma, G. 2020. "Indian poultry industry: challenges, problems and opportunities." *The Poultry Punch: English Monthly Magazine*, October 4, 2020. www.thepoultrypunch.com.

Shih, J.C.H. 1993. "Recent development in poultry waste digestion and feather utilization: a review." *Poultry Science* 72: 1617–1620. http://doi.org/10.3382/ps.0721617

Shih, J.C.H., and M.B. Kozink. 1980. "Ultrafiltration treatment of poultry processing wastewater and recovery of a nutritional by-product." *Poultry Science* 59: 247–252. http://doi.org/10.3382/ps.0590247

Simpson, T.W. 1991. "Agronomic use of poultry industry waste." *Poultry Science* 70: 1126–1131. http://doi.org/10.3382/ps.0701126

Singh, P., T. Mondal, R. Sharma, N. Mahalakshmi, and M. Gupta. 2018. "Poultry Waste management." *International Journal of Current Microbiology and Applied Sciences* 7(8): 701–712. http://doi.org/10.20546/ijcmas.2018.708.076

Smith, T.C., W.A. Gebreyes, M.J. Abley, A.L. Harper, B.M. Forshey, M.J. Male, and H. Wayne Martin, et al. 2013. "Methicillin-resistant *Staphylococcus aureus* in pigs and farm workers on conventional and antibiotic-free swine farms in the USA." PLoS One 8(5). http://doi.org/10.1371/journal.pone.0063704

Stadelman, W. 2000. *Encyclopedia of Food Science and Technology*. John Wiley and Sons, pp. 593–599.

Swan, J.E. 1992. "Animal by-product processing." In Y.H. Hui (Ed.), *Encyclopedia Food Science Technology*, vol. 4. John Wiley & Sons, pp. 42–49.

Tangboriboon, N., R. Kunanuruksapong, and A. Sirivat. 2012. "Preparation and properties of calcium oxide from eggshells via calcination." *Materials Science-Poland* 30(4): 313–322. http://doi.org/10.2478/s13536-012-0055-7

Taupe, N.C., D. Lynch, R. Wnetrzak, M. Kwapinska, W. Kwapinski, and J.J. Leahy. 2016. "Updraft gasification of poultry litter at farm-scale: a case study." *Waste Management* 50: 324–333. http://doi.org/10.1016/j.wasman.2016.02.036

Tensingh, G.P., S. Ezhilvalavan, A. Bharathidhasan, and A. Sundaresan. "Production of energy from poultry waste in India: an overview." In *Proceeding of World's Poultry Congress-2016*. http://krishikosh.egranth.ac.in/handle/1/5810003642.

Thyagarajan, D., M. Barathi, and R. Sakthivadivu. 2013. "Scope of poultry waste utilization." *IOSR Journal of Agriculture and Veterinary Science* 6(5): 29–35.

Tiwary, E., and R. Gupta. 2012. "Rapid conversion of chicken feather to feather meal using dimeric keratinase from *Bacillus licheniformis* ER-15." *Journal of Bioprocessing and Biotechniques* 2: 4. http://doi.org/10.4172/2155-9821.1000123

Tsai, W.T., K.J. Hsien, H.C. Hsu, C.M. Lin, K.Y. Lin, and C.H. Chiu. 2008a. "Utilization of ground eggshell waste as an adsorbent for the removal of dyes from aqueous solution." *Bioresource Technology* 99: 1623–1629. http://doi.org/10.1016/j.biortech.2007.04.010

Tsai, W.T., J.M. Yang, H.C. Hsu, C.M. Lin, K.Y. Lin, and C.H. Chiu. 2008b "Development and characterization of mesoporosity in eggshell ground by planetary ball milling." *Microporous and Mesoporous Materials* 111: 379–386. http://doi.org/10.1016/j.micromeso.2007.08.010

Turnell, J.R., R.D. Faulkner, and G.N. Hinch. 2007. "Recent advances in Australian broiler litter utilisation." *World's Poultry Science Journal* 63: 223–231. http://doi.org/10.1017/S0043933907001420

U.S. Environmental Protection Agency (USEPA). 2013. "Literature review of contaminants in livestock and poultry manure and implications for water quality." EPA 820-R-13-002. https://nepis.epa.gov/Exe/ZyPURL.cgi?Dockey=P100H2NI.txt (accessed 8.22.17).

Ubando, A.T., C.B. Felix, and W.H. Chen. 2020. "Biorefineries in circular bioeconomy: a comprehensive review." *Bioresource Technology* 299: 122585. http://doi.org/10.1016/j.biortech.2019.122585

USDA. 2014. "Livestock and poultry: world markets and trade." Foreign Agricultural Service USDA. http://apps.fas.usda.gov.

Van Boeckel, T.P., C. Brower, M. Gilbert, B.T. Grenfell, S.A. Levin, T.P. Robinson, and R. Laxminarayan. 2015. "Global trends in antimicrobial use in food animals." *Proceedings of the National Academy of Sciences of the United States of America* 112(18): 5649–5654. http://doi.org/10.1073/pnas.150314111

Verma, N., V. Kumar, and M.C. Bansal. 2012. "Utilization of egg shell waste in cellulase production by *Neurospora crassa* under wheat bran-based solid state fermentation." *Polish Journal of Environmental Studies* 21: 491–497.

Verraes, C., S.V. Boxstael, E.V. Meervenne, E.V. Coillie, P. Butaye, B. Catry, M.A. Schaetzen, et al. 2013. "Antimicrobial resistance in the food chain: a review." *International Journal of Environmental Research and Public Health* 10(7): 2643–2669. http://doi.org/10.3390/ijerph10072643

Webb, A.R., and F.R. Hawkes. 1985a. "Laboratory scale anaerobic digestion of poultry litter: gas yield-loading rate relationship." *Agricultural Waste* 13: 31–49. http://doi.org/10.1016/0141-4607(85)90010-1

Webb, A.R., and F.R. Hawkes. 1985b. "The anaerobic digestion of poultry manure: variation of gas yield with influent concentration and ammonium–nitrogen levels." *Agricultural Waste* 14: 135–156. http://doi.org/10.1016/S0141-4607(85)80025-1

Williams, P.T. 1999. *Waste Treatment and Disposal.* Wiley. http://doi.org/10.1002/0470012668

Yetilmezsoy, K., I. Turkdogan-Aydinol, A. Gunay, and I. Ozis. 2011. "Post treatment of poultry slaughterhouse wastewater and appraisal of the economic outcome." *Environmental Engineering and Management Journal* 10(11): 1635–1645. http://doi.org/10.30638/eemj.2011.225

Zhao, N., J. Lehmann, and F. You. 2020. "Poultry waste valorization via pyrolysis technologies: economic and environmental life cycle optimization for sustainable bioenergy systems." *ACS Sustainable Chemistry & Engineering* 8(11): 4633–4646. http://doi.org/10.1021/acssuschemeng.0c00704

Zhao, W., E.V.D. Voet, Y. Zhang, and G. Huppes. 2009. "Life cycle assessment of municipal solid waste management with regard to greenhouse gas emissions: case study of Tianjin, China." *Science of The Total Environment* 407(5): 1517–1526.

5 Byproducts of a Microalgal Biorefinery as a Resource for a Circular Bioeconomy

Madonna Roy and Kaustubha Mohanty

CONTENTS

5.1 INTRODUCTION

The global economy is highly dependent on nonrenewable energy, such as fossil reserves, to maintain pace with technological and energy demand, which eventually imposes alarming environmental pressure. Exploration and implementation of a renewable energy resource is necessary to overcome this energy demand, along with global warming resulting from the combustion of fossil fuels.

DOI: 10.1201/9781003335740-5

Among several other renewable energy resources, microalgae came to the limelight owing to its metabolic versatility (Suali and Sarbatly 2012). In the bioeconomy context, microalgae are known as a promising player owing to their increased carbon dioxide sequestration ability, high growth rate, ability to provide feedstock for biofuels (biogas, biodiesel, bioethanol, etc.), and marketable value-added products (β-carotene, polyunsaturated fatty acid [PUFA], polysaccharide, biopeptides, biopolymers, antioxidants, and pigments, etc.) (S. V. Mohan et al. 2019). However, the commercialization of microalgae-based products is a challenge, as at the current stage the production process suffers from several limitations during the upstream and downstream processing (Roy and Mohanty 2019). The production stage involves a high cost and energy for cultivating, harvesting, and drying microalgal biomass, making the process unsustainable.

The valorization of microalgal biomass to a single end product is unsustainable and limits the economic viability. Therefore the transition from a linear end-of-use to circular-based bioeconomy is required to be implemented for strengthening the current socioeconomic condition and reducing the environmental impacts (Kalmykova, Sadagopan, and Rosado 2018). A microalgae-based biorefinery approach allows to produce food, feed, energy, and valuable products within a single framework. The commercial bioenergy production process can be self-sustained and efficient if the waste generated during the upstream and downstream processes is reduced, recycled, and reused (Venkata Mohan et al. 2016). The final step in the circularity of a microalgal bioeconomy is represented by the opportunities of further exploiting microalgal biomass residuals after their primary use for the production of high value-added products. These microalgal residuals are present either in the liquid or solid form. This chapter presents a framework to further improve the economic viability of the biorefinery system by reducing, recycling, and reusing these byproducts.

5.2 MICROALGAL BIOREFINERY AND THE CIRCULAR BIOECONOMY

The bioprocessing of a microalgal biomass involves the extraction of various biomolecules such as lipids, proteins, and carbohydrates through various extraction processes. However, the exploitation of a single product is not economical and produces undesirable waste. This problem can be solved by the biorefinery approach, where byproducts and wastes produced from biomass conversion pathways are converted to marketable products by a cascade extraction approach (Gifuni et al. 2019). A microalgal biorefinery is a platform where microalgal growth facilities and biomass conversion processes (biochemical, thermochemical, and chemical) are integrated to offer multifaceted benefits from CO_2 mitigation to the production of biochemicals, bioenergy, and value-added products (Figure 5.1) (Rajak, Jacob, and Kim 2020). The biorefinery approach allows us to profitably exploit a microalgal biomass by producing multiple products with minimum waste. Previously, the performance of a biorefinery system was evaluated in terms of either environmental or economic impact. However, recently Tuazon et al. (2013) proposed that the sustainability assessment of biorefineries can be analyzed based on the triple bottom line framework, which consists of the environmental, economic, and social aspects (Figure 5.2).

In the latter parts of the 2000s, the concept of a bioeconomy gained the attention of researchers and industry practitioners (Dieckhoff, El-Cichakli, and Patermann 2015). The German Bioeconomy Council defined bioeconomy as the knowledge-based production and utilization of renewable resources to mitigate climate change while providing value-added products, processes, and job opportunities in all economic sectors within the framework of an economically sustainable system (S. V. Mohan et al. 2019). To accomplish these goals set for the bioeconomy concept, the biorefinery approach plays a significant role. The circular bioeconomy is defined as the sustainable cascade processing of organic residues into products that can be either shared, reused, recycled, remanufactured, or released safely into the environment through nutrient cycles (Carrez et al. 2017).

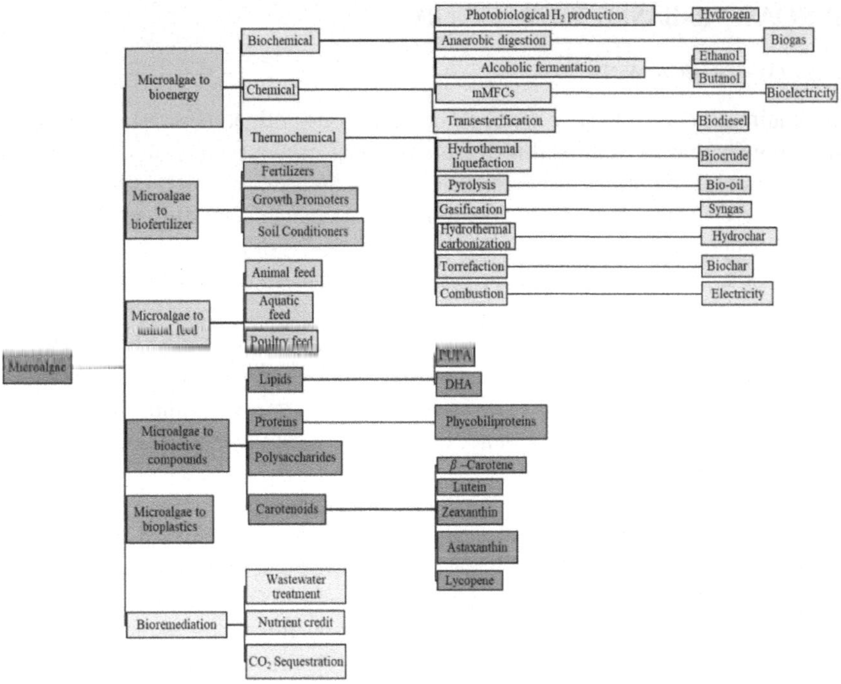

FIGURE 5.1 A complete spectrum of microalgal products and environmental benefits from freshwater microalgae.

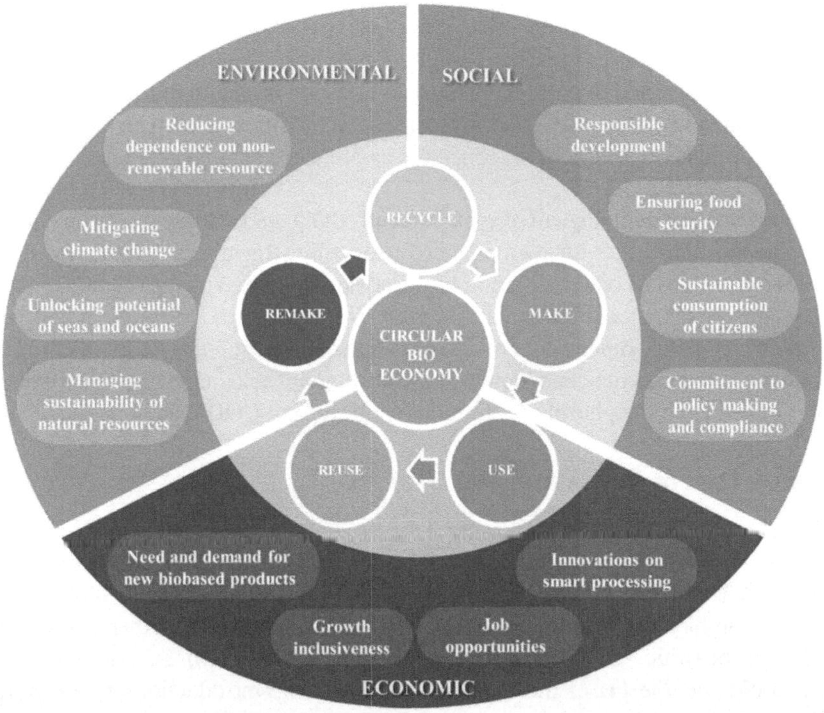

FIGURE 5.2 The triple bottom line as an assessment tool for a circular bioeconomy. Adapted from (Ubando et al. 2020).

5.3 MICROALGAE-BASED BIOREFINERY

5.3.1 Wastewater Treatment

Microalgae require nitrogen, phosphorous, and carbon, along with numerous trace elements for sustaining their growth. The conventional fertilizers used for growing microalgae are not sustainable for industrial microalgal biorefineries. Freshwater microalgae are generally cultivated in a freshwater medium. However, as the demand for freshwater is increasing across the world, it is not feasible to cultivate microalgae in freshwater for large-scale production of a biomass. Therefore, alternative nutrient and water sources are required for making the process economically viable. Microalgae can be cultured in wastewater, as it is a reservoir of various nutrients (S. Mishra and Mohanty 2019). Moreover, potential microalgae isolated from wastewater have higher acclimatization ability to the existing stressful environment and are a crucial player in phycoremediation of wastewater (Kumar and Singh 2019). Valorization of wastewater and the spent medium for cultivating microalgae reduces the water and nutrient demand, thereby reducing the freshwater and nutritional cost footprint of a microalgal biorefinery.

A considerable amount of energy is consumed during the production of synthetic nutrients for microalgal cultivation along with the release of greenhouse gases (GHGs) in the atmosphere (Johnson, Palou-Rivera, and Frank 2013). It has been reported that the production of synthetic nutrients may constitute about 50% of the total energy consumed and GHGs emitted, making the overall process unsustainable (Lam and Lee 2012). Maximum recycling of the spent medium along with the valorization of wastewater can reduce the demand for synthetic nutrients. Moreover, it has been reported that when the nitrogen and phosphorus content in the medium is depleted, microalgae tend to accumulate more lipids in their cells (Sathish and Sims 2012). Zhu et al. (2013) found that a nitrogen- and phosphorus-deprived medium increased the lipid content of *Chlorella zofingiensis* by 15% to 20% compared to the nutrient-replete medium. Moreover, the spent medium was recycled and reused twice after adding sufficient nutrients for cultivating *C. zofingiensis* without any reduction in biomass productivity. Thus, incorporation of nutrient stress during cultivation not only enhances the lipid content but also reduces the consumption of nutrients. The integrated approach for phycoremediation of wastewater and microalgal biomass production is able to minimize the total nitrogen and freshwater demand by 94% and 90%, respectively. In addition, the diversion of waste streams to algal biomass production will help to reduce the existing biomass production cost by about 90% (Kumar and Singh 2019).

Waste streams are rich in biological oxygen demand (BOD) and chemical oxygen demand (COD) and are thus potential substitutes for synthetic media for cultivating microalgae. Cultivation in waste streams serves two purposes: bioremediation of wastewater by reducing the organic load and simultaneous production of a biomass by using the nutrients from waste streams (Kumar and Singh 2019). Mishra and Mohanty (2019) reported that the cultivation of *Monoraphidium* sp. in domestic sewage wastewater not only reduced the dependency on freshwater but also helped in the bioremediation of sewage water by reducing the phosphate, ammonia, nitrate, and COD load.

5.3.2 Carbon Dioxide Mitigation

Atmospheric CO_2 concentration is increasing because of fossil fuel combustion, thereby causing global warming. Photosynthetic organisms have made a significant contribution to the global carbon cycle through CO_2 fixation. Microalgal cell factories have the ability to biologically fix and recycle CO_2 into the biomass through photosynthesis, which could be further used for producing bioenergy and various value-added products (Jacob-Lopes et al. 2020). As microalgae grow, consuming carbon dioxide, the pH of the medium increases. The introduction of CO_2 in the system increases the carbonic acid concentration, thereby lowering the pH. This approach helps to maintain the optimum pH of the medium for maximum microalgal growth while delivering enough CO_2 for sustaining growth (Wilson et al. 2014). Synthetic CO_2 has been widely used during microalgal cultivation for producing biomass, but this involves additional cost, making the process economically unsustainable.

Microalgae have developed a CO_2-concentrating mechanism (CCM), which enables the microalgal cells to carry out photosynthesis efficiently by acquiring inorganic carbon from very low atmospheric CO_2 concentration (Singh and Dhar 2019). Moreover, microalgae have the ability to utilize hazardous gases such as CO_2, NO_X, and SO_X from flue gas as nutrients for producing a biomass (Yen et al. 2015). The combined approach of SO_X, NO_X, and CO_2 removal using microalgae and biomass production using these flue gases as nutrients represents a sustainable biorefinery. The simple cellular structures and rapid growth rate of microalgae enable them to have 10- to 50-fold higher CO_2-fixing efficiency than terrestrial plants (Singh and Dhar 2019). A wide range of microalgal species such as *Chlamydomonas*, *Botryococcus*, *Chlorella*, and *Scenedesmus* has been used for CO_2 sequestration (Basu et al. 2013; Ho et al. 2014; Chiu et al. 2008; Yoo et al. 2010).

5.3.3 Microalgae to Bioenergy

Three conversion technologies are used to transform a microalgal biomass to bioenergy, namely biochemical, thermochemical, and chemical conversions. The sustainability of the bioenergy conversion process is based on the following key parameters: biomass composition, reaction conditions, bioenergy yield, and energy efficiency (S. Mishra, Roy, and Mohanty 2019).

5.3.3.1 Biochemical Conversion Processes and Products

5.3.3.1.1 Photobiological Process for Hydrogen Production

Biohydrogen is considered a potential biofuel, as it yields high energy (122 KJ g^{-1}) with energy density three-fold higher in comparison to hydrocarbon fuels (S. Mishra, Roy, and Mohanty 2019). Biohydrogen can be produced through (a) direct biophotolysis, (b) indirect biophotolysis, and (c) two-stage dark/photo fermentation. The hydrogenase enzyme is harbored in some microalgal strains belonging to the genera *Chlorella, Botryococcus, Chlorococcum, Synechocystis, Tetraspora, Anabaena, Scenedesmus, Nostoc, Chlamydomonas*, etc., which makes them promising candidates for producing biohydrogen (Khetkorn et al. 2017).

Biohydrogen generation through direct biophotolysis has been witnessed only in green microalgae. This process utilizes abundantly available water and light and is not involved in any GHG emissions. Direct biophotolysis produces molecular hydrogen through water splitting activity of photosystem II (PSII) in the absence of oxygen. The hydrogenase enzyme produced by green algae catalyzes this mechanism (Winkler et al. 2009). In this process, H_2 molecules are formed for a very short duration as H_2, and O_2 (H_2:O_2 = 2:1) molecules are produced simultaneously, resulting in water formation. The water molecule, on the other hand, inhibits the enzymatic reaction of hydrogenase and suppresses the gene expression of [FeFe]-hydrogenases.

Indirect biophotolysis is a process that permanently eradicates the problem of [FeFe]-hydrogenase inhibition by O_2 during hydrogen evolution. Indirect biophotolysis is a two-stage process, where the algal cells are initially allowed to reproduce and store carbohydrates under normal photosynthetic conditions. In the final stage, also referred to as anaerobic dark fermentation, the stored carbohydrates are used as a substrate by the nitrogenase enzyme to produce hydrogen. Indirect biophotolysis is advantageous over direct biophotolysis, as the former process utilizes water as the electron donor, carbohydrate as the carbon source, and most importantly, O_2 inhibition is insignificant during the process of H_2 synthesis (Brentner, Peccia, and Zimmerman 2010). However, as the conversion of light energy to hydrogen involves several steps, the conversion efficiency was found to be reduced (Pilon, Berberoğlu, and Kandilian 2011).

Up to now, the commercialization of biohydrogen produced from microalgae via direct or indirect biophotolysis has not been possible due to several bottlenecks. The low H_2 yield, high sensitivity of [FeFe]-hydrogenases towards O_2, the requirement of a carbon source by fermentative bacteria, the high cost of photobioreactors, and the technical issues in genetic engineering limit the commercialization of biohydrogen. Therefore, two-stage dark/photo fermentation is applied to overcome these limitations. In this process, initially, microalgae produces biohydrogen through direct biophotolysis, and later the fermentative bacteria utilizes the microalgal biomass as a substrate to produce hydrogen via dark fermentation (Nagarajan et al. 2017).

5.3.3.1.2 Anaerobic Digestion for Biogas Production

Microalgal biomass can be converted into biogas through a biochemical process known as anaerobic digestion. Anaerobic digestion involves a sequence of processes: (a) hydrolysis, (b) acidogenesis, (c) acetogenesis, and (d) methanogenesis, where the product of the first phase serves as a substrate for the next phase. In the early stage of anaerobic digestion, complex organic molecules are hydrolyzed to soluble organic compounds. During acidogenesis, the acidogenic bacteria degrade the hydrolyzed compounds from the first stage, producing NH_3, volatile fatty acids, and CO_2. Acetogenesis is the third stage, where the acetogens digest the simple molecules produced in the acidogenesis phase to produce CH_3COOH, CO_2, and H_2. Methanogenesis is the final stage of anaerobic digestion, where the methanogenic bacteria convert the intermediate products from the acetogenesis phase to CH_4 (55%-75%), CO_2 (25%-45%), and H_2O (Suganya et al. 2016). However, an excessive product from one phase might impair the entire process by accumulating intermediates, modifying the culture pH, and decreasing the process efficiency. Nitrogen-rich microalgal species produce a high amount of NH_3, thereby reducing the digestion process. In order to overcome this issue, a codigestion process is being practiced, where organic waste with low N_2 and high C is added to the microalgal biomass.

5.3.3.1.3 Alcoholic Fermentation for Bioethanol Production

Alcoholic fermentation is a biological process, where the microorganisms convert the microalgal polysaccharides to C_2H_5OH and CO_2 under anaerobic conditions. Pretreatment of the microalgal biomass is required before the fermentation process to facilitate the release of intracellular starch from the rigid cell walls. The starch is then converted to simple sugars through a saccharification process, so that microorganisms can ferment it. The most widely used pretreatment methods are physical or mechanical (e.g., sonication, pyrolysis, milling, grinding), biological (e.g., enzymatic treatments), and chemical (e.g., thermal, acid, or alkali treatments). Distillation is carried out to remove water and other impurities from the crude alcohol. The concentrated bioethanol (approximately 95%) is finally withdrawn and condensed to a liquid form, which can be blended with fossil fuel (Suganya et al. 2016). Microalgal species such as *C. vulgaris* are rich in starch content (approximately 37% dry weight), for which 65% bioethanol conversion efficiency was achieved, making them a potential candidate for bioethanol production (Hirano et al. 1997).

5.3.3.1.4 Microalgae-Based Microbial Fuel Cell for Bioelectricity Production

A microalgae-based microbial fuel cell (mMFC) is a sustainable electrochemical system with advanced technology that came to the limelight owing to its capability of treating waste, harnessing energy, and cultivating algae simultaneously under a single platform. In this bioelectric system, microalgae play a crucial role in converting light and biochemical energy to bioelectrical energy through oxidation and reduction reactions. Oxidation of organic substrates generates electrons and protons at the anode chamber. The terminal electron acceptor (TEA) in the cathode chamber serves as a driving force, transferring the electrons (e^-) via an external circuit to the cathode. At the cathode chamber, the transferred e^- are harnessed for bioelectricity generation, whereas the protons (H^+) are carried to the cathode via an ion exchange membrane that separates the anode and cathode chamber (Ishii et al. 2013). At the cathode, the oxygen molecules, which are imparted either from the atmosphere or by a catalyst such as microalgae, are reduced by the e^-, which then combines with H^+ to form water (Kakarla and Min 2014). In combination with bioelectricity production, the microalgae in the cathode chamber also help to fix CO_2, N, and P. Researchers are trying to develop a hybrid technology by integrating systems with similar principles for simultaneous production of microalgae and bioenergy along with waste management.

5.3.3.2 Thermochemical Conversion

5.3.3.2.1 Hydrothermal Liquefaction for Liquid Biofuel Production

Hydrothermal liquefaction (HTL) is a thermal depolymerization process by which wet microalgal biomass is converted to biocrude. The reaction is operated at temperatures between 250°C and 370°C and high pressures of 5 MPa to 20 MPa, with 15% to 20% of microalgal biomass in the feed, where water acts as a catalyst (W.-H. Chen et al. 2015). Depending on the operating conditions, the water in the microalgal biomass becomes either subcritical or supercritical and serves as a solvent, reactant, and catalyst to promote the conversion of biomass to biocrude. High pressure maintains the water at a subcritical state, where the water breaks down the biomolecules in the microalgal biomass to small reactive compounds, which are then repolymerized to a range of products. The biomolecules undergo a series of reactions such as isomerization, depolymerization, and repolymerization, which are influenced by various factors such as temperature, pressure, retention time, feedstock composition, feedstock loading, and catalyst (Peterson et al. 2008). The HTL process produces biocrude oil (10%-73%) as the main product, whereas a gaseous mixture (8%-20%), ash (0.2%-0.5%), and nutrient-rich aqueous phase are produced as the byproducts (Shuping et al. 2010). Biocrude is chemically composed of ketones, acids, alkenes, esters, and nitrogen-containing heterocycles, and it has a calorific value between 30 and 50 MJKg, which is comparable to fossil fuels (Shuping et al. 2010). The gaseous mixture comprises carbon dioxide, hydrogen, methane, nitrogen, and other C_2-C_3 compounds. HTL technique have gained huge attention, as the process is able to convert wet biomass (up to 90% moisture) to biocrude, thus eliminating the predrying step of the biomass. HTL of microalgal biomass has been performed with *Spirulina, Chlorella, Scenedesmus, Nannochloropsis, Dunaliella,* and *Botryococcus braunii* (Biller et al. 2012; Jena, Das, and Kastner 2011; W.-H. Chen et al. 2014; Shuping et al. 2010; Dote et al. 1994).

5.3.3.2.2 Hydrothermal Carbonization for Hydrochar Production

Hydrothermal carbonization (HTC) is a thermochemical process that converts wet biomass to hydrochar. The HTC process is similar to HTL; however, in this process the reaction is carried out at relatively low temperature (180°C-250°C) and pressure (2 MPa-10 MPa), which lowers the O_2 and H_2 content of the biomass via dehydration and decarboxylation (Funke and Ziegler 2010). Hydrothermal parameters like reaction temperature and residence time govern the process. The hydrochar is highly hydrophobic and friable in nature, and so it is easily separable from the aqueous product. Besides bioenergy production, hydrochar has been extensively used for wastewater treatment, soil amelioration, and carbon sequestration (Saidur et al. 2011).

5.3.3.2.3 Pyrolysis for Bio-Oil Production

Pyrolysis is a thermochemical process where biomass is thermally cracked under relatively mild conditions: atmospheric pressure, temperature (200°C-750°C), and inert atmosphere. Pyrolysis of microalgal biomass produces solid (charcoal), liquid (bio-oil), and gaseous products based on the operational conditions. During pyrolysis, microalgal biomolecules (proteins and lipids) undergo thermal cracking and depolymerization, which results in the production of fatty oxygenate– and nitrogenous compound–rich bio-oil. The stability of the bio-oil produced through the pyrolysis of microalgal biomass is relatively more than that produced from lignocellulosic biomass (D. Mohan, Pittman, and Steele 2006). Feedstock properties and operational parameters such as heating rate, temperature, residence time, particle size, gas flow rate, and reactor design are the key factors that influence the bio-oil yield (R. K. Mishra and Mohanty 2018). Based on the pyrolysis temperature, the process is divided into fast and slow pyrolysis. Fast pyrolysis is carried out at a higher heating rate (10°Cs^{-1}-600°Cs^{-1}) with shorter residence time, which results in the production of bio-oil (19%-57%) and biochar. On the other hand, slow pyrolysis occurs at a lower heating rate (5°Cmin^{-1}-10°Cmin^{-1}) with longer residence time producing bio-oils, methane, carbon dioxide, and a considerable amount of biochar (W.-H. Chen et al. 2015). Microwave and catalytic pyrolysis are two other types of pyrolysis. Microwave pyrolysis uniformly heats the biomass using microwave heat and temperature at 800°C, whereas catalytic pyrolysis is operated at around 300°C in the presence of a catalyst. Bio-oils can be directly employed for bioenergy production through combustion or can be converted to liquid biofuels. Microalgal species such as *Chlorella* sp., *Tetraselmis* sp., and *Nannochloropsis* sp. have been widely used for bio-oil production through pyrolysis (Grierson et al. 2009; Pan et al. 2010).

5.3.3.2.4 Torrefaction for Solid Biofuel Production

Torrefaction is a pretreatment process of the microalgal biomass for generating bioenergy at an industrial scale through processes like direct combustion. Torrefaction degrades organic biomass in an inert atmosphere at around 200°C to 300°C and 1 atmospheric pressure. Torrefaction enhances the feedstock properties by reducing the moisture content, increasing the energy density, reducing the O/C ratio, increasing the heating value, and improving the ignitability and reactivity of the processed biofuel (W.-H. Chen and Kuo 2011). Torrefaction dehydrates the microalgal biomass and thermally decomposes the biomolecules (proteins and carbohydrates), causing partial carbonization. The torrefaction efficiency depends on the reaction temperature and the holding time. Based on the reaction temperature, torrefaction can be light (200°C-235°C), mild (235°C-275°C), or severe (275°C-300°C) (Briens, Piskorz, and Berruti 2008). Recent research on oxidative and nonoxidative torrefaction of microalgal biomass and residue suggests that a high-energy-density solid biofuel can be successfully produced for industrial applications (C. Zhang et al. 2019).

5.3.3.2.5 Gasification for Syngas Production

Gasification is a thermochemical process where carbonaceous materials are converted to carbon monoxide (8%-53%), hydrogen (6%-55%), methane (2%-26%), and other gases known as syngas. Syngas can be produced either by conventional gasification methods or by supercritical water gasification methods. During conventional gasification, dry biomass is processed at around 800°C to 1000°C under 1bar to 10bar pressure. Supercritical water gasification is carried out at around 400°C to 500°C and 24MPa to 36MPa pressure using wet biomass, which results in the production of methane and hydrogen. During supercritical water gasification, microalgal biomolecules are decomposed by supercritical water, and the C-C bond is broken down to smaller gaseous molecules. Generally, a high reaction temperature, high residence time, and high water density with lower feedstock load favor high syngas yield. The syngas produced from microalgae has low calorific value, and so it can be directly used as biofuels for running engines and turbines. Moreover, syngas also finds its application for producing methanol and dimethyl ether through the Fischer-Tropsch process (W.-H. Chen et al. 2012). Gasification produces tars, char, and ashes as a byproduct during the process.

5.3.3.2.6 Combustion for Energy Production

Combustion is a chemical process where fuel reacts with oxygen in a furnace, boiler, or steam turbine at above 800°C producing carbon dioxide, water, and heat energy. Combustion directly burns the biomass in the presence of air and converts the chemical energy stored in the biomass to heat energy (Goyal, Seal, and Saxena 2008). The moisture content of the feedstock should be <50% for the combustion process (McKendry 2002). Biomass with low energy density and high ash content is considered a desirable feedstock for the combustion process. Net energy conversion efficiency varies from 20% to 40% in the case of biomass combustion plants, whereas higher conversion efficiency can be achieved in larger plants (>100 MW) or when microalgal biomass is co-combusted with coal and other waste biomass (Suganya et al. 2016).

5.3.3.3 Chemical Conversion

5.3.3.3.1 Transesterification for Biodiesel Production

Transesterification is a chemical conversion process in which triglycerides react with alcohol to produce fatty acid alkyl esters (biodiesel) and glycerol. Methanol and ethanol are the most widely used alcohols during the transesterification reaction due to their low cost and availability. The transesterification process can be either catalytic (acid or base catalyst) or noncatalytic (Suganya et al. 2016). Based on the catalyst phase, acid or base catalyst is further divided as a homogeneous, heterogeneous, and enzymatic catalyst. Transesterification using a homogenous catalyst is time consuming; results in saponification; and requires separation of oil, glycerol, and catalyst after the process, making the overall process unsustainable. Heterogeneous catalysts can easily convert feedstocks with high free fatty acid to alkyl esters; moreover, the catalyst can be easily removed from the reaction mixer and reused. Enzymatic transesterification avoids the prelipid extraction step, as the process directly converts the wet microalgal biomass to fatty acid alkyl esters and glycerol under low reaction temperature, making the process sustainable and energy efficient. However, enzymatic transesterification is a time-consuming process. Recent studies have shown that supercritical and microwave-assisted transesterification of wet microalgal biomass can produce substantial amounts of alkyl esters (S. Mishra, Roy, and Mohanty 2019).

5.3.4 Microalgae to Biofertilizer

Soil loses its essential nutrients (nitrogen and phosphorus) with time due to frequent cultivation in arable soil. Fertilizers play a significant role by improving soil fertility to obtain maximum production. However, the intensive application of synthetic fertilizers pollutes the soil (Rai and Shukla 2020). The application of biofertilizers as an alternative to synthetic fertilizers for enhancing soil quality is gaining importance, as biofertilizers are economical and sustainable. Biofertilizers comprise microorganisms (algae, bacteria, fungi) either in the live or powdered form. Microalgal biomass can be used as a biofertilizer, as it is rich in nitrogen (N), phosphorus (P), and potassium (K). Microalgal biofertilizer releases nutrients in a gradual way depending on the nutrient demand of the plants, thus preventing nutrient loss as compared to the synthetic fertilizers (Mulbry, Kondrad, and Pizarro 2007). Microalgal biofertilizers also contain micronutrients such as calcium, manganese, iron, and zinc in higher quantities as compared to synthetic fertilizers. Each of these micronutrients has a vital role in the plant's growth: calcium is required for the development of new roots; manganese is involved in chlorophyll synthesis; iron is a primary component of a structural enzyme; and zinc plays an important role in enzyme synthesis (Wuang et al. 2016). In addition to nutrients, microalgae contain plant growth–promoting substances (auxins, amino acids, polyamines, betaines, and gibberellins) and antifungal compounds, which further help in plant growth (Spolaore et al. 2006). Microalgae also aid in stimulating plant growth hormones and β-glucans (Vigani 2020). Microalgal biomass serves as a potential biopesticide by protecting plants against pathogens and various abiotic stresses (Jacob-Lopes et al. 2020).

5.3.5 Microalgae to Feed

Microalgae are considered a potential food source for animals because of their high nutritional content. Microalgal biomasses are either mixed with other feed materials or used alone in the form of pellets, powder, or paste. It has been found that about 75% of the microalgal biomass produced annually is used for producing microalgae-based supplements for animal and human consumption (Chacón-Lee and González-Mariño 2010). Microalgae holds the potential to substitute commonly used animal feed, thereby alleviating the competition among food, feed, and bioenergy (Madeira et al. 2017). Microalgal biomass is highly nutritious as:

1. The protein content in microalgae varies from 8% to 71%. They have a balanced amino acid profile, which helps in the growth and development of animals (Wu 2014).

2. They are rich in lipid content (16%-75%) with a significant amount of PUFAs, docosahexaenoic acid, arachidonic acid, eicosapentaenoic acid, and g-linolenic acid (Safinaz and Ragaa 2013).
3. The carbohydrate content in microalgae varies from 8% to 57%. Microalgal carbohydrates (starch, sugars, polysaccharides, and cellulose) are rich in dietary fiber, which enhances the digestibility of animals. Since carbohydrates make up the largest portion of animal feed and energy sources, these features of microalgal carbohydrates make them a notable feed ingredient (Madeira et al. 2017).
4. They are rich in pigments, vitamins, and bioactive compounds (Navarro et al. 2016).

However, microalgae as a feed for animals have certain limitations. The microalgal cell composition varies with species and growth conditions. The sulfur-containing amino acids, such as methionine and cysteine, are usually found in small concentrations and thus need to be supplemented externally. For some animals, the rigid cell wall of algae is indigestible (Madeira et al. 2017). Therefore, either mechanical or heat pretreatment should be implemented on the microalgal biomass to rupture the cell wall, so that confined nutrients become available for the animals. Microalgae also serve as a nutritious feed for aquaculture. Microalgae of the genera *Phaeodactylum*, *Tetraselmis*, *Isochrysis*, *Nannochloropsis*, and *Haematococcus* are widely used as aquaculture feed (Kleivdal, Chauton, and Reitan 2013). Proteins and PUFAs that are abundantly available in microalgae are an essential nutrient component of aquaculture feed. It has been reported that about 10% of the microalgal biomass produced is being used in the poultry industry (Madeira et al. 2017). In a comparative study between conventional and microalgae-based poultry feed, microalgal feed showed increased disease tolerance, improved yolk color, and low cholesterol (Fredriksson, Elwinger, and Pickova 2006). However, some essential criteria need to be fulfilled when using microalgae as feed, such as (a) animal acceptance; (b) product nutrition and digestibility; (c) absence of toxins; (d) impact on animal's weight gain; and (e) product odor and taste (Koutra et al. 2020).

5.3.6 Microalgae to Bioactive Compounds

The microalgal biomass is enriched with a wide variety of bioactive compounds, which have potential applications in various sectors, such as food, pharmaceutical, and cosmetic industries. Within the context of the integrated biorefinery approach, bioactive compounds play a significant role in boosting the economics, owing to their high market value (Koller, Muhr, and Braunegg 2014). Novel extraction methods such as microwave-assisted extraction, supercritical fluid extraction, enzyme-assisted extraction, ultrasound-assisted extraction, and pressurized liquid extraction enable the extraction of bioactive compounds without their degradation. The bioactive compounds derived from microalgae have antibacterial, antitumor, antimalarial, antifungal, antioxidant, antiviral, and anti-inflammatory properties and so have gained research interest in curing many diseases (Ummalyma, Sahoo, and Pandey 2020). Some of the bioactive compounds are discussed in the following sections.

5.3.6.1 Lipids

Microalgal lipids may contribute about 30% to 50% of their total weight (Jacob-Lopes et al. 2020). However, at extreme environmental conditions, microalgae can accumulate lipids as high as 70% to 90% of their dry weight (Koyande et al. 2019). The higher lipid productivity of microalgae as compared to other lipid-containing raw materials makes microalgae a potential feedstock for the production of bioenergy and bioactive compounds. The lipid fraction of microalgae primarily consists of PUFAs such as omega-3 fatty acids (eicosapentaenoic acid and docosahexaenoic acid) and omega-6 fatty acids (γ-linolenic acid and arachidonic acid) (Priyadarshani and Rath 2012). PUFAs play a vital role for the proper functioning of the human body. Due to their antioxidative, antibacterial, and antifungal properties, PUFAs are widely used in nutritional, pharmaceutical, and food industries (Borowitzka 2013). Microalgal lipids can be extracted from the cell using various novel techniques. Among various lipid extraction techniques, the supercritical carbon dioxide extraction method is considered a potential technology, as it can be implemented for large-scale lipid extraction. This extraction technique is nontoxic, has high selectivity for acylglycerols, and yields solvent-free lipids. However, the high cost and energy involved during this extraction process make this technique economically unsustainable (Halim, Danquah, and Webley 2012). Sterols are another interesting group of lipids with many attractive properties. Phytosterols extracted from microalgae find wide application in pharmaceutical (production of therapeutic steroids), nutraceutical (anticholesterol additives in functional foods and anticancer properties), and cosmetic industries (Francavilla et al. 2012).

Among all the applications of microalgal lipids, the most promising one is biodiesel production. Microalgae have the potential to reduce the dependency on fossil fuels and mitigate GHG emissions. Many biomass conversion techniques such as catalytic cracking, microemulsification, and transesterification can be used to produce biodiesel (fatty acid alkyl esters). Transesterification is the most widely used method for producing biodiesel, as this process is relatively inexpensive and produces high-quality diesel compared to other methods (Faried et al. 2017).

5.3.6.2 Proteins

Microalgae are a rich source of protein, accounting for 50% to 70% of the total composition. Microalgae are considered a promising protein source, as they have a high protein content with a favorable amino acid profile comprising essential amino acids (leucine, valine, lysine, isoleucine, and branched-chain amino acids). According to the World Health Organization (WHO), these essential amino acids are present in microalgal protein in an adequate amount for human requirements (Jacob-Lopes et al. 2020). Among various microalgal species, *Spirulina* is known for its single-cell protein content and nutraceutical properties. *Spirulina* is being produced commercially by many industries; for example, Parry Nutraceuticals (India) manufactures *Parry Organic Spirulina,* which contains 60% protein and is known as a green food (Michalak and Chojnacka 2015).

Bioactive proteins are one of the most important products of the microalgal biorefinery. Phycobiliproteins are a group of water-soluble proteins that are generally present in cyanobacteria and certain algae such as cryptomonads, glaucocystophytes, rhodophytes, cyanelles, etc. (Ibañez et al. 2012). Based on the spectral properties and structures, these hydrophilic phycobiliproteins are classified into three major types, namely phycoerythrin, phycocyanin, and allophycocyanin (Jacob-Lopes et al. 2020). In most of the microalgal species, these phycobiliproteins are arranged in specialized aggregated structures known as phycobilisomes. These phycobilisomes, in conjunction with chlorophyll a (Chl a), optimize their light-harvesting functions. Phycobiliproteins have antioxidant, anti-inflammatory, neuroprotective, antiviral, antitumor, and hepatoprotective properties and so find wide applications in pharmaceutical, cosmetic, and food industries. These pigments are also commercially used as natural dyes and fluorescent agents (Bhagavathy, Sumathi, and Bell 2011). It has been reported that phycobiliproteins can be efficiently extracted from microalgae using a microwave-assisted extraction technique (Juin et al. 2015).

5.3.6.3 Polysaccharides

Polysaccharides are macromolecules of polymeric carbohydrates with complex structures (Chanda, Merghoub, and EL Arroussi 2019). Polysaccharides vary significantly, both biochemically and structurally, from one species to another. Microalgae excrete extracellular polysaccharides (EPSs) consisting of polysaccharides, lipids, proteins, and nucleic and uronic acids into their medium. EPSs are being exploited for various industrial applications. Microalgal species such as *Dunaliella, Chlorella, Ankistrodesmus,* and *Botryococcus* excrete EPSs (Venkata Mohan et al. 2020). The highly viscous EPSs are widely used as flocculants, emulsifiers, and surfactants. They are also used in medical applications as antitumor, antiviral, anticoagulant, antioxidant, and anti-inflammatory agents. Kaplan, Christiaen, and Arad (1987) reported that EPS extracted from *Chlorella* species has good metal ion chelating properties, which can be used in the production of new formulations such as carriers of microelements in biofertilizers. Polysaccharides are categorized as soluble and insoluble fibers based on their solubility. The soluble fibers such as fucans, laminarans, and alginates are obtained from brown algae (*Phaeophyta*), whereas the insoluble fibers, also known as sulfated galactans (agars, xylans, starch, and floridean), are extracted from red algae (*Rhodophyta*) (Kraan 2012). Green algae contains sulfuric acid polysaccharides, xylans, and sulfated galactans (Mišurcová, Orsavová, and Vávra Ambrožová 2015).

From an economic viewpoint, microalgal polysaccharides are the most important bioactive compounds extracted from microalgae. Green extraction techniques such as microwave-assisted and enzyme-assisted extraction methods have been successfully applied for the recovery of polysaccharides (B. Wang et al. 2011; Ko et al. 2012). Microalgal polysaccharides are widely being used in food and pharmaceutical industries, as these compounds possess antitumor, antioxidative, antiviral, and antibacterial properties. Fucoidan has antitumor, antithrombotic, antioxidant, and anticoagulant properties and is used as a dietary complement (O'sullivan et al. 2011). Agar is a combination of agaropectin and agarose. This bioactive compound is commercially used as a gelling agent to change the viscous properties of food products. Carrageenan is a water-soluble sulfated polysaccharide that is extensively used as a stabilizing and emulsifying agent in various food products such as puddings, jams, jellies, chocolates, ice creams, etc. It is also used as an anticoagulant, antiviral, and antitumor agent (Aslam et al. 2020). Moreover, it has been reported that carrageenan gel extracted from *Chondrus cripus* has the potential to prevent husk the diffusion of human immunodeficiency virus and some viruses that are linked to sexually transmitted infections (Carlucci et al. 1997).

5.3.6.4 Carotenoids

Carotenoids are lipophilic pigments that are photosynthetic accessory pigments in plants. Carotenoids are further classified as β-carotene, lutein, zeaxanthin, astaxanthin, and lycopene. Microalgal carotenoids are considered to be involved in the reduction of risk associated with certain cancers, macular degeneration, cataract formation, and cardiovascular diseases (Jacob-Lopes et al. 2020). In 2017, the global demand for carotenoids was estimated at US$1577 million and is projected to reach US$2098 million by 2025, registering a compound annual growth rate of 3.6% from 2018 to 2025 (Ferreira and Gouveia 2020).

β-Carotene is a red-orange colored pigment found in cyanobacteria and microalgae (Malik, Tokkas, and Goyal 2012). Among all the microalgal species, *Dunaliella* sp. has the highest β-carotene–producing potential (100 g kg-1 dry cell weight) (Coesel et al. 2008). β-Carotene is used as an antiaging, anticancer, antioxidant, and coloring agent (Harvie 2014). Lutein is gaining commercial importance due to its potential nutraceutical properties. The productivity of microalgal lutein depends on various environmental factors such as pH, salinity, temperature, light intensity, and nitrogen content (Guedes, Amaro, and Malcata 2011). Microalgal species such as *Scenedesmus almeriensis*, *Chlorella zofingiensis*, *Neospongiococcus gelatinosum*, *Chlorella protothecoides*, and *Chlorococcum citriforme* have been widely used for extracting lutein (Jacob-Lopes et al. 2020). Zeaxanthin is a yellow-colored carotenoid that is extensively used in cosmetic, pharmaceutical, and food industries. Zeaxanthin can also be used as a supplement for improving the visual performance of humans (Kvansakul et al. 2006). Astaxanthin is a keto-carotenoid belonging to the xanthophyll family. It is a blood red–colored pigment, which is naturally produced in certain microalgal species such as *Haematococcus pluvialis* and *C. zofingiensis* (J. Liu et al. 2014). Astaxanthin is a potent antioxidant with antiaging, anti-inflammatory, and sun-protection properties. Lycopene has received commercial importance due to its antibiotic, antioxidative, anticarcinogenic, and antiatherogenic properties (Friedman 2013). Renju et al. (2014) reported that lycopene extracted from *Chlorella marina* showed a high antioxidative and anti-inflammatory effect on an atherogenic rat.

5.3.7 Microalgae to Bioplastic

Increased urbanization and growing populations have resulted in the accumulation of a substantial amount of plastic waste. It has been reported that about 80% of plastics are being accumulated as waste in the environment, causing an alarming hazard (Karan et al. 2019). Bioplastics derived from biopolymers can substitute the conventional petrochemical-based plastics. Bioplastics are eco-friendly and reduce the reliance on fossil reserves. Biopolymers (polyhydroxyalkanoates, cellulose acetate, starch, polylactic acids, and polyvinyl chloride) derived from food crops are conventionally used to produce bioplastics (Ummalyma, Sahoo, and Pandey 2020). However, bioplastics produced from food crops are not sustainable. Microalgae can be exploited as a potential source for bioplastic production, as they are rich in carbohydrate polymers and protein (Karan et al. 2019). Moreover, microalgae-based bioplastics are highly compatible and biodegradable. It has been reported that microalgal species such as *Spirulina* and *Chlorella* are being successfully used to produce bioplastics and thermostable plastic bends (K. Wang et al. 2016). Das et al. (2018) reported that a deoiled microalgal biomass obtained after biodiesel production from *Chlorella pyrenoidosa* could be used to produce bioplastic primarily polyhydroxybutyrate (PHB) through chemical treatment. Bioplastics produced from microalgae are a byproduct in the microalgal biorefinery, which helps to generate revenue along with bioenergy and coproducts as part of a circular bioeconomy. For example, *Synechocystis salina* can be exploited as a feedstock for the coproduction of PHB, feed, pigments, biogas, and biofertilizers (Meixner et al. 2018).

5.4 THE CIRCULAR BIOECONOMY APPROACH

Although the microalgal biorefinery has gained huge attention for its potential applications, it lacks commercial applicability due to the unsustainability of the process. The techno-economic challenges during industrial-scale cultivation and downstream processing lead to the high cost of the biofuel. A microalgal biorefinery generates a considerable amount of byproducts in the form of liquid effluents and solid residues. A sustainable biorefinery can be achieved if these byproducts are reduced, recycled, and reused (Figure 5.3). However, very limited studies have been conducted for the valorization of solid and liquid byproducts. Table 5.1 summarizes all the solid and liquid byproducts that have been processed further for the production of bioenergy and value-added products.

5.4.1 Processing of Liquid Waste Streams

A large volume of liquid waste is produced during the upstream and downstream processing of a microalgal biorefinery. Therefore, in order to minimize the water footprint and generate renewable energy, the recycling and processing of such liquid waste are important.

5.4.1.1 Effluent from Microalgal Culture

The microalgal biomass to culture medium ratio is small; hence, a considerable amount of wastewater is generated from microalgal culture cultivated in a photobioreactor and open raceway pond. It has been estimated that to produce 1 kg of algal biomass, approximately 1 metric ton of water is consumed (Guieysse, Béchet, and Shilton 2013). Therefore, the recycling of the spent medium is necessary to meet the water demand. The chemical composition of the spent medium is mostly influenced by three factors: culture medium, microalgal species, and harvesting technique. Among all three factors, the harvesting process has a significant impact on the quality of the spent medium. Microalgae harvesting using chemical flocculants may alter the color and chemical composition of the spent medium, thereby inhibiting the reusability of the medium (Roy and Mohanty 2019). Biomass harvesting by pH shift is commonly applied, as it removes the risk of chemical contamination; however, the spent medium needs to be neutralized prior to recycling. Biological harvesting

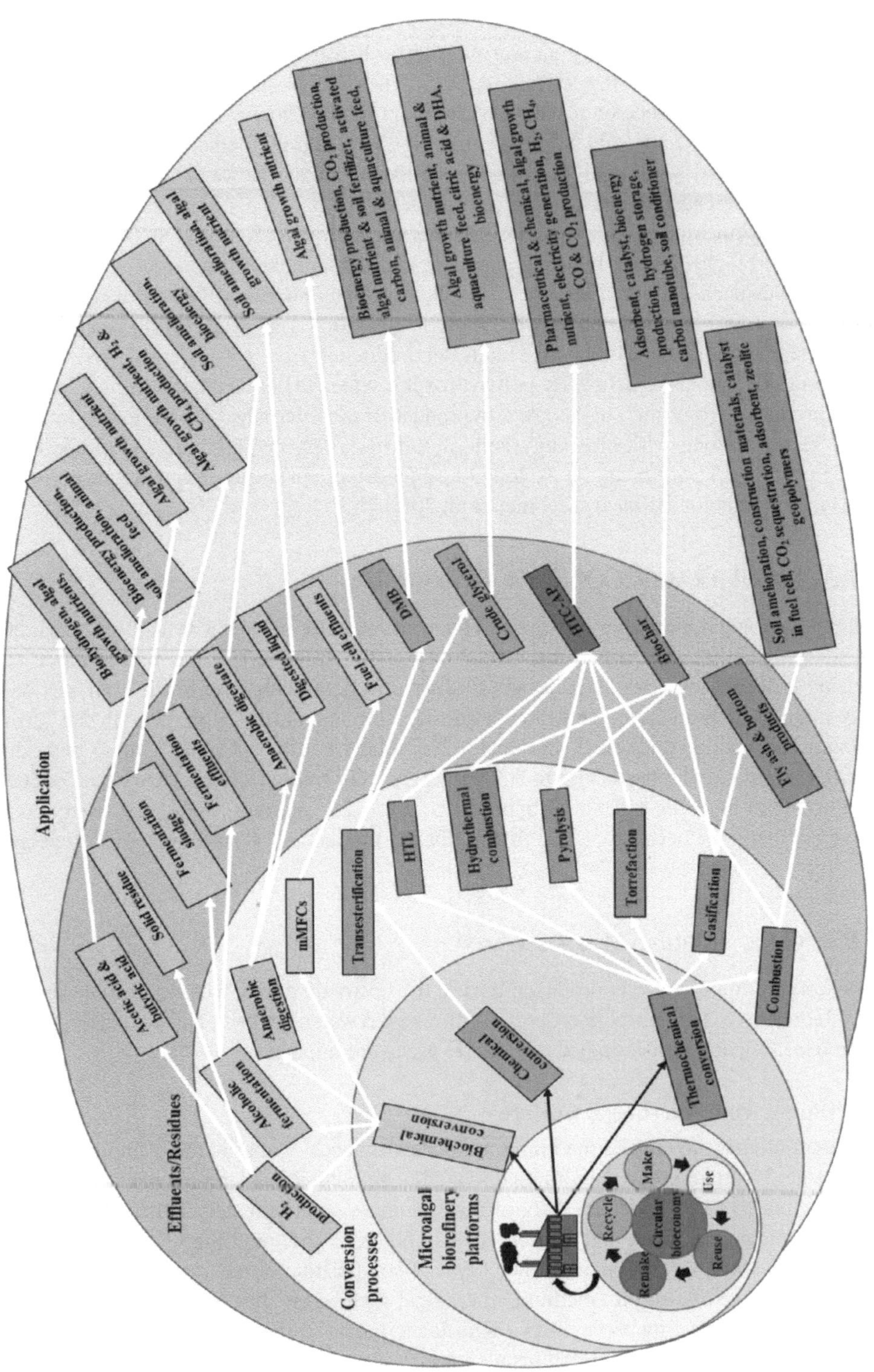

FIGURE 5.3 The biorefinery framework on a circular bioeconomy depicting the applications of liquid effluents and solid residues.

TABLE 5.1
Recycling of Solid Residues and Liquid Effluents for the Production of Bioenergy and Value-Added Products

Source of Residue	Microalgal Species	Application	Features	References
Liquid residue				
Biochemical	*Chlorella pyrenoidosa*	Microalgae cultivation	69.37% N and 60.22% P were released into digested liquid	(L. Yang et al. 2018)
	Gelidium amansii	Methane	Methane yield 283 L kg^{-1} VS	(J.-H. Park et al. 2012)
Thermochemical	*C. sorokiniana*	Nutrient recycling	Contains high concentration of phosphate, organic nitrogen, and polysaccharides	(L. Chen et al. 2017)
Transesterification	-	Hydrogen, syngas, butanol	Releases high amount of crude glycerol	(F. Yang, Hanna, and Sun 2012)
	Chlorella minutissima	Microalgal biomass and lipid production	Glycerol is a rich source of carbon	(Katiyar et al. 2017)
Solid residue				
Biochemical	*Nannochloropsis* sp.	Crude biodiesel	Fatty acid ethyl ester ~14 wt%	(Rahman et al. 2019)
	Chlamydomonas mexicana	Electrical energy	Energy output 1.82 kJ L^{-1}, power density 88.6 mW m^{-2}	(Song et al. 2019)
	Chlorella sp.	Hydrogen and methane	Hydrogen yield 12.5 mL g^{-1} VS, methane yield 81 mL g^{-1} VS	(Lunprom et al. 2019)
	Gelidium amansii	Methane	Methane yield 239 L kg^{-1} VS	(J.-H. Park et al. 2012)
Thermochemical	*Chlorella vulgaris*	Wastewater treatment	COD, NO_3, NH_3, and PO_4 were reduced by 53.3%, 58.6%, 94.7%, and 59.7%	Arun et al. 2018)
	Spirulina sp.	Soil amendment	Contains 11.82% C, 1.81% H,1.41% N, and 0.61% S	(Barreiro et al. 2013)
Transesterification	*Scenedesmus dimorphus*	Bioethanol	Fermentation of DMB produced 0.26 g g^{-1} of bioethanol	(Chng, Chan, and Lee 2016)
	Chlorella sp.	Activated carbon	Activated carbon with 840 m^2 g^{-1} and 0.46 cm^2 g^{-1} surface area and total pore volume, respectively	(Chang, Tsai, and Li 2015)
	Botryococcus braunii	Biogas	AD of glycerol:DMB = 10:90 produced 448 mL CH_4 g^{-1} VS	(Neumann et al. 2015)
	Nannochloropsis oceanica, Desmodesmus sp.	Dietary source	DMB of these two species served as an iron and protein source for an anemic pig	(Manor et al. 2017)
	Chlorella pyrenoidosa	Methane	28.38% energy recovered as methane from DMB	(L. Yang et al. 2018)

Source: Adapted from (S. Mishra, Roy, and Mohanty 2019)

techniques are chemical-free processes, but there is a high chance of microbial contamination. Therefore, appropriate pretreatment of the spent medium is necessary before recycling it for algal cultivation. Physical pretreatments such as centrifugation, autosedimentation, and membrane filtration are generally employed for the purification of the spent medium. Recently, researchers have come up with various pretreatment methods such as chlorination, ozonation, filtration, and sterilization.

It is necessary to understand the physiology of the microalgae grown before reusing the spent medium. Some microalgae release biomolecules known as allelochemicals in the medium during their growth. These allelochemicals may either inhibit or enhance the growth of microalgae cultured in the spent medium (Bacellar Mendes and Vermelho 2013). Several biotic and abiotic factors stimulate the microalgae to produce allelochemicals such as fatty acids, phenolic compounds, peptides, alkaloids, and aldehydes. The growth of *Nannochloropsis* sp. was found to be significantly reduced in the spent medium as compared to the fresh medium. *Nannochloropsis* sp. shed their cell wall in the medium during cell division; these cell walls in the spent medium reduced the productivity of microalgae (Rodolfi et al. 2003). On the other hand, the growth rate and lipid productivity of *Tetradesmus obliquus* was relatively high when cultured in the spent medium recovered after the harvesting process. The calcium carbonate released from the bioflocculant in the harvested medium might have served as a carbon source, thereby enhancing the growth rate and lipid productivity (Roy and Mohanty 2020). Thus, proper utilization of the spent medium will help to reduce the water footprint for microalgal cultivation.

5.4.1.2 Effluent from Biochemical Conversion

Anaerobic digestion of the microalgal biomass produces a significant amount of digestate as a byproduct. The anaerobic digestate can be divided into a solid and liquid fraction. Liquid digestate contributes to a major fraction of digestate (80%-90% by mass) and is rich in nitrogen (629.70 mg L^{-1}) and phosphorus (54.00 mg L^{-1}) content (Fuchs and Drosg 2013; L. Yang et al. 2018). The conventional application of liquid digestate as a biofertilizer for soil amelioration may lead to eutrophication. Alternatively, microalgae can offer a solution for treating liquid digestate. It has been reported that microalgae can efficiently utilize the nutrients from the liquid digestate while providing high-value biomass for biorefinery applications (Xia and Murphy 2016). The economic analysis estimated that liquid digestate could reduce nutrient cost by 41.77% (L. Yang et al. 2018).

Volatile fatty acids (VFAs) such as acetic acid and butyric acid are the major byproducts produced during hydrogen production via dark fermentation (H. Yang, Guo, and Liu 2010). However, if these effluents are not treated properly, it may cause a serious threat to the environment. These dark fermentative effluents, which are mostly composed of organic acids, serve as a potential substrate for photosynthetic nonsulfur bacteria to produce biohydrogen (Fang, Liu, and Zhang 2005). Moreover, heterotrophic and mixotrophic microalgae can efficiently utilize these organic acids as an inexpensive carbon and energy source for their sustainability. Mixotrophic growth with VFAs as a carbon source maximized the growth and lipid productivity of *Chlamydomonas reinhardtii* (Moon et al. 2013).

Fermentation effluents are major byproducts of the microalgal fermentation process. These effluents contain a considerable amount of mineral elements that can be efficiently used for cell metabolism of microalgae. *Schizochytrium* sp. was able to produce 22.23 g L^{-1} of docosahexaenoic acid (DHA) by replacing 20% of freshwater with fermentation effluent (Yin et al. 2018). The fermentation effluent is also a rich source of carbon and other nutrients, which can serve as a potential feedstock for hydrogen and methane production via anaerobic digestion (De Paoli et al. 2011).

5.4.1.3 Effluent from Thermochemical Conversion

A large fraction of wastewater is generated during HTC of wet microalgal biomass. During the HTC process, approximately 20% to 50% of the organic compounds are transferred into the aqueous phase (HTC-AP) (S. Mishra, Roy, and Mohanty 2019). The high concentration of organic compounds such as nitrogen, phosphorus, polysaccharides, phenols, proteins, and organic acids in the HTC-AP makes it a potential feedstock for various applications (Y. Chen et al. 2017). Moreover, as the HTC-AP is composed of phenolic and aromatic compounds, it must be appropriately treated before being disposed of. Fernandez et al. (2018) codigested 30% (v/v), HTC-AP with manure for simultaneously degrading salts and organic compounds while producing biogas.

The nutrient-rich HTC-AP can be recycled for cultivating microalgae. Du et al. (2012) reported that the biomass productivity of microalga *Chlorella vulgaris* was much higher in HTC-AP as compared to in BG-11 medium. In another study, *Chlorella sorokiniana and C. vulgaris* were able to efficiently utilize polysaccharides in HTC-AP for its growth (L. Chen et al. 2017). Phenol and carboxylic acid recovered from HTC-AP using nanofiltration technology can be used in the chemical and pharmaceutical industries (X. Zhang et al. 2018). Monosaccharide-rich HTC-AP can be employed for producing bioethanol. Moreover, HTC-AP can serve as a promising feedstock for bioenergy production through microbial fuel cells. In another study, the organic carbon-rich HTC-AP was hydrothermally gasified to produce hydrogen. Moreover, the aqueous phase recovered after gasification was found to be rich in nutrients that could be recycled for microalgal cultivation (Cherad et al. 2016).

5.4.1.4 Effluent from Transesterification

Biodiesel production through the transesterification process generates approximately 10% (w/w) glycerol as the primary byproduct. It has been reported that about 50% to 80% glycerol is generated from the transesterification process, which means that 66% of the world's glycerol is recovered from the biodiesel industries (Kaur et al. 2020). Crude glycerol is mainly composed of catalysts, alcohol, soap, organic matter, salts, and feedstock impurities (F. Yang, Hanna, and Sun 2012). With the increasing biodiesel production, enormous amounts of glycerol are being produced. However, the purification of this low-grade glycerol involves a high cost; therefore, the disposal of crude glycerol has become a serious environmental problem. Thus, valorization of crude glycerol to value-added products will simultaneously reduce the environmental issues and make the biodiesel industry economically sustainable.

Glycerol, also known as glycerin, has great opportunities for various applications, as it is a chemical with high commercial value. Crude glycerol recovered after the transesterification process can be recycled for microalgal cultivation, as it is rich in organic carbon. S. Xu et al. (2019) could achieve a bioethanol yield of 5.58 g L^{-1} from microalga *Scenedesmus obliquus* grown in wastewater enriched with crude glycerol. Crude glycerol is a promising substrate for DHA production (Ethier et al. 2011). Crude glycerol can be copyrolyzed with microalgal biomass for enhanced bio-oil production (S. Wang et al. 2019). Crude glycerol is also a promising feedstock for PUFAs, biohydrogen, biopolymer, bioethanol, and n-butanol production (Garlapati, Shankar, and Budhiraja 2016). Since the glycerol kinase enzyme in the animal liver can convert glycerol to glucose, crude glycerol can be used as animal feed (Kerr, Dozier, and Bregendahl 2007). Moreover, hydrogenolysis of crude glycerol produces 1,2,-propanediol, which finds its application in pharmaceuticals, polyester resins, foods, functional fluids, detergents, and animal feed (Monteiro et al. 2018). Commercially important chemicals such as propionic acid, citric acid, succinic acid, glyceric acid, lactic acid, acetate, 1,3-propanediol, etc., can be produced from crude glycerol, which can contribute to the bioeconomy of the system (Kaur et al. 2020).

5.4.2 Processing of Solid Residue

The primary processing of microalgal biomass for the production of bioenergy and value-added products generates a considerable amount of solid residues. The residual biomass obtained after primary application generally retains about 70% of its energy level along with various nutrients such as lipids, proteins, and carbohydrates. These solid residues can be valorized through various processes for multiple product recovery to achieve a zero-waste microalgal biorefinery. However, it is important to choose an appropriate pathway for the valorization of solid waste to prevent downstream energy consumption.

5.4.2.1 Residue from Biochemical Conversion

The solid residue obtained after microalgal fermentation is generally composed of about 50% organic matter such as lipids, proteins, carbohydrates, polysaccharides, volatile solids, and biodegradable compounds, which can be a promising substrate for further applications (Lunprom et al. 2019). The residual biomass obtained after the fermentation process contains a considerable amount of volatile solids, which can be further processed for bioenergy production. The residual biomass obtained after anaerobic fermentation of *Chlorella* sp. was found to be a potential feedstock for biohythane production post acid-thermal pretreatment (Lunprom et al. 2019). Fermentation residue can also be used as a feedstock for a microbial fuel cell (MFC) to harvest electrical energy (Song et al. 2019). Moreover, the integration of fermentation with the MFC system will provide a platform for the simultaneous treatment of fermentation residue and energy recovery. In addition to bioenergy production, a fermentation residue with high inorganic salts and nitrogen content can partly substitute the yeast extract used in the traditional culture medium for DHA production. Yin et al. (2018) could achieve a DHA yield of 27.10 g L^{-1} from *Schizochytrium* sp. by replacing 80% of the yeast extract with fermentation residue. Protein- and amino acid–enriched microalgal fermentation residues can also substitute nitrate in the culture medium. In addition to providing a nitrogen source, these protein-rich fermentation residues contribute to a carbon "heterotrophic boost" in subsequent cultivation, thereby enhancing overall biomass and lipid productivity (Gu et al. 2015). Park et al. (2012) stated that anaerobic digestion of fermentation residue recovered energy 2.24 times higher than that of bioethanol produced in the primary process.

Anaerobic digestion of the microalgal biomass partially extracts hydrogen, carbon, and oxygen for biogas production; however, a considerable amount of organic matter, water, nitrogen, phosphorous, and micronutrients is left in the solid digestate. Recovery of phosphorus from solid digestate to produce value-added products is gaining importance in anaerobic digestion industries to make the overall process economically sustainable. Phosphorous can be recovered from the solid digestate using physiochemical, biological, and thermochemical methods (F. Xu et al. 2018). Ideally, the recovered phosphorus can be used as an organic fertilizer for soil amelioration. However, with an expansion of biogas plants, land application alone of solid digestate can become problematic with negative impacts such as biological hazards, GHG emissions, eutrophication, etc. (Fabbri and Torri 2016). Consequently, various possibilities for the valorization of solid digestate should be developed for achieving economic and environmental sustainability.

In a typical anaerobic digestion process, approximately 33% to 50% of the energy is converted and more than half of the energy is retained in the solid digestate, making it a promising feedstock for bioenergy production (Hossain et al. 2016). Recently, anaerobic digestion was coupled with pyrolysis for simultaneous biogas production and conversion of a solid digestate to syngas, bio-oil, and biochar (Fabbri and Torri 2016). In another study, a typical anaerobic digestion process was integrated with a gasification system in a resource closed-loop approach, where the solid digestate was converted to bioenergy and soil fertilizer (Antoniou et al. 2019).

The lipid-rich solid residue obtained after photobiological hydrogen production can be transesterified to produce biodiesel. The deoiled residue can be further used as animal feed or anaerobically digested to produce biogas (Torri et al. 2011). Moreover, the integration of photobiological hydrogen production with pyrolysis will result in the simultaneous production of hydrogen, bio-oil, and biochar. The biochar retains a considerable amount of ash and nitrogen, which could be used for soil amelioration, thereby contributing to the abatement of GHGs (Pan et al. 2010). A hydrogen-extracted residual biomass with relatively high nitrogen content favors its application as a feedstock for animal feed compared to bioenergy production (Rashid et al. 2013).

5.4.2.2 Residue from Thermochemical Conversion

The solid residue generated from an ideal HTL system is <10 wt.% of the feedstock. The solid residue obtained from the HTL process has a high ash content. It also contains hydrogen, sulfur, nitrogen, and several other minerals that are transferred from the feedstock (S. Mishra, Roy, and Mohanty 2019). These solid residues with adequate nutrients and minerals are a promising candidate as biofertilizers for soil amendment. Xu et al. (2019) stated that mineral-rich solid residues might exhibit catalytic properties during the HTL process, thereby enhancing the reaction mechanism.

However, biocrude contamination with mineral elements present in the solid residue may interfere during the downstream processing by reducing the biocrude quality (R. Liu et al. 2018). The biochar obtained from the HTL process can be activated and utilized as an adsorbent for treating various wastewater streams (Arun et al. 2018).

Biochar, also known as black carbon, is the solid residue produced during pyrolysis, hydrothermal carbonization, and gasification processes. Though biochar is primarily composed of carbon, it also contains oxygen, nitrogen, hydrogen, sulfur, and ash (N. Liu et al. 2015). Biochar is generally characterized by a large surface area, porous structure, and surface functional groups, because of which it not only finds its application for bioenergy production but is also widely used as an adsorbent for environmental remediation, as a catalyst to enhance reaction, as fuel cell, and as a soil conditioner (Cha et al. 2016). Recently, it has been reported that biochar also has the potential to substitute synthetic carbon nanotubes for use in supercapacitors (Gupta et al. 2015). An activated biochar with a large surface area and micropore volume can also be used for hydrogen storage (Cha et al. 2016).

Fly ash and bottom products are released as solid byproducts during the combustion and gasification processes of a microalgal biomass. Fly ash is mostly pozzolanic and <100 micro in size and is widely used in construction materials as a binder; in catalysis; in polymers as fillers; in wastewater treatment as an adsorbent; in zeolites; and in soil amelioration (Gollakota, Volli, and Shu 2019). It has been reported that co-combustion of a biomass with coal produces ash with low pozzolanic activity, low illite content, high bulk density, and high anhydrite and hematite content, making the co-combustion fly ash favorable for agricultural applications as compared to applications in construction industries (Belviso 2018). Fly ash can also be used as an adsorbent for capturing CO_2, as a CO_2 storage medium through mineralization, and as a catalyst support or catalyst for CO_2 utilization processes (Dindi et al. 2019). Recent research is focused on green utilization of fly ash that surpasses the conventional fly ash applications. The green applications of fly ash include the synthesis of geopolymers to substitute cement, silica aerogels as an insulating material, and carbon nanotubes for electronic devices (Gollakota, Volli, and Shu 2019).

5.4.2.3 Residue from the Chemical Conversion

Large-scale production of biodiesel leads to the generation of a huge amount of deoiled microalgal biomass (DMB) as a byproduct. Moreover, Chisti (2007) assumed that if 25% lipid could be extracted from a microalgal biomass, then to produce 1 metric ton of biodiesel, triple the amount of DMB will be generated. Therefore, researchers are trying to utilize DMB for various applications to make biodiesel production economically sustainable through a zero-waste biorefinery approach. Carbohydrates, proteins, and minerals are the primary compositions of residual DMB (Maurya et al. 2016). DMB with a high carbon:nitrogen ratio (C:N) acts as a potential feedstock for bioethanol, biohydrogen, and biomethane production, while DMB with a low C:N ratio is rich in protein content and can be used as a biofertilizer, fermentation medium, or an animal feed supplement (S. Mishra, Roy, and Mohanty 2019). However, the chemical composition and application of DMB depend on the microalgal species, growth, harvesting, lipid extraction, and pretreatment conditions (Rashid, Rehman, and Han 2013). Harvesting and lipid extraction through chemical flocculation and organic solvents, respectively, impair the composition of DMB, making it toxic and unfit for animal feed.

DMB serves as a promising feedstock for biomethane production through anaerobic digestion. However, DMB with a high protein content is not favorable for methane production, as the excess ammonia inhibits the fermentation process (Sialve, Bernet, and Bernard 2009). To overcome this problem, researchers have implemented the codigestion approach, where DMB with a low C:N ratio is codigested with high carbon-containing waste to increase the overall C:N ratio (S. Park and Li 2012). In the case of biomethane production from DMB, the energy output to input ratio is found to be high in comparison to bioethanol, biohydrogen, and biodiesel production processes (Maurya et al. 2016). It has been reported that biohydrogen production from pretreated DMB (thermal, alkaline, acid, thermos-alkaline) was greater compared to untreated DMB (Subhash and Mohan 2014; Z. Yang et al. 2011). Thermo-alkaline pretreatment of DMB resulted in a three-fold increase in biohydrogen production compared to untreated DMB (Z. Yang et al. 2011). Apart from protein, carbohydrate is the second largest component of DMB, which can be utilized for bioethanol production via saccharification and microbial fermentation after pretreatment processes. Moreover, Geun Goo et al. (2013) reported that EPS extracted from deoiled microalgae *Dunaliella tertiolecta* can be exploited for producing glucose and bioethanol on an industrial scale. Carbohydrate-rich DMB can also serve as a promising candidate for maltodextrin production (Lam, Tan, and Lee 2014).

DMB can be used as a feedstock for the production of bio-oil, biobitumen, and a carbon-based catalyst through HTL (Rolland et al. 2020; Fu et al. 2013; Vardon et al. 2012). DMB obtained from various microalgal species such as *Nannochloropsis*, *Chlamydomonas*, *D. tertiolecta*, and *Chlorella* has been widely used for the production of bio-oil and biochar through pyrolysis (Maurya et al. 2016). In fact, pyrolysis of a lipid-extracted biomass consumes less energy compared to a fresh microalgal biomass. Syngas can be efficiently produced by the gasification of carbonaceous materials derived from DMB (Watanabe et al. 2015).

In addition to bioenergy production, the DMB serves as a promising feedstock for the production of activated carbon. Gokhale et al. (2008) stated that the biosorption efficiency of DMB is more compared to a fresh microalgal biomass. This can be attributed to the large surface area of DMB, as the ruptured microalgal cells of DMB expose more active sites for biosorption. The DMB of various microalgal species such as *Nannochloris oculata*, *Pseudochoricystis ellipsoidea*, *Microspora* sp., *Spirulina platensis*, *Chlorella vulgaris*, and *Scenedesmus dimorphus* have been utilized for the biosorption of dyes and heavy metals released from industrial effluents (Gokhale, Jyoti, and Lele 2008; Maurya et al. 2016).

The DMB also serves as a highly nutritious growth medium for microorganisms owing to its high nitrogen and phosphorus content (Mishra and Mohanty 2019). Gu et al. (2015) reported that the biomass and lipid yield of *Scenedesmus acutus* was enhanced when DMB was used as the sole nitrogen source for cultivation. The protein-rich DMB has the potential to substitute the commercial animal and aquaculture feed (Gatrell et al. 2014). The DMB is a rich resource for producing biofertilizers. The DMB can be used as an alternative to chemical fertilizers, as it comprises nitrogen, phosphorus, and minerals in a substantial amount, which are the primary components of a fertilizer (S. C. P. Mishra et al. 2014). Bohutskyi et al. (2015) reported that DMB of *Auxenochlorella protothecoides* is a rich source of nitrogen, phosphorus, magnesium, calcium, sulfur, manganese, and iron, which can be used as a potential biofertilizer, thereby reducing 45% of the fertilizer cost.

5.5 CHALLENGES AND FUTURE PERSPECTIVE

Rapid industrialization and its increasing environmental impact enforce a shift from a linear economy to a circular economy (S. V. Mohan et al. 2019). The biorefinery approach promotes the utilization of green resources for the production of bioenergy and other value-added products. This holistic approach not only mitigates GHG emissions and energy demand but also offers job opportunities in various industrial sectors. The wastes and byproducts generated from the microalgal biorefinery can be integrated with the circular economy via bioeconomy processes. Circularity allows efficient use of resources for attaining a sustainable bioeconomy. Thus, a circular economy enhances the bioeconomy of a system.

However, it is important to maintain a balance between the GHGs emitted during the primary product generation with the GHGs emitted during the secondary product generation. Appropriate integration of conversion technologies for achieving maximum circularity is one of the greatest challenges (Schieb et al. 2015). Toxic chemicals might be accumulated during the production of primary products, thereby inhibiting the circularity of the system. Therefore, the primary emphasis should not be on not only maximizing circularity but also on optimizing the overall outcomes for environmental and economical sustainability. The feasibility of a zero-waste biorefinery approach at a pilot scale is another challenge—for now, this cascading approach is restricted to only the lab scale. Further, the recovery of solid and liquid byproducts after the primary processing of the biomass and its subsequent valorization in a pilot scale is also a challenge that needs to be addressed.

A futuristic zero-waste biorefinery approach in which a circular bioeconomy is fueled by the byproducts recovered from the primary processing of microalgal biomass is suggested for a sustainable biorefinery. Moreover, integration of a microalgal biorefinery with wastewater treatment and flue gas carbon dioxide mitigation helps to reduce environmental pollution. Integration of various feedstocks such as lignocellulosic biomass and its residues, aquatic waste, agricultural waste, food waste, sewage waste, and industrial gases is suggested, as this approach will help to maintain an uninterrupted plant operation (S. Mishra, Roy, and Mohanty 2019; J. Wang et al. 2019). Furthermore, as this approach is still in its infancy, it is necessary to develop trust and communication among lay people, governments, industries, and research and development (R&D) sectors. Finally, the proper techno-economic analysis should be carried out to determine the feasibility of this biorefinery approach in a pilot scale.

5.6 CONCLUSION

The microalgal biomass is a powerhouse of products with commercial potential and is considered a doorway to a multibillion dollar industry. Since single-product recovery is not sustainable, the biorefinery approach is highly recommended. A microalgal biorefinery allows complete valorization of the biomass by exploiting all the valuable parts that are present. A biorefinery approach makes the overall system economically sustainable by reducing the production cost and generating extra revenue. The economic performance of a microalgal biorefinery system can be further improved by substituting freshwater with sewage water and utilizing flue gas carbon dioxide as a carbon source for microalgal cultivation. The economic viability of this biorefinery system can be achieved by making a transition from a linear economy to a circular economy through a zero-waste biorefinery approach, where the byproducts recovered after primary processing of the biomass are further processed by integrating appropriate conversion technologies to achieve maximum circularity.

REFERENCES

Antoniou, N, F Monlau, C Sambusiti, E Ficara, A Barakat, and A Zabaniotou. 2019. "Contribution to Circular Economy Options of Mixed Agricultural Wastes Management: Coupling Anaerobic Digestion with Gasification for Enhanced Energy and Material Recovery." *Journal of Cleaner Production* 209: 505–514. https://doi.org/10.1016/j.jclepro.2018.10.055.

Arun, Jayaseelan, Padmanabhan Varshini, P Kamath Prithvinath, Venkataramani Priyadarshini, and Kannappan Panchamoorthy Gopinath. 2018. "Enrichment of Bio-Oil after Hydrothermal Liquefaction (HTL) of Microalgae C. Vulgaris Grown in Wastewater: Bio-Char and Post HTL Wastewater Utilization Studies." *Bioresource Technology* 261 (March): 182–187. https://doi.org/10.1016/j.biortech.2018.04.029.

Aslam, Ambreen, Tahir Fazal, Qamar uz Zaman, Ali Shan, Fahad Rehman, Javed Iqbal, Naim Rashid, and Muhammad Saif Ur Rehman. 2020. "Biorefinery of Microalgae for Nonfuel Products." In *Microalgae Cultivation for Biofuels Production*, 197–209. Elsevier.

Bacellar Mendes, Leonardo Brantes, and Alane Beatriz Vermelho. 2013. "Allelopathy as a Potential Strategy to Improve Microalgae Cultivation." *Biotechnology for Biofuels* 6 (1): 152. https://doi.org/10.1186/1754-6834-6-152.

Barreiro, Diego López, Wolter Prins, Frederik Ronsse, and Wim Brilman. 2013. "Hydrothermal Liquefaction (HTL) of Microalgae for Biofuel Production: State of the Art Review and Future Prospects." *Biomass and Bioenergy* 53: 113–127.

Basu, Samarpita, Abhijit Sarma Roy, Kaustubha Mohanty, and Aloke K Ghoshal. 2013. "Enhanced CO2 Sequestration by a Novel Microalga: Scenedesmus Obliquus SA1 Isolated from Bio-Diversity Hotspot Region of Assam, India." *Bioresource Technology* 143: 369–377. https://doi.org/https://doi.org/10.1016/j.biortech.2013.06.010.

Belviso, Claudia. 2018. "State-of-the-Art Applications of Fly Ash from Coal and Biomass: A Focus on Zeolite Synthesis Processes and Issues." *Progress in Energy and Combustion Science* 65: 109–135. https://doi.org/10.1016/j.pecs.2017.10.004.

Bhagavathy, S, P Sumathi, and I Jancy Sherene Bell. 2011. "Green Algae Chlorococcum Humicola-a New Source of Bioactive Compounds with Antimicrobial Activity." *Asian Pacific Journal of Tropical Biomedicine* 1 (1): S1–7.

Biller, P, A B Ross, S C Skill, A Lea-Langton, B Balasundaram, C Hall, R Riley, and C A Llewellyn. 2012. "Nutrient Recycling of Aqueous Phase for Microalgae Cultivation from the Hydrothermal Liquefaction Process." *Algal Research* 1 (1): 70–76. https://doi.org/https://doi.org/10.1016/j.algal.2012.02.002.

Bohutskyi, Pavlo, Ben Ketter, Steven Chow, Kameron J Adams, Michael J Betenbaugh, F C Thomas Allnutt, and Edward J Bouwer. 2015. "Anaerobic Digestion of Lipid-Extracted Auxenochlorella Protothecoides Biomass for Methane Generation and Nutrient Recovery." *Bioresource Technology* 183: 229–239. https://doi.org/https://doi.org/10.1016/j.biortech.2015.02.012.

Borowitzka, Michael A. 2013. "High-Value Products from Microalgae—their Development and Commercialisation." *Journal of Applied Phycology* 25 (3): 743–756.

Brentner, Laura B, Jordan Peccia, and Julie B Zimmerman. 2010. "Challenges in Developing Biohydrogen as a Sustainable Energy Source: Implications for a Research Agenda." *Environmental Science & Technology* 44 (7): 2243–2254. https://doi.org/10.1021/es9030613.

Briens, Cedric, Jan Piskorz, and Franco Berruti. 2008. "Biomass Valorization for Fuel and Chemicals Production—A Review." *International Journal of Chemical Reactor Engineering* 6 (1). https://doi.org/https://doi.org/10.2202/1542-6580.1674.

Carlucci, Maria J, Carlos A Pujol, Marina Ciancia, Miguel D Noseda, Maria C Matulewicz, Elsa B Damonte, and Alberto S Cerezo. 1997. "Antiherpetic and Anticoagulant Properties of Carrageenans from the Red Seaweed Gigartina Skottsbergii and Their Cyclized Derivatives: Correlation between Structure and Biological Activity." *International Journal of Biological Macromolecules* 20 (2): 97–105.

Carrez, Dirk, Michael Carus, Michel Griffon, Jirina Jilkova, Anikó Juhász, Lene Lange, Robert Mavsar, Tiina Pursula, and Ortega Consuelo Varela. 2017. "Expert Group Report: Review of the EU Bioeconomy Strategy and Its Action Plan." European Commission.

Cha, Jin Sun, Sung Hoon Park, Sang-Chul Jung, Changkook Ryu, Jong-Ki Jeon, Min-Chul Shin, and Young-Kwon Park. 2016. "Production and Utilization of Biochar: A Review." *Journal of Industrial and Engineering Chemistry* 40: 1–15.

Chacón-Lee, T L, and Gloria Eugenia González-Mariño. 2010. "Microalgae for 'Healthy' Foods—Possibilities and Challenges." *Comprehensive Reviews in Food Science and Food Safety* 9 (6): 655–675.

Chanda, Mutale-joan, Nawal Merghoub, and Hicham EL Arroussi. 2019. "Microalgae Polysaccharides: The New Sustainable Bioactive Products for the Development of Plant Bio-Stimulants?" *World Journal of Microbiology and Biotechnology* 35 (11): 177. https://doi.org/10.1007/s11274-019-2745-3.

Chang, Yuan-Ming, Wen-Tien Tsai, and Ming-Hsuan Li. 2015. "Characterization of Activated Carbon Prepared from Chlorella-Based Algal Residue." *Bioresource Technology* 184: 344–348. https://doi.org/https://doi.org/10.1016/j.biortech.2014.09.131.

Chen, Limei, Tao Zhu, Jose Salomon Martinez Fernandez, Shulin Chen, and Demao Li. 2017. "Recycling Nutrients from a Sequential Hydrothermal Liquefaction Process for Microalgae Culture." *Algal Research* 27: 311–317. https://doi.org/https://doi.org/10.1016/j.algal.2017.09.023.

Chen, Wei-Hsin, Ming-Yueh Huang, Jo-Shu Chang, and Chun-Yen Chen. 2014. "Thermal Decomposition Dynamics and Severity of Microalgae Residues in Torrefaction." *Bioresource Technology* 169: 258–264. https://doi.org/https://doi.org/10.1016/j.biortech.2014.06.086.

Chen, Wei-Hsin, and Po-Chih Kuo. 2011. "Torrefaction and Co-Torrefaction Characterization of Hemicellulose, Cellulose and Lignin as Well as Torrefaction of Some Basic Constituents in Biomass." *Energy* 36 (2): 803–811. https://doi.org/https://doi.org/10.1016/j.energy.2010.12.036.

Chen, Wei-Hsin, Bo-Jhih Lin, Ming-Yueh Huang, and Jo-Shu Chang. 2015. "Thermochemical Conversion of Microalgal Biomass into Biofuels: A Review." *Bioresource Technology* 184 (May): 314–327. https://doi.org/10.1016/j.biortech.2014.11.050.

Chen, Wei-Hsin, Bo-Jhih Lin, How-Ming Lee, and Men-Han Huang. 2012. "One-Step Synthesis of Dimethyl Ether from the Gas Mixture Containing CO2 with High Space Velocity." *Applied Energy* 98: 92–101. https://doi.org/https://doi.org/10.1016/j.apenergy.2012.02.082.

Chen, Yu, Nannan Zhao, Yulong Wu, Kejing Wu, Xiuyun Wu, Ji Liu, and Mingde Yang. 2017. "Distributions of Organic Compounds to the Products from Hydrothermal Liquefaction of Microalgae." *Environmental Progress & Sustainable Energy* 36 (1): 259–268.

Cherad, Ramzi, J A Onwudili, P Biller, P T Williams, and A B Ross. 2016. "Hydrogen Production from the Catalytic Supercritical Water Gasification of Process Water Generated from Hydrothermal Liquefaction of Microalgae." *Fuel* 166: 24–28. https://doi.org/https://doi.org/10.1016/j.fuel.2015.10.088.

Chisti, Y. 2007. "Biodiesel from Microalgae." *Biotechnology Advances* 25 (3). https://doi.org/10.1016/j.biotechadv.2007.02.001.

Chiu, Sheng-Yi, Chien-Ya Kao, Chiun-Hsun Chen, Tang-Ching Kuan, Seow-Chin Ong, and Chih-Sheng Lin. 2008. "Reduction of CO2 by a High-Density Culture of Chlorella Sp. in a Semicontinuous Photobioreactor." *Bioresource Technology* 99 (9): 3389–3396.

Chng, Lee Muei, Derek J C Chan, and Keat Teong Lee. 2016. "Sustainable Production of Bioethanol Using Lipid-Extracted Biomass from Scenedesmus Dimorphus." *Journal of Cleaner Production* 130: 68–73.

Coesel, Sacha Nicole, Alexandra Cordeiro Baumgartner, Licia Marlene Teles, Ana Alexandra Ramos, Nuno Miguel Henriques, Leonor Cancela, and João Carlos Serafim Varela. 2008. "Nutrient Limitation Is the Main Regulatory Factor for Carotenoid Accumulation and for Psy and Pds Steady State Transcript Levels in Dunaliella Salina (Chlorophyta) Exposed to High Light and Salt Stress." *Marine Biotechnology* 10 (5): 602.

Das, Sruthy K, Asha Sathish, and John Stanley. 2018. "Production of Biofuel and Bioplastic from Chlorella Pyrenoidosa." *Materials Today: Proceedings* 5 (8): 16774–16781.

Dieckhoff, P, B El-Cichakli, and C Patermann. 2015. "Bioeconomy Policy Synopsis and Analysis of Strategies in the G7." German Bioeconomy Council, Berlin.

Dindi, Abdallah, Dang Viet Quang, Lourdes F Vega, Enas Nashef, and Mohammad R M Abu-Zahra. 2019. "Applications of Fly Ash for CO2 Capture, Utilization, and Storage." *Journal of CO2 Utilization* 29: 82–102. https://doi.org/https://doi.org/10.1016/j.jcou.2018.11.011.

Dote, Yutaka, Shigeki Sawayama, Seiichi Inoue, Tomoaki Minowa, and Shin-ya Yokoyama. 1994. "Recovery of Liquid Fuel from Hydrocarbon-Rich Microalgae by Thermochemical Liquefaction." *Fuel* 73 (12): 1855–1857. https://doi.org/https://doi.org/10.1016/0016-2361(94)90211-9.

Du, Zhenyi, Bing Hu, Aimin Shi, Xiaochen Ma, Yanling Cheng, Paul Chen, Yuhuan Liu, Xiangyang Lin, and Roger Ruan. 2012. "Cultivation of a Microalga Chlorella Vulgaris Using Recycled Aqueous Phase Nutrients from Hydrothermal Carbonization Process." *Bioresource Technology* 126: 354–357.

Ethier, Shannon, Kevin Woisard, David Vaughan, and Zhiyou Wen. 2011. "Continuous Culture of the Microalgae Schizochytrium Limacinum on Biodiesel-Derived Crude Glycerol for Producing Docosahexaenoic Acid." *Bioresource Technology* 102 (1): 88–93. https://doi.org/https://doi.org/10.1016/j.biortech.2010.05.021.

Fabbri, Daniele, and Cristian Torri. 2016. "Linking Pyrolysis and Anaerobic Digestion (Py-AD) for the Conversion of Lignocellulosic Biomass." *Current Opinion in Biotechnology* 38 (March): 167–173. https://doi.org/10.1016/j.copbio.2016.02.004.

Fang, Herbert H P, Hong Liu, and Tong Zhang. 2005. "Phototrophic Hydrogen Production from Acetate and Butyrate in Wastewater." *International Journal of Hydrogen Energy* 30 (7): 785–793.

Faried, M, M Samer, E Abdelsalam, R S Yousef, Y A Attia, and A S Ali. 2017. "Biodiesel Production from Microalgae: Processes, Technologies and Recent Advancements." *Renewable and Sustainable Energy Reviews* 79: 893–913.

Fernandez, Sebastian, Keerthi Srinivas, Andrew J Schmidt, Marie S Swita, and Birgitte K Ahring. 2018. "Anaerobic Digestion of Organic Fraction from Hydrothermal Liquefied Algae Wastewater Byproduct." *Bioresource Technology* 247: 250–258.

Ferreira, Alice, and Luisa Gouveia. 2020. "Microalgal Biorefineries." In *Handbook of Microalgae-Based Processes and Products*, 771–798. Elsevier.

Francavilla, Matteo, M Colaianna, M Zotti, M G Morgese, P Trotta, Paolo Tucci, Stefania Schiavone, V Cuomo, and Luigia Trabace. 2012. "Extraction, Characterization and in Vivo Neuromodulatory Activity of Phytosterols from Microalga Dunaliella Tertiolecta." *Current Medicinal Chemistry* 19 (18): 3058–3067.

Fredriksson, Sofia, Klas Elwinger, and Jana Pickova. 2006. "Fatty Acid and Carotenoid Composition of Egg Yolk as an Effect of Microalgae Addition to Feed Formula for Laying Hens." *Food Chemistry* 99 (3): 530–537.

Friedman, Mendel. 2013. "Anticarcinogenic, Cardioprotective, and Other Health Benefits of Tomato Compounds Lycopene, α-Tomatine, and Tomatidine in Pure Form and in Fresh and Processed Tomatoes." *Journal of Agricultural and Food Chemistry* 61 (40): 9534–9550.

Fu, Xiaobo, Dianhong Li, Jie Chen, Yuanming Zhang, Weiya Huang, Yi Zhu, Jun Yang, and Chengwu Zhang. 2013. "A Microalgae Residue Based Carbon Solid Acid Catalyst for Biodiesel Production." *Bioresource Technology* 146: 767–770. https://doi.org/https://doi.org/10.1016/j.biortech.2013.07.117.

Fuchs, W, and B Drosg. 2013. "Assessment of the State of the Art of Technologies for the Processing of Digestate Residue from Anaerobic Digesters." *Water Science and Technology* 67 (9): 1984–1993.

Funke, Axel, and Felix Ziegler. 2010. "Hydrothermal Carbonization of Biomass: A Summary and Discussion of Chemical Mechanisms for Process Engineering." *Biofuels, Bioproducts and Biorefining* 4 (2): 160–177.

Garlapati, Vijay Kumar, Uttara Shankar, and Amrita Budhiraja. 2016. "Bioconversion Technologies of Crude Glycerol to Value Added Industrial Products." *Biotechnology Reports* 9: 9–14. https://doi.org/https://doi.org/10.1016/j.btre.2015.11.002.

Gatrell, S, K Lum, J Kim, and X G Lei. 2014. "Potential of Defatted Microalgae from the Biofuel Industry as an Ingredient to Replace Corn and Soybean Meal in Swine and Poultry Diets." *Journal of Animal Science* 92 (4): 1306–1314.

Geun Goo, Bon, Gu Baek, Doo Jin Choi, Yong Il Park, Andriy Synytsya, Roman Bleha, Dong Ho Seong, Choul-Gyun Lee, and Jae Kweon Park. 2013. "Characterization of a Renewable Extracellular Polysaccharide from Defatted Microalgae Dunaliella Tertiolecta." *Bioresource Technology* 129: 343–350. https://doi.org/https://doi.org/10.1016/j.biortech.2012.11.077.

Gifuni, Imma, Antonino Pollio, Carl Safi, Antonio Marzocchella, and Giuseppe Olivieri. 2019. "Current Bottlenecks and Challenges of the Microalgal Biorefinery." *Trends in Biotechnology* 37 (3): 242–252. https://doi.org/10.1016/j.tibtech.2018.09.006.

Gokhale, S V, K K Jyoti, and S S Lele. 2008. "Kinetic and Equilibrium Modeling of Chromium (VI) Biosorption on Fresh and Spent Spirulina Platensis/Chlorella Vulgaris Biomass." *Bioresource Technology* 99 (9): 3600–3608. https://doi.org/https://doi.org/10.1016/j.biortech.2007.07.039.

Gollakota, Anjani R K, Vikranth Volli, and Chi-Min Shu. 2019. "Progressive Utilisation Prospects of Coal Fly Ash: A Review." *Science of The Total Environment* 672: 951–989. https://doi.org/https://doi.org/10.1016/j.scitotenv.2019.03.337.

Goyal, H B, Diptendu Seal, and R C Saxena. 2008. "Bio-Fuels from Thermochemical Conversion of Renewable Resources: A Review." *Renewable and Sustainable Energy Reviews* 12 (2): 504–517.

Grierson, Scott, Vladimir Strezov, Gary Ellem, Ross Mcgregor, and Joe Herbertson. 2009. "Thermal Characterisation of Microalgae under Slow Pyrolysis Conditions." *Journal of Analytical and Applied Pyrolysis* 85 (1): 118–123. https://doi.org/https://doi.org/10.1016/j.jaap.2008.10.003.

Gu, Huiya, Nick Nagle, Philip T Pienkos, and Matthew C Posewitz. 2015. "Nitrogen Recycling from Fuel-Extracted Algal Biomass: Residuals as the Sole Nitrogen Source for Culturing Scenedesmus Acutus." *Bioresource Technology* 184: 153–160.

Guedes, Ana Catarina, Helena M Amaro, and Francisco Xavier Malcata. 2011. "Microalgae as Sources of Carotenoids." *Marine Drugs* 9 (4): 625–644.

Guieysse, Benoit, Quentin Béchet, and Andy Shilton. 2013. "Variability and Uncertainty in Water Demand and Water Footprint Assessments of Fresh Algae Cultivation Based on Case Studies from Five Climatic Regions." *Bioresource Technology* 128: 317–323.

Gupta, Rakesh Kumar, Mukul Dubey, Parashu Kharel, Zhengrong Gu, and Qi Hua Fan. 2015. "Biochar Activated by Oxygen Plasma for Supercapacitors." *Journal of Power Sources* 274: 1300–1305. https://doi.org/https://doi.org/10.1016/j.jpowsour.2014.10.169.

Halim, R, M K Danquah, and P A Webley. 2012. "Extraction of Oil from Microalgae for Biodiesel Production: A Review." *Biotechnology Advances* 30 (3). https://doi.org/10.1016/j.biotechadv.2012.01.001.

Harvie, Michelle. 2014. "Nutritional Supplements and Cancer: Potential Benefits and Proven Harms." *American Society of Clinical Oncology Educational Book* 34 (1): e478–e486.

Hirano, Atsushi, Ryohei Ueda, Shin Hirayama, and Yasuyuki Ogushi. 1997. "CO2 Fixation and Ethanol Production with Microalgal Photosynthesis and Intracellular Anaerobic Fermentation." *Energy* 22 (2): 137–142. https://doi.org/https://doi.org/10.1016/S0360-5442(96)00123-5.

Ho, Shih-Hsin, Akihito Nakanishi, Xiaoting Ye, Jo-Shu Chang, Kiyotaka Hara, Tomohisa Hasunuma, and Akihiko Kondo. 2014. "Optimizing Biodiesel Production in Marine Chlamydomonassp. JSC4 through Metabolic Profiling and an Innovative Salinity-Gradient Strategy." *Biotechnology for Biofuels* 7 (1): 97.

Hossain, A K, C Serrano, J B Brammer, A Omran, F Ahmed, D I Smith, and P A Davies. 2016. "Combustion of Fuel Blends Containing Digestate Pyrolysis Oil in a Multi-Cylinder Compression Ignition Engine." *Fuel* 171: 18–28. https://doi.org/https://doi.org/10.1016/j.fuel.2015.12.012.

Ibañez, E, M Herrero, J A Mendiola, and M Castro-Puyana. 2012. "Marine Bioactive Compounds: Sources, Characterization and Applications." *Extraction and Characterization of Bioactive Compounds with Health Benefits from Marine Resources: Macro and Micro Algae, Cyanobacteria, and Invertebrates. Madrid: Springer*, 55–98.

Ishii, Shun'ichi, Shino Suzuki, Trina M Norden-Krichmar, Angela Wu, Yuko Yamanaka, Kenneth H Nealson, and Orianna Bretschger. 2013. "Identifying the Microbial Communities and Operational Conditions for Optimized Wastewater Treatment in Microbial Fuel Cells." *Water Research* 47 (19): 7120–7130. https://doi.org/https://doi.org/10.1016/j.watres.2013.07.048.

Jacob-Lopes, Eduardo, Mariana Manzoni Maroneze, Maria Isabel Queiroz, and Leila Queiroz Zepka. 2020. *Handbook of Microalgae-Based Processes and Products: Fundamentals and Advances in Energy, Food, Feed, Fertilizer, and Bioactive Compounds.* Academic Press.

Jena, Umakanta, K C Das, and J R Kastner. 2011. "Effect of Operating Conditions of Thermochemical Liquefaction on Biocrude Production from Spirulina Platensis." *Bioresource Technology* 102 (10): 6221–6229. https://doi.org/https://doi.org/10.1016/j.biortech.2011.02.057.

Johnson, Michael C, Ignasi Palou-Rivera, and Edward D Frank. 2013. "Energy Consumption during the Manufacture of Nutrients for Algae Cultivation." *Algal Research* 2 (4): 426–436.

Juin, Camille, Jean-René Chérouvrier, Valérie Thiéry, Anne-Laure Gagez, Jean-Baptiste Bérard, Nicolas Joguet, Raymond Kaas, Jean-Paul Cadoret, and Laurent Picot. 2015. "Microwave-Assisted Extraction of Phycobiliproteins from Porphyridium Purpureum." *Applied Biochemistry and Biotechnology* 175 (1): 1–15.

Kakarla, Ramesh, and Booki Min. 2014. "Evaluation of Microbial Fuel Cell Operation Using Algae as an Oxygen Supplier: Carbon Paper Cathode vs. Carbon Brush Cathode." *Bioprocess and Biosystems Engineering* 37 (12): 2453–2461.

Kalmykova, Yuliya, Madumita Sadagopan, and Leonardo Rosado. 2018. "Circular Economy–From Review of Theories and Practices to Development of Implementation Tools." *Resources, Conservation and Recycling* 135: 190–201.

Kaplan, Drora, Daniel Christiaen, and Shoshana (Malis) Arad. 1987. "Chelating Properties of Extracellular Polysaccharides from Chlorella Spp." *Applied and Environmental Microbiology* 53 (12): 2953–2956. https://aem.asm.org/content/53/12/2953.

Karan, Hakan, Christiane Funk, Martin Grabert, Melanie Oey, and Ben Hankamer. 2019. "Green Bioplastics as Part of a Circular Bioeconomy." *Trends in Plant Science* 24 (3): 237–249. https://doi.org/https://doi.org/10.1016/j.tplants.2018.11.010.

Katiyar, Richa, B R Gurjar, Randhir K Bharti, Amit Kumar, Shalini Biswas, and Vikas Pruthi. 2017. "Heterotrophic Cultivation of Microalgae in Photobioreactor Using Low Cost Crude Glycerol for Enhanced Biodiesel Production." *Renewable Energy* 113: 1359–1365. https://doi.org/https://doi.org/10.1016/j.renene.2017.06.100.

Kaur, Jaspreet, Anil Kumar Sarma, Mithilesh Kumar Jha, and Poonam Gera. 2020. "Valorisation of Crude Glycerol to Value-Added Products: Perspectives of Process Technology, Economics and Environmental Issues." *Biotechnology Reports* 27: e00487. https://doi.org/https://doi.org/10.1016/j.btre.2020.e00487.

Kerr, Brian J, William A Dozier, and K Bregendahl. 2007. "' Nutritional Value of Crude Glycerin for Nonruminants,' 68th Minnesota Nutrition Conference and University of Minnesota Research Update Session: Modern Concepts in Livestock Production for 2007, Proceedings, September 18-19-2007, Minneapolis, Minnesota."

Khetkorn, Wanthanee, Rajesh P Rastogi, Aran Incharoensakdi, Peter Lindblad, Datta Madamwar, Ashok Pandey, and Christian Larroche. 2017. "Microalgal Hydrogen Production – A Review." *Bioresource Technology* 243: 1194–1206. https://doi.org/https://doi.org/10.1016/j.biortech.2017.07.085.

Kleivdal, Hans, Matilde S Chauton, and Kjell Inge Reitan. 2013. "Industrial Production of Marine Microalgae as a Source of EPA and DHA Rich Raw Material in Fish Feed–Basis, Knowledge Status and Possibilities." *Bergen, SINTEF.*

Ko, Seok-Chun, Seung-Hong Lee, Ginnae Ahn, Kil-Nam Kim, Seon-Heui Cha, Se-Kwon Kim, Byong-Tae Jeon, Pyo-Jam Park, Ki-Wan Lee, and You-Jin Jeon. 2012. "Effect of Enzyme-Assisted Extract of Sargassum Coreanum on Induction of Apoptosis in HL-60 Tumor Cells." *Journal of Applied Phycology* 24 (4): 675–684.

Koller, Martin, Alexander Muhr, and Gerhart Braunegg. 2014. "Microalgae as Versatile Cellular Factories for Valued Products." *Algal Research* 6: 52–63.

Koutra, Eleni, Panagiota Tsafrakidou, Myrsini Sakarika, and Michael Kornaros. 2020. "Microalgal Biorefinery." In *Microalgae Cultivation for Biofuels Production*, 163–185. Elsevier.

Koyande, Apurav Krishna, Pau-Loke Show, Ruixin Guo, Bencan Tang, Chiaki Ogino, and Jo-Shu Chang. 2019. "Bio-Processing of Algal Bio-Refinery: A Review on Current Advances and Future Perspectives." *Bioengineered* 10 (1): 574–592.

Kraan, Stefan. 2012. "Algal Polysaccharides, Novel Applications and Outlook." In *Carbohydrates-Comprehensive Studies on Glycobiology and Glycotechnology*. IntechOpen.

Kumar, Dipesh, and Bhaskar Singh. 2019. "Algal Biorefinery: An Integrated Approach for Sustainable Biodiesel Production." *Biomass and Bioenergy* 131: 105398. https://doi.org/https://doi.org/10.1016/j.biombioe.2019.105398.

Kvansakul, Jessica, Marisa Rodriguez-Carmona, David F Edgar, Felix M Barker, Wolfgang Köpcke, Wolfgang Schalch, and John L Barbur. 2006. "Supplementation with the Carotenoids Lutein or Zeaxanthin Improves Human Visual Performance." *Ophthalmic and Physiological Optics* 26 (4): 362–371.

Lam, Man Kee, and Keat Teong Lee. 2012. "Microalgae Biofuels: A Critical Review of Issues, Problems and the Way Forward." *Biotechnology Advances* 30 (3): 673–690.

Lam, Man Kee, Inn Shi Tan, and Keat Teong Lee. 2014. "Utilizing Lipid-Extracted Microalgae Biomass Residues for Maltodextrin Production." *Chemical Engineering Journal* 235: 224–230. https://doi.org/https://doi.org/10.1016/j.cej.2013.09.023.

Liu, Jin, Zheng Sun, Henri Gerken, Zheng Liu, Yue Jiang, and Feng Chen. 2014. "Chlorella Zofingiensis as an Alternative Microalgal Producer of Astaxanthin: Biology and Industrial Potential." *Marine Drugs* 12 (6): 3487–3515.

Liu, Na, Alberto Bento Charrua, Chih-Huang Weng, Xiaoling Yuan, and Feng Ding. 2015. "Characterization of Biochars Derived from Agriculture Wastes and Their Adsorptive Removal of Atrazine from Aqueous Solution: A Comparative Study." *Bioresource Technology* 198: 55–62. https://doi.org/https://doi.org/10.1016/j.biortech.2015.08.129.

Liu, Renlong, Wenying Tian, Shengyan Kong, Yanghao Meng, Huansheng Wang, and Jinglai Zhang. 2018. "Effects of Inorganic and Organic Acid Pretreatments on the Hydrothermal Liquefaction of Municipal Secondary Sludge." *Energy Conversion and Management* 174 (June): 661–667. https://doi.org/10.1016/j.enconman.2018.08.058.

Lunprom, Siriporn, Orawan Phanduang, Apilak Salakkam, Qiang Liao, Tsuyoshi Imai, and Alissara Reungsang. 2019. "Bio-Hythane Production from Residual Biomass of Chlorella Sp. Biomass through a Two-Stage Anaerobic Digestion." *International Journal of Hydrogen Energy* 44 (6): 3339–3346. https://doi.org/10.1016/j.ijhydene.2018.09.064.

Madeira, Marta S, Carlos Cardoso, Paula A Lopes, Diogo Coelho, Cláudia Afonso, Narcisa M Bandarra, and José A M Prates. 2017. "Microalgae as Feed Ingredients for Livestock Production and Meat Quality: A Review." *Livestock Science* 205: 111–121.

Malik, Kamla, Jayanti Tokkas, and Sneh Goyal. 2012. "Microbial Pigments: A Review." *International Journal of Microbiology Resource Technology* 1 (4): 361–365.

Manor, Meghan L, Jonggun Kim, Theodore J Derksen, Rebecca L Schwartz, Carol A Roneker, Rohil Sahai Bhatnagar, and Xin Gen Lei. 2017. "Defatted Microalgae Serve as a Dual Dietary Source of Highly Bioavailable Iron and Protein in an Anemic Pig Model." *Algal Research* 26: 409–414. https://doi.org/https://doi.org/10.1016/j.algal.2017.07.018.

Maurya, Rahulkumar, Chetan Paliwal, Tonmoy Ghosh, Imran Pancha, Kaumeel Chokshi, Madhusree Mitra, Arup Ghosh, and Sandhya Mishra. 2016. "Applications of De-Oiled Microalgal Biomass towards Development of Sustainable Biorefinery." *Bioresource Technology* 214: 787–796. https://doi.org/https://doi.org/10.1016/j.biortech.2016.04.115.

McKendry, Peter. 2002. "Energy Production from Biomass (Part 2): Conversion Technologies." *Bioresource Technology* 83 (1): 47–54.

Meixner, K, A Kovalcik, E Sykacek, M Gruber-Brunhumer, W Zeilinger, K Markl, C Haas, I Fritz, N Mundigler, and F Stelzer. 2018. "Cyanobacteria Biorefinery—Production of Poly (3-Hydroxybutyrate) with Synechocystis Salina and Utilisation of Residual Biomass." *Journal of Biotechnology* 265: 46–53.

Michalak, Izabela, and Katarzyna Chojnacka. 2015. "Algae as Production Systems of Bioactive Compounds." *Engineering in Life Sciences* 15 (2): 160–176. https://doi.org/10.1002/elsc.201400191.

Mishra, Ranjeet Kumar, and Kaustubha Mohanty. 2018. "Pyrolysis Kinetics and Thermal Behavior of Waste Sawdust Biomass Using Thermogravimetric Analysis." *Bioresource Technology* 251 (October 2017): 63–74. https://doi.org/10.1016/j.biortech.2017.12.029.

Mishra, Sandhya Chandrika Prasad, Pushpito Kumar Ghosh, Mahesh Ramniklal Gandhi, Sourish Bhattacharya, Subarna Maiti, Sumesh Chandra Upadhyay, Arup Ghosh, Rachapudi Badari Narayana Prasad, Sanjit Kanjilal, and Sanjiv Kumar Mishra. 2014. "Engine Worthy Fatty Acid Methyl Ester (Biodiesel) from Naturally Occuring Marine Microalgal Mats and Marine Microalgae Cultured in Open Salt Pans Together with Value Addition of Co-Products." Google Patents.

Mishra, Sanjeev, and Kaustubha Mohanty. 2019. "Comprehensive Characterization of Microalgal Isolates and Lipid-Extracted Biomass as Zero-Waste Bioenergy Feedstock: An Integrated Bioremediation and Biorefinery Approach." *Bioresource Technology* 273: 177–184. https://doi.org/https://doi.org/10.1016/j.biortech.2018.11.012.

Mishra, Sanjeev, Madonna Roy, and Kaustubha Mohanty. 2019. "Microalgal Bioenergy Production under Zero-Waste Biorefinery Approach: Recent Advances and Future Perspectives." *Bioresource Technology* 292: 122008. https://doi.org/https://doi.org/10.1016/j.biortech.2019.122008.

Mišurcová, Ladislava, Jana Orsavová, and Jarmila Vávra Ambrožová. 2015. "Algal Polysaccharides and Health." *Polysaccharides: Bioactivity and Biotechnology.*

Mohan, Dinesh, Charles U Pittman, and Philip H Steele. 2006. "Pyrolysis of Wood/Biomass for Bio-Oil: A Critical Review." *Energy & Fuels* 20 (3): 848–889. https://doi.org/10.1021/ef0502397.

Mohan, S Venkata, Shikha Dahiya, K Amulya, Ranaprathap Katakojwala, and T K Vanitha. 2019. "Can Circular Bioeconomy Be Fueled by Waste Biorefineries — A Closer Look." *Bioresource Technology Reports* 7: 100277. https://doi.org/https://doi.org/10.1016/j.biteb.2019.100277.

Monteiro, Marcos Roberto, Cristie Luis Kugelmeier, Rafael Sanaiotte Pinheiro, Mario Otávio Batalha, and Aldara da Silva César. 2018. "Glycerol from Biodiesel Production: Technological Paths for Sustainability." *Renewable and Sustainable Energy Reviews* 88: 109–122.

Moon, Myounghoon, Chul Woong Kim, Won-Kun Park, Gursong Yoo, Yoon-E Choi, and Ji-Won Yang. 2013. "Mixotrophic Growth with Acetate or Volatile Fatty Acids Maximizes Growth and Lipid Production in Chlamydomonas Reinhardtii." *Algal Research* 2 (4): 352–357. https://doi.org/https://doi.org/10.1016/j.algal.2013.09.003.

Mulbry, Walter, Shannon Kondrad, and Carolina Pizarro. 2007. "Biofertilizers from Algal Treatment of Dairy and Swine Manure Effluents." *Journal of Vegetable Science* 12 (4): 107–125. https://doi.org/10.1300/J484v12n04_08.

Nagarajan, Dillirani, Duu-Jong Lee, Akihiko Kondo, and Jo-Shu Chang. 2017. "Recent Insights into Biohydrogen Production by Microalgae: From Biophotolysis to Dark Fermentation." *Bioresource Technology* 227: 373–387. https://doi.org/https://doi.org/10.1016/j.biortech.2016.12.104.

Navarro, Francisco, Eduardo Forján, María Vázquez, Zaida Montero, Elisabeth Bermejo, Miguel Ángel Castaño, Alberto Toimil, Enrique Chagüaceda, Miguel Ángel García-Sevillano, and Marisa Sánchez. 2016. "Microalgae as a Safe Food Source for Animals: Nutritional Characteristics of the Acidophilic Microalga Coccomyxa Onubensis." *Food & Nutrition Research* 60 (1): 30472.

Neumann, P, A Torres, F G Fermoso, R Borja, and D Jeison. 2015. "Anaerobic Co-Digestion of Lipid-Spent Microalgae with Waste Activated Sludge and Glycerol in Batch Mode." *International Biodeterioration & Biodegradation* 100: 85–88. https://doi.org/https://doi.org/10.1016/j.ibiod.2015.01.020.

O'sullivan, A M, Y C O'callaghan, M N O'grady, B Queguineur, D Hanniffy, D J Troy, J P Kerry, and N M O'brien. 2011. "In Vitro and Cellular Antioxidant Activities of Seaweed Extracts Prepared from Five Brown Seaweeds Harvested in Spring from the West Coast of Ireland." *Food Chemistry* 126 (3): 1064–1070.

Pan, Pan, Changwei Hu, Wenyan Yang, Yuesong Li, Linlin Dong, Liangfang Zhu, Dongmei Tong, Renwei Qing, and Yong Fan. 2010. "The Direct Pyrolysis and Catalytic Pyrolysis of Nannochloropsis Sp. Residue for Renewable Bio-Oils." *Bioresource Technology* 101 (12): 4593–4599. https://doi.org/https://doi.org/10.1016/j.biortech.2010.01.070.

Paoli, F De, A Bauer, C Leonhartsberger, B Amon, and T Amon. 2011. "Utilization of By-Products from Ethanol Production as Substrate for Biogas Production." *Bioresource Technology* 102 (11): 6621–6624.

Park, Jeong-Hoon, Jeong-Jun Yoon, Hee-Deung Park, Dong Jung Lim, and Sang-Hyoun Kim. 2012. "Anaerobic Digestibility of Algal Bioethanol Residue." *Bioresource Technology* 113: 78–82. https://doi.org/https://doi.org/10.1016/j.biortech.2011.12.123.

Park, Stephen, and Yebo Li. 2012. "Evaluation of Methane Production and Macronutrient Degradation in the Anaerobic Co-Digestion of Algae Biomass Residue and Lipid Waste." *Bioresource Technology* 111: 42–48.

Peterson, Andrew A, Frédéric Vogel, Russell P Lachance, Morgan Fröling, Michael J Antal, Jr., and Jefferson W Tester. 2008. "Thermochemical Biofuel Production in Hydrothermal Media: A Review of Sub- and Supercritical Water Technologies." *Energy & Environmental Science* 1 (1): 32–65. https://doi.org/10.1039/B810100K.

Pilon, Laurent, Halil Berberoğlu, and Razmig Kandilian. 2011. "Radiation Transfer in Photobiological Carbon Dioxide Fixation and Fuel Production by Microalgae." *Journal of Quantitative Spectroscopy and Radiative Transfer* 112 (17): 2639–2660. https://doi.org/https://doi.org/10.1016/j.jqsrt.2011.07.004.

Priyadarshani, Indira, and Biswajit Rath. 2012. "Commercial and Industrial Applications of Micro Algae–A Review." *Journal of Algal Biomass Utilization* 3 (4): 89–100.

Rahman, Quazi Mahzabin, Bo Zhang, Lijun Wang, Gail Joseph, and Abolghasem Shahbazi. 2019. "A Combined Fermentation and Ethanol-Assisted Liquefaction Process to Produce Biofuel from Nannochloropsis Sp." *Fuel* 238: 159–165. https://doi.org/https://doi.org/10.1016/j.fuel.2018.10.116.

Rai, Swapnil, and Nidhi Shukla. 2020. "Biofertilizer: An Alternative of Synthetic Fertilizers." *Plant Archives* 20 (2): 1374–1379.

Rajak, Rajiv Chandra, Samuel Jacob, and Beom Soo Kim. 2020. "A Holistic Zero Waste Biorefinery Approach for Macroalgal Biomass Utilization: A Review." *Science of The Total Environment* 716: 137067. https://doi.org/10.1016/j.scitotenv.2020.137067.

Rashid, Naim, Muhammad Saif Ur Rehman, and Jong-In Han. 2013. "Recycling and Reuse of Spent Microalgal Biomass for Sustainable Biofuels." *Biochemical Engineering Journal* 75: 101–107.

Renju, G L, G Muraleedhara Kurup, and C H Saritha Kumari. 2014. "Effect of Lycopene from Chlorella Marina on High Cholesterol-Induced Oxidative Damage and Inflammation in Rats." *Inflammopharmacology* 22 (1): 45–54.

Rodolfi, Liliana, Graziella Chini Zittelli, Laura Barsanti, Giovanna Rosati, and Mario R Tredici. 2003. "Growth Medium Recycling in Nannochloropsis Sp. Mass Cultivation." *Biomolecular Engineering* 20 (4): 243–248. https://doi.org/https://doi.org/10.1016/S1389-0344(03)00063-7.

Rolland, Antoine, Alain Sarda, Gaël Colomines, Yannick Madec, Emmanuel Chailleux, and Eric Leroy. 2020. "Biobased Bitumen Analogue Formation during Hydrothermal Treatment of Microalgae Residues, Part 1: Influence of Reaction Enthalpy on the Process." *Journal of Analytical and Applied Pyrolysis* 151: 104921. https://doi.org/https://doi.org/10.1016/j.jaap.2020.104921.

Roy, Madonna, and Kaustubha Mohanty. 2019. "A Comprehensive Review on Microalgal Harvesting Strategies: Current Status and Future Prospects." *Algal Research* 44: 101683. https://doi.org/https://doi.org/10.1016/j.algal.2019.101683.

Roy, Madonna, and Kaustubha Mohanty. 2020. "Valorization of Waste Eggshell-Derived Bioflocculant for Harvesting T. Obliquus: Process Optimization, Kinetic Studies and Recyclability of the Spent Medium for Circular Bioeconomy." *Bioresource Technology* 307: 123205. https://doi.org/https://doi.org/10.1016/j.biortech.2020.123205.

Safinaz, A F, and A H Ragaa. 2013. "Effect of Some Red Marine Algae as Biofertilizers on Growth of Maize (Zea Mayz L.) Plants." *International Food Research Journal* 20 (4): 1629.

Saidur, Rahman, E A Abdelaziz, Ayhan Demirbas, M S Hossain, and Saad Mekhilef. 2011. "A Review on Biomass as a Fuel for Boilers." *Renewable and Sustainable Energy Reviews* 15 (5): 2262–2289.

Sathish, Ashik, and Ronald C Sims. 2012. "Biodiesel from Mixed Culture Algae via a Wet Lipid Extraction Procedure." *Bioresource Technology* 118: 643–647.

Schieb, Pierre-Alain, Honorine Lescieux-Katir, Maryline Thénot, and Barbara Clément-Larosière. 2015. "An Original Business Model: The Integrated Biorefinery." In *Biorefinery 2030: Future Prospects for the Bioeconomy*, 25–66. Springer. https://doi.org/10.1007/978-3-662-47374-0_2.

Shuping, Zou, Wu Yulong, Yang Mingde, Imdad Kaleem, Li Chun, and Junmao Tong. 2010. "Production and Characterization of Bio-Oil from Hydrothermal Liquefaction of Microalgae Dunaliella Tertiolecta Cake." *Energy* 35 (12): 5406–5411. https://doi.org/https://doi.org/10.1016/j.energy.2010.07.013.

Sialve, Bruno, Nicolas Bernet, and Olivier Bernard. 2009. "Anaerobic Digestion of Microalgae as a Necessary Step to Make Microalgal Biodiesel Sustainable." *Biotechnology Advances* 27 (4): 409–416.

Singh, Jyoti, and Dolly Wattal Dhar. 2019. "Overview of Carbon Capture Technology: Microalgal Biorefinery Concept and State-of-the-Art." *Frontiers in Marine Science* 6: 29.

Song, Young Eun, Marwa M El-Dalatony, Changman Kim, Mayur B Kurade, Byong-Hun Jeon, and Jung Rae Kim. 2019. "Harvest of Electrical Energy from Fermented Microalgal Residue Using a Microbial Fuel Cell." *International Journal of Hydrogen Energy* 44 (4): 2372–2379. https://doi.org/https://doi.org/10.1016/j.ijhydene.2018.08.151.

Spolaore, P, C Joannis-Cassan, Elie Duran, and A Isambert. 2006. "Commercial Applications of Microalgae." *Journal of Bioscience and Bioengineering* 101 2: 87–96.

Suali, Emma, and Rosalam Sarbatly. 2012. "Conversion of Microalgae to Biofuel." *Renewable and Sustainable Energy Reviews* 16 (6): 4316–4342.

Subhash, G Venkata, and S Venkata Mohan. 2014. "Deoiled Algal Cake as Feedstock for Dark Fermentative Biohydrogen Production: An Integrated Biorefinery Approach." *International Journal of Hydrogen Energy* 39 (18): 9573–9579.

Suganya, T, M Varman, H H Masjuki, and S Renganathan. 2016. "Macroalgae and Microalgae as a Potential Source for Commercial Applications along with Biofuels Production: A Biorefinery Approach." *Renewable and Sustainable Energy Reviews* 55: 909–941. https://doi.org/https://doi.org/10.1016/j.rser.2015.11.026.

Torri, Cristian, Chiara Samorì, Alessio Adamiano, Daniele Fabbri, Cecilia Faraloni, and Giuseppe Torzillo. 2011. "Preliminary Investigation on the Production of Fuels and Bio-Char from Chlamydomonas Reinhardtii Biomass Residue after Bio-Hydrogen Production." *Bioresource Technology* 102 (18): 8707–8713. https://doi.org/10.1016/j.biortech.2011.01.064.

Tuazon, D, G D Corder, and B C McLellan. 2013. "Sustainable Development: A Review of Theoretical Contributions." *International Journal Sustainable Future for Human Security* 1 (1): 40–48.

Ubando, Aristotle T, Charles B Felix, and Wei-Hsin Chen. 2020. "Biorefineries in Circular Bioeconomy: A Comprehensive Review." *Bioresource Technology* 299: 122585. https://doi.org/https://doi.org/10.1016/j.biortech.2019.122585.

Ummalyma, Sabeela Beevi, Dinabandhu Sahoo, and Ashok Pandey. 2020. "Microalgal Biorefineries for Industrial Products." In *Microalgae Cultivation for Biofuels Production*, 187–195. Elsevier.

Vardon, Derek R, Brajendra K Sharma, Grant V Blazina, Kishore Rajagopalan, and Timothy J Strathmann. 2012. "Thermochemical Conversion of Raw and Defatted Algal Biomass via Hydrothermal Liquefaction and Slow Pyrolysis." *Bioresource Technology* 109: 178–187. https://doi.org/https://doi.org/10.1016/j.biortech.2012.01.008.

Venkata Mohan, S, Manupati Hemalatha, Debkumar Chakraborty, Sulogna Chatterjee, Palle Ranadheer, and Rajesh Kona. 2020. "Algal Biorefinery Models with Self-Sustainable Closed Loop Approach: Trends and Prospective for Blue-Bioeconomy." *Bioresource Technology* 295: 122128. https://doi.org/https://doi.org/10.1016/j.biortech.2019.122128.

Venkata Mohan, S, G N Nikhil, P Chiranjeevi, C Nagendranatha Reddy, M V Rohit, A Naresh Kumar, and Omprakash Sarkar. 2016. "Waste Biorefinery Models towards Sustainable Circular Bioeconomy: Critical Review and Future Perspectives." *Bioresource Technology* 215: 2–12. https://doi.org/10.1016/j.biortech.2016.03.130.

Vigani, Mauro. 2020. "The Bioeconomy of Microalgae-Based Processes and Products." In *Handbook of Microalgae-Based Processes and Products*, 799–821. Elsevier.

Wang, Bin, Guo Zhong Tong, You Le Qu, and Li Li. 2011. "Microwave-Assisted Extraction and in Vitro Antioxidant Evaluation of Polysaccharides from Enteromorpha Prolifera." In *Applied Mechanics and Materials*, 79: 204–209. Trans Tech Publ.

Wang, Jingjing, Xiaowei Peng, Xinfei Chen, and Xiaoqian Ma. 2019. "Co-Liquefaction of Low-Lipid Microalgae and Starch-Rich Biomass Waste: The Interaction Effect on Product Distribution and Composition." *Journal of Analytical and Applied Pyrolysis* 139 (February): 250–257. https://doi.org/10.1016/j.jaap.2019.02.013.

Wang, K, A Mandal, E Ayton, R Hunt, M A Zeller, and S Sharma. 2016. "Modification of Protein Rich Algal-Biomass to Form Bioplastics and Odor Removal." In *Protein Byproducts*, 107–117. Elsevier.

Wang, Shuang, Hao Shang, Abd El-Fatah Abomohra, and Qian Wang. 2019. "One-Step Conversion of Microalgae to Alcohols and Esters through Co-Pyrolysis with Biodiesel-Derived Glycerol." *Energy Conversion and Management* 198: 111792. https://doi.org/https://doi.org/10.1016/j.enconman.2019.111792.

Watanabe, Hideo, Dalin Li, Yoshinao Nakagawa, Keiichi Tomishige, and Makoto M Watanabe. 2015. "Catalytic Gasification of Oil-Extracted Residue Biomass of Botryococcus Braunii." *Bioresource Technology* 191: 452–459.

M. H. Wilson, J. Groppo, A Placido, S Graham, S A Morton, E Santillan-Jimenez, A Shea, M Crocker, C Crofcheck, and R Andrews. 2014. "CO_2 Recycling Using Microalgae for the Production of Fuels." *Applied Petrochemical Research* 4 (1): 41–53. https://doi.org/10.1007/s13203-014-0052-3.

Winkler, Martin, Sebastian Kuhlgert, Michael Hippler, and Thomas Happe. 2009. "Characterization of the Key Step for Light-Driven Hydrogen Evolution in Green Algae." *Journal of Biological Chemistry* 284 (52): 36620–36627.

Wu, Guoyao. 2014. "Dietary Requirements of Synthesizable Amino Acids by Animals: A Paradigm Shift in Protein Nutrition." *Journal of Animal Science and Biotechnology* 5 (1): 34.

Wuang, Shy Chyi, Mar Cho Khin, Pei Qiang Danny Chua, and Yanpei Darren Luo. 2016. "Use of Spirulina Biomass Produced from Treatment of Aquaculture Wastewater as Agricultural Fertilizers." *Algal Research* 15: 59–64. https://doi.org/https://doi.org/10.1016/j.algal.2016.02.009.

Xia, Ao, and Jerry D Murphy. 2016. "Microalgal Cultivation in Treating Liquid Digestate from Biogas Systems." *Trends in Biotechnology* 34 (4): 264–275. https://doi.org/https://doi.org/10.1016/j.tibtech.2015.12.010.

Xu, Donghai, Yang Wang, Guike Lin, Shuwei Guo, Shuzhong Wang, and Zhiqiang Wu. 2019. "Co-Hydrothermal Liquefaction of Microalgae and Sewage Sludge in Subcritical Water: Ash Effects on Bio-Oil Production." *Renewable Energy* 138: 1143–1151. https://doi.org/10.1016/j.renene.2019.02.020.

Xu, Fuqing, Adam Khalaf, Johnathon Sheets, Xumeng Ge, Harold Keener, and Yebo Li. 2018. "Phosphorus Removal and Recovery From Anaerobic Digestion Residues." In *Advances in Bioenergy*, edited by Yebo Li and Xumeng Ge, 3: 77–136. Elsevier. https://doi.org/10.1016/bs.aibe.2018.02.003.

Xu, Shannan, Mahdy Elsayed, Gehan A Ismail, Chunhou Li, Shuang Wang, and Abd El-Fatah Abomohra. 2019. "Evaluation of Bioethanol and Biodiesel Production from Scenedesmus Obliquus Grown in Biodiesel Waste Glycerol: A Sequential Integrated Route for Enhanced Energy Recovery." *Energy Conversion and Management* 197: 111907. https://doi.org/https://doi.org/10.1016/j.enconman.2019.111907.

Yang, Fangxia, Milford A Hanna, and Runcang Sun. 2012. "Value-Added Uses for Crude Glycerol--a Byproduct of Biodiesel Production." *Biotechnology for Biofuels* 5 (1): 13. https://doi.org/10.1186/1754-6834-5-13.

Yang, Honghui, Liejin Guo, and Fei Liu. 2010. "Enhanced Bio-Hydrogen Production from Corncob by a Two-Step Process: Dark- and Photo-Fermentation." *Bioresource Technology* 101 (6): 2049–2052. https://doi.org/https://doi.org/10.1016/j.biortech.2009.10.078.

Yang, Libin, Xiaobo Tan, Buchun Si, Fangchao Zhao, Huaqiang Chu, Xuefei Zhou, and Yalei Zhang. 2018. "Nutrients Recycling and Energy Evaluation in a Closed Microalgal Biofuel Production System." *Algal Research* 33: 399–405. https://doi.org/https://doi.org/10.1016/j.algal.2018.06.009.

Yang, Zhiman, Rongbo Guo, Xiaohui Xu, Xiaolei Fan, and Shengjun Luo. 2011. "Fermentative Hydrogen Production from Lipid-Extracted Microalgal Biomass Residues." *Applied Energy* 88 (10): 3468–3472.

Yen, Hong-Wei, Shih-Hsin Ho, Chun-Yen Chen, and Jo-Shu Chang. 2015. "CO2, NOx and SOx Removal from Flue Gas via Microalgae Cultivation: A Critical Review." *Biotechnology Journal* 10 (6): 829–839. https://doi.org/10.1002/biot.201400707.

Yin, Feng Wei, Dong Sheng Guo, Lu Jing Ren, Xiao Jun Ji, and He Huang. 2018. "Development of a Method for the Valorization of Fermentation Wastewater and Algal-Residue Extract in Docosahexaenoic Acid Production by Schizochytrium Sp." *Bioresource Technology* 266 (May): 482–487. https://doi.org/10.1016/j.biortech.2018.06.109.

Yoo, Chan, So-Young Jun, Jae-Yon Lee, Chi-Yong Ahn, and Hee-Mock Oh. 2010. "Selection of Microalgae for Lipid Production under High Levels Carbon Dioxide." *Bioresource Technology* 101 (1): S71–74.

Zhang, Congyu, Chengyu Wang, Guoliang Cao, Wei Hsin Chen, and Shih Hsin Ho. 2019. "Comparison and Characterization of Property Variation of Microalgal Biomass with Non-Oxidative and Oxidative Torrefaction." *Fuel* 246 (March): 375–385. https://doi.org/10.1016/j.fuel.2019.02.139.

Zhang, Xuesong, John Scott, Brajendra K Sharma, and Nandakishore Rajagopalan. 2018. "Advanced Treatment of Hydrothermal Liquefaction Wastewater with Nanofiltration to Recover Carboxylic Acids." *Environmental Science: Water Research & Technology* 4 (4): 520–528.

Zhu, L D, J Takala, E Hiltunen, and Z M Wang. 2013. "Recycling Harvest Water to Cultivate Chlorella Zofingiensis under Nutrient Limitation for Biodiesel Production." *Bioresource Technology* 144: 14–20. https://doi.org/https://doi.org/10.1016/j.biortech.2013.06.061.

6 Sewer Mining in a Circular Economy

Drivers, Challenges, and Opportunities

Long D. Nghiem and Andrzej Listowski

CONTENTS

6.1 INTRODUCTION

Water scarcity is often a consequence of natural hydrological variability. In recent years, population growth and urbanization have further aggravated the water shortage problem. Some cities and regions with abundant water resources have begun to experience severe shortages of clean water. In this context, large-scale water reuse projects have been successfully implemented in many countries throughout the world, including Australia, Israel, Namibia, Singapore, the United States, and many European countries [1-3]. Due to unfavorable public perception (although not scientifically supported) of the toilet-to-tap image, most of these projects are classified as indirect potable water reuse [4, 5]. In other words, the reclaimed water is not directly used for potable water consumption. Instead, it is used to supplement existing water resources in underground aquifers or above-ground reservoirs. Only in the rare instance of Namibia is the reclaimed water directly used for potable purposes [2].

Today, indirect water reuse is a well-established practice in terms of treatment technologies, operational experience, legal framework, and public acceptance [3, 6]. It is, however, costly and energy intensive. Wastewater is collected and treated at a centralized location, then transferred to the storage facility that can be very far away, before additional purification for reuse purposes. Building on the success of many indirect water reuse projects over the last few decades, an alternative known as sewer mining has gradually emerged.

Sewer mining is not a new concept, as the first sewer mining project can be dated back to more than a decade ago. It has, however, not received adequate scientific and public attention compared to

DOI: 10.1201/9781003335740-6

the significant potential contribution toward a circular economy, especially in a growing and highly population urban center. The definition of sewer mining varies from one jurisdiction to another. Since the concept was first described in Australia where there are arguably the largest number of well-documented sewer mining projects, in this chapter, the Australian definition of sewer mining is adapted. Sewer mining is the process of tapping into the sewer to extract wastewater, which is treated and used on-site for activities of high economic value. Residuals, including excess wastewater, sludge, and any other by-products from the treatment process, are returned back to the sewer.

Compared to traditional water reuse, a notable advantage of sewer mining is on-site and on-demand production and utilization of water. This means the reclaimed water can be allocated to activities of highest economic value at the most critical time, since the water source (i.e., wastewater) is always available and long-distance water transfer is avoided. The overarching aim of sewer mining is to optimize water resource yields. Although it is at the heart of the circular water economy, to date, sewer mining has only established a niche market within the water reuse sector. Despite several successful sewer mining projects, especially in Sydney, sewer mining has only accounted for 0.2% of the total municipal water supply in the region. There remain multiple barriers to further development of sewer mining as a mainstream practice to address water scarcity. Some of these are intrinsic to water use. Some of these are unique to sewer mining and thus have not been seen in traditional centralized water use projects.

This chapter first discusses the evolution of sewer mining within the context of a circular water economy. Successful sewer mining projects in Sydney are then examined as case studies to delineate current challenges in terms of technology, economic viability, planning, and public perception. The analysis and discussion in this chapter are expected to result in more sewer mining applications in other urban centers around the world.

6.2 THE EVOLUTION OF SEWER MINING

Since the 2000s drought in Australia, also known as the Millennium drought [7], new and large-scale developments within Australia have had to incorporate minimum requirements for energy and water efficiency within the design and planning. This requirement is also endorsed by the Green Building Council Australia through the Green Star certifications, as well as many other authorities throughout Australia. The concept of sewer mining can therefore be dated back to the early 2000s. It was conceptualized through international research collaborations. One of such collaborations is the European Union (EU)–funded project "Integrated concepts for reuse of upgraded wastewater" commonly known as AQUAREC, of which Australia was an international non-European partner [8]. The Millennium drought in Australia and the highly urbanized demography together have resulted in significant research and development (R&D) work and subsequently many sewer mining projects across the country. In the Sydney Central Business District, there are almost 10 current or planned sewer mining projects to cater for the anticipated increase in both wastewater volume and water supply demand (Figure 6.1).

The term "sewer mining" was first defined in a research brief to the Australian Parliament in 2005 in response to the Millennium drought in Australia [9]. The first sewer mining project was in fact part of the Water Reclamation and Management Scheme (WRAMS) at Sydney Olympic Park [10]. WRAMS was built for the Sydney Olympic Games in 2000 when water use was still in its infancy before the concept of sewer mining was clearly defined and recognized. Since its conception, WRAMS has been maintained and upgraded to become an essential infrastructure component of the modern Sydney Olympic Park precinct that currently serves as a large-scale entertainment venue (hosting over 10 million visitors each year), a major commercial and business center with over 15,000 employees, and a residential suburb for over 30,000 people. Since WRAMS, there have been many other sewer mining projects in Sydney alone. To cover the breadth of the constraints and potentials, the Barangaroo and the Pennant Hill sewer mining schemes are selected for further discussion in this chapter (Table 6.1).

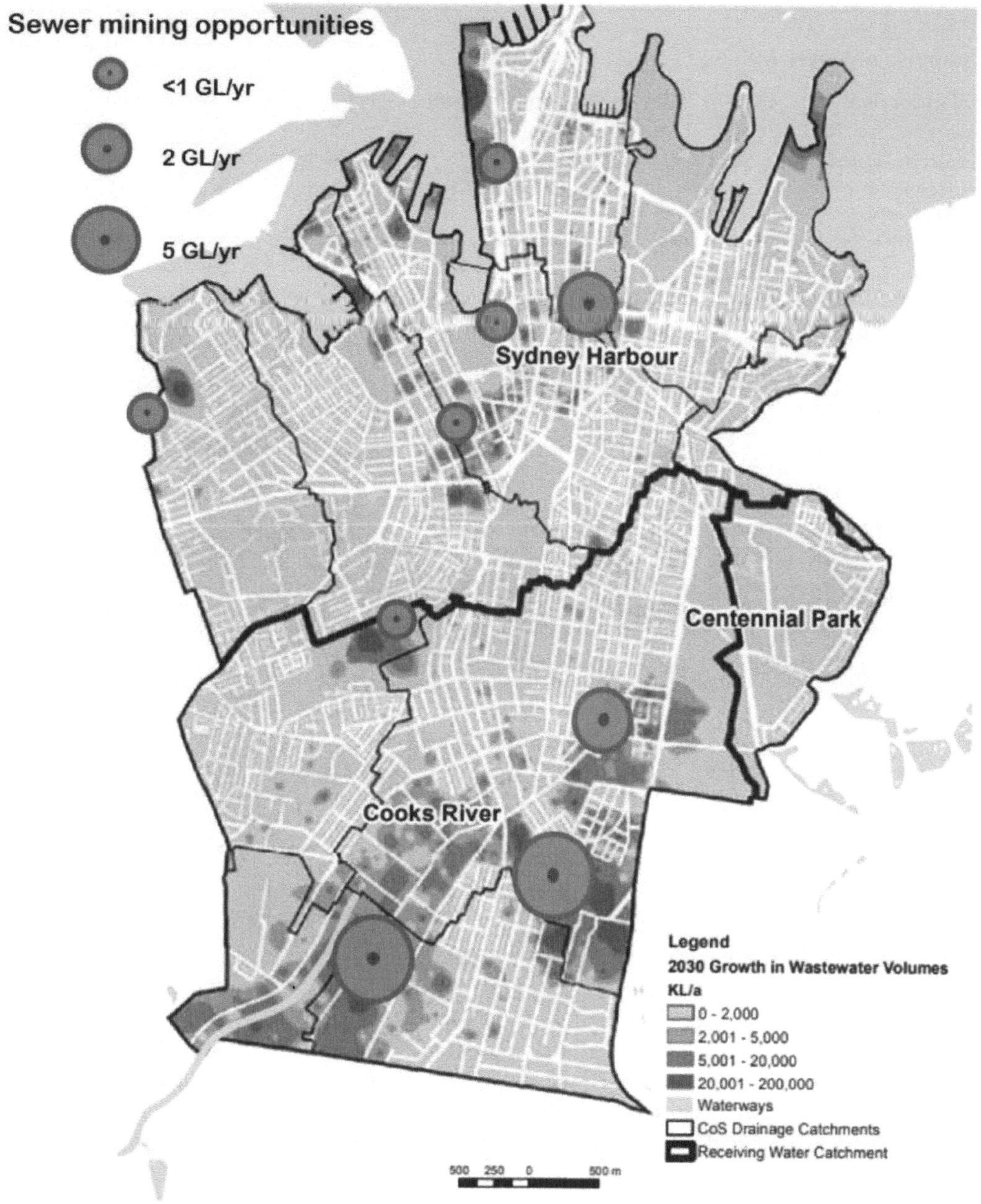

FIGURE 6.1 The potential of sewer mining in the Sydney Central Business District and the projected growth in wastewater volume by 2030.

Each of those sewer mining projects in Table 6.1 is unique in their own way. The main building of the WRAMS scheme that houses the MF and RO plants also serves as an educational center to educate the public about water reuse. The building is therefore made of transparent glass for public inspection (Figure 6.2). The Pennant Hill project was born out of necessity for the survival of the Gold Club. The Barangaroo project is cleverly hidden away in the basement of one of the most expensive real estate areas in the world (Figure 6.2).

TABLE 6.1
Examples of Sewer Mining Projects in Sydney

Project Location	Capacity (ML/yr)	Treatment Processes	Commission Date	Reclaimed Water Usage
Sydney Olympic Park	500	SBR-MF-RO-Cl_2	2000	Irrigation, toilet flushing, etc.
Pennant Hill Gold Club	100	MBR-UV	2008	Irrigation
Barangaroo	200	MBR-RO-UV/Cl_2	2018	Irrigation and toilet flushing

Note: SBR = Sequencing Batch Reactor; MF = Microfiltration; RO = Reverse Osmosis; UV = Ultraviolet treatment; Cl_2 = Chlorination.

FIGURE 6.2 (A) The Sydney Olympic Park – Sewer Mining Plant and (B) The Underground Sewer Mining Plant at Barangaroo.

6.3 CASE STUDIES OF SEWER MINING PROJECTS IN SYDNEY

6.3.1 Sydney Olympic Park

Sydney Olympic Park is about 13 km west of the Sydney Central Business District. The current land of Sydney Olympic Park, approximately 6.6 km^2, was previously an abandoned industrial estate, used as a limestone quarry, abattoir, and even disposal site for industrial waste. When Sydney was selected to host the 2000 Olympic and Paralympic Games, the site was completely transformed to become a state-of-the-art sporting and event venue. To achieve the "green" Olympic objective, many innovative and visionary environmentally features were incorporated into the design and planning of the Sydney Olympic Park precinct. These include WRAMS.

WRAMS is Australia's first large-scale urban water reuse scheme and one of the few pioneering projects in the world. Initially, wastewater from the site is treated and reused for toilet flushing in sporting venues, irrigation, and other nonpotable uses. The treatment train consists of a 2 ML/d sequencing biological reactor, microfiltration (MF), and reverse osmosis (RO). Due to limited space and proximity to the existing sewer network, residuals from the treatment process, including sludge, MF backwashing waste, and RO, are disposed to the sewer. It is noted that wastewater is only collected from Sydney Olympic Park. Thus, the overall design of WRAM only partially meets the definition of a sewer mining project.

WRAMS currently provides 500 ML of high-quality reused water each year to the Sydney Olympic Park precinct, effectively offsetting 50% of potable water consumption. WRAMS also has a network of digital water meters for accurate water accounting within the scheme. The combined capital and operational investment cost of reused water production and distribution by WRAMS was about 50% of the normal potable water supply. Since 50% of the potable water demand is self-sustained and most wastewater is treated on-site, the water infrastructure cost of the Sydney Olympic Park precinct is significantly reduced. This is in the context of a sporting and entertainment venue with extreme variation in water usage. As an example, the Stadium Australia can hold 110,000 spectators. At the interval during a sporting game, many of these spectators would visit the washroom, resulting in a surge in water demand and wastewater discharge.

6.3.2 Pennant Hills Golf Club

The Pennant Hills sewer mining project was a direct response to water restriction during the Millennium drought in Australia. With 23 hectares of course ground, the Pennant Hills Golf Club faced a major challenge when the ongoing drought triggered the ban on using potable water for irrigation. To overcome this, an on-site sewer mining plant utilizing a ZeeWeed membrane bioreactor (MBR) was constructed and commissioned by Suez in 2008.

Up to 100 ML of reclaimed water each year that would have previously been potable water is used for irrigation of the course grounds (Suez 2008). The economic benefit in this case is beyond potable water saving. Under a water restriction, without sewer mining, the club would have to be closed, with millions of dollars of lost revenue and infrastructure damage. Reclaimed water from sewer mining has also resulted in a reduction of the fertilizer used by the club due to high nitrogen levels found in the water. This has added to the cost/benefit of utilizing recycled wastewater for irrigation. Pennant Hills is the first golf club in Australia to utilize reclaimed water from sewer mining for irrigation. Following its success after commissioning, many other golf courses throughout Australia have also commissioned sewer mining for irrigation.

6.3.3 Barangaroo

Barangaroo is located in central Sydney, on the northwestern side of the Central Business District. It overlooks the Sydney Harbour Bridge from the southern end. In 2009, the state government of New South Wales established the Barangaroo Delivery Authority to redevelop the precinct from shipping and stevedoring facilities to provide more commercial office space and recreational areas. The redevelopment is expected to be completed in 2023 to become a mixed-use precinct for high-end commercial real estate, entertainment, and residential purposes, including the establishment of the Barangaroo Reserve and other public domains. The high-rise complex consists of six towers. The first three office towers (known as International Towers Sydney) were completed in 2017 to include 320,000 m^2 gross floor area at a height of over 210 m. Barangaroo Reserve is an open parkland of 6 ha with direct access to Sydney Harbour at the northern end of Barangaroo. The park has been opened to the public since 22 August 2015. It has a reconstructed, naturalistic headland based around the pre-1836 shoreline. Given the large landscaping area and the development density of the precinct, the water demand is very significant. To address this challenge, a sewer mining plant has been built and installed in the basement of the International Towers Sydney.

A comprehensive train of wastewater treatment and reuse processes is used at Barangaroo. It includes two parallel MBRs for biological treatment of wastewater, RO for salinity removal and polishing, and chlorine and ultraviolet (UV) disinfection. Raw wastewater is extracted from a large sewer main that runs across the basement of the International Towers Sydney. The Barangaroo sewer mining plant was commissioned in 2018. It is licensed to produce 200 ML of reclaimed water for toilet flushing and landscaping purposes. To date, the plant has always maximized their licensing condition, and most of the reclaimed water is produced during the summer when the demand for irrigation is high.

Unlike WRAMS and the Pennant Hills Gold Club, space is a major constraint at Barangaroo. Thus, the entire plant was designed for space optimization. Since wastewater treatment is in an enclosed basement of a building, there have been some significant and additional capital and operational costs associated with safety provision for the plant operators, ventilation, and incident contingency response. High wastewater salinity is another notable challenge to the Barangaroo plant, since the sewer network is within Sydney Harbour and is below seawater level. A conductivity sensor is used at the inlet to detect excessive seawater intrusion at high tide and temporarily close the intake valve.

6.4 CHALLENGES AND OPPORTUNITIES

Sewer mining has already been established at various sites across Sydney, in Australia, and in many other countries in the world. Nevertheless, there remain numerous technological, economic, social, and planning issues for further consideration. Some of these issues are impediments to further application of sewer mining. Some can also be potential opportunities to further strengthen the case for more sewer mining in the urban environment.

6.4.1 Technology

In terms of treatment technology, the basis of sewer mining is similar to all other water reuse schemes. It is, however, challenged by the need for more compact infrastructure and more automation in operation. In addition to the brief discussion in Section 2, a sewer mining scheme usually consists of all components of a typical large-scale water reuse facility. These components include screening, a balancing tank, biological treatment, additional polishing steps (usually MF and/or RO), disinfection, and a reclaimed water reservoir. The treatment technologies are mature and reliable. To date, there have been very few issues associated with reclaimed water quality from all sewer mining projects in Sydney and elsewhere in the world [3].

The fundamental shift toward decentralized sewer mining can also entail new challenges and opportunities (Table 6.2). The challenge for space optimization has facilitated the deployment of modern and highly compact treatment technologies for sewer mining. Indeed, almost all current sewer mining projects, including the three case studies discussed here, have deployed membrane technology as core treatment processes. There is also a trade-off between space and treatment redundancy, which is the hallmark of large-scale wastewater treatment. For example, the Pennant Hill sewer mining project depends on a single MBR. As a result, during maintenance and unexpected operational disruptions, reclaimed water may not be available. When these incidents occur, potable water is temporarily used instead of reclaimed water.

Due to the lack of economies of scale, the unit operation cost of sewer mining can be substantially higher than that of a large-scale water reuse project. This is particularly prevalent in Australia, where the labor cost is high. This impediment can be mitigated by remote monitoring and flexible operational arrangement. All three sewer mining case studies discussed in this chapter are fully

TABLE 6.2
Major Technical Challenges and Opportunities Associated with Sewer Mining

Challenge	Solution/Opportunity	Remark
Space limitation	Deployment of modern and compact treatment technologies	Membrane separation is core to most sewer mining projects
Space limitation	Trade-off between space and treatment redundancy	Potable water can be used during downtime
High operational cost	Automation and remote monitoring to reduce labor cost	All sewer mining plants are remotely monitored and fully automated; information sharing with the public
High operational cost	Redistributing technical skills of the crew of operators	Operators are responsible for several sewer mining schemes at the same time

automated and can be remotely monitored. They are assigned to a crew of operators who are in charge of several adjacent sewer mining projects. With the recent progress in information technology, sensing, and automation, further cost reduction can be expected from the integration of the digital infrastructure to sewer mining. The ability to share information with the public through the Internet of Things can also improve public support for sewer mining.

6.4.2 Economic Considerations

All water reuse projects, whether they are traditional large scale or sewer mining, are market driven. Economic consideration of a water reuse project is governed by its context. It is argued here that economic viability of a water reuse project cannot always be evaluated based solely on the cost of producing reclaimed water and the economic value of water use. Most (if not all) recent sewer mining projects in Australia are still in operation. By contrast, many large-scale water reuse projects in Australia have either been abandoned or mothballed due to inadequate and unstable demand for reclaimed water [3].

The cost of reclaimed water production by sewer mining varies drastically. However, in many instances there are additional incentives (some of which may not be able to be converted to financial outcomes) to support sewer mining (Figure 6.3). A detailed life cycle cost analysis is usually proprietary and thus rarely available in full. Thus, the following discussion is mostly based on personal conversations between the authors and their industry partners or from the authors themselves.

The three case studies described in Section 3 have been specifically selected to demonstrate the full range of the economic imperative of sewer mining. As noted earlier, a major economic fundamental of sewer mining is the ability to produce on-site and on-demand reclaimed water, as well as the reliable demand for reclaimed water. WRAMS is a well-planned and well-executed sewer mining project to satisfy the highly variable water demand at Sydney Olympic Park. The current cost of reclaimed water supply (including treatment and distribution) is less than 1.3 $\$/m^3$, which is

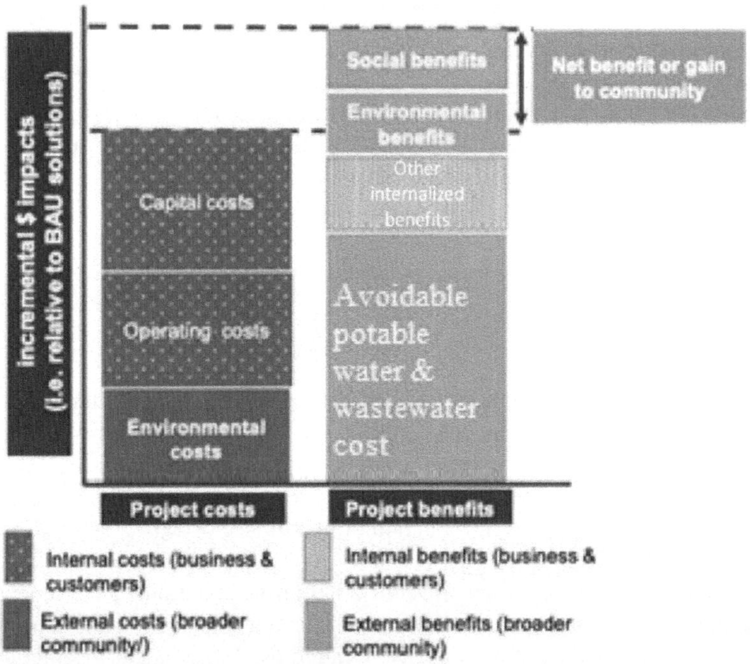

FIGURE 6.3 Net benefits of sewer mining to the community.

about half of the cost of a potable water supply. WRAMS has been a financially sustainable scheme since its inception in 2000. Annual profit can be used to amortize the cost of capital investment and off-set the cost of public visit to the WRAMS water recycling center.

The Pennant Hill sewer mining project is a response to water restriction that, although not permanent, is expected to occur frequently in the Australian climatic condition. Benefits from the reliable access to irrigation water outweigh the cost of reclaimed water production. The cost of reclaimed water production is also insignificant compared to other operational costs by the Golf Club and revenue from membership fees. Thus, a traditional economic analysis cannot be applied in this case.

Capital and operational costs of the Barangaroo sewer mining scheme are marginally higher than simply paying for potable water. The key driver in this case is the collective determination of the asset owners, their tenants, and other stakeholders to demonstrate the value of a circular water economy. In other words, they are willing to internalize negative externalities of using (partially subsidized) potable water. The Barangaroo sewer mining scheme is also designed to satisfy the water sustainability requirement to achieve a 6-star green building rating. This is an important intangible asset, although it cannot be readily converted to a financial value.

6.4.3 Social and Environmental Considerations

Over time, the demand for potable water in urban areas will continue to increase driven by population growth, urbanization, and improved living standards. The ever-growing increase in potable water demand can lead to an extended period of water restriction, especially in arid regions of the world. Water reuse via sewer mining also results in a reduction in the discharge of wastewater to the natural environment.

The social benefits of sewer mining include the preservation of potable water, leading to public confidence, and the ability to introduce more irrigated green space in urban areas, thus adding to city aesthetics and decreased urban heat island effects. These benefits attribute to an increased quality of life for urban and city dwellers.

One of the biggest challenges for any reclaimed or recycled water operation is public acceptance. This includes both community and individual acceptance of reclaimed water. For individuals, the main concerns stemmed from personal choices such as the use and source of the water, trust, and perceptions on water quality. Unlike large-scale water reuse, most sewer mining schemes have received favorable public acceptance. This is because the use of reclaimed water is clearly defined and is exclusively for nonhuman contact and nonpotable applications such as irrigation and toilet flushing. In addition, sewer mining schemes are usually located in areas of well-informed individuals [11]. Initiatives to provide educational visits to the public to learn about sewer mining have proven to be very effective in gaining trust and confidence from the public [11].

6.4.4 Regulations and Planning for Sewer Mining

The water and wastewater treatment market is, by definition, a natural monopoly. In other words, water supply and wastewater treatment are controlled by a single utility. This has created a major hurdle for the private sector to participate in sewer mining projects. In the past, the regulatory and legislative framework in Australia has been slow to keep up with the rapid pace of technological developments, particularly in the introduction of private-sector participation and allowance for competition within the water industry [12].

In Australia, to balance the monopoly power of water utilities, the state government has set up an independent authority to regulate price and commercial arrangements for water supply and services. In the state of New South Wales, this authority is called the Independent Pricing and Regulatory Tribunal (IPART). For simplicity, the Greater Sydney Region is used as an example to illustrate the importance of a suitable regulatory framework. Recognizing the social and environmental values of sewer mining schemes, the New South Wales government, through IPART and other agencies, has introduced a set of guidelines and legislation to encourage the involvement of the private sector in sewer mining. Indeed, after the WRAMS scheme, which was financially supported by the government, most recent sewer mining schemes are directly funded by the private sector. These guidelines and legislation mandate Sydney Water (who provides water and wastewater services to the Greater Sydney Region) to negotiate and consider all applications for sewer mining. An approval to extract wastewater from the sewer can then be granted to the applicant by Sydney Water. The return of wastewater residuals to the sewer is also regulated by a trade waste agreement between Sydney Water and the applicant.

The extraction of wastewater and return of high-solid-content residuals to the sewer can alter fundamental design parameters of the sewer network, such as the self-cleansing velocity, the rate of solid deposition, and the intensity of sewer odor. Many of these impacts have not been fully understood. In recognizing these potential impacts, the authority can set a sewer mining exclusion zone. The sewer mining exclusion zone is usually set in areas of low population density where wastewater flow is low (and thus the risk of sewer blockage is high) or where there is an existing sewer mining scheme. Surprisingly, to date, there has been very little research work to set objective parameters to define these sewer mining exclusion zones.

6.5 CONCLUSION

Sewer mining is an innovative concept consistent with a circular water economy. Key treatment technologies for sewer mining are mature and readily available. These technologies are adapted for small and compact applications to suit the requirement of sewer mining. The small scale of sewer mining allows for the private sector to become water suppliers and for consumers to consider an alternative source of water for nonpotable use. In regard to cost-effectiveness, it is important to consider that the cost of reclaimed water is not always lower than that of a traditional potable water supply; however, this does not mean that potential schemes should be discredited, as benefits exist that extend beyond that of the economy.

This chapter demonstrates the value of sewer mining within an urban center. In Sydney, with several active schemes contributing 1.2 billion liters of reclaimed water each year, sewer mining has been successfully established as a credible practice to extend the economic and environmental value of the water flow through the city. For further adoption of sewer mining, there must be a shift in public perception of wastewater as a by-product of society to a valuable resource. The consideration of sewer mining as a requirement to new developments should also be introduced to allow for purpose-built schemes and to offset future water consumption in highly urbanized areas. The Greater Sydney Region has established a legal template for implementing sewer mining. It is by no means perfect, and further work both in research and in policy development is necessary to support additional sewer mining schemes and to replicate the success in Sydney in other cities.

REFERENCES

1. Bixio, D., B. De Heyder, H. Cikurel, M. Muston, V. Miska, D. Joksimovic, A.I. Schäfer, A. Ravazzini, A. Aharoni, D. Savic, and C. Thoeye. 2005. Municipal wastewater reclamation: Where do we stand? An overview of treatment technology and management practice. *Water Science and Technology: Water Supply* **5**(1): p. 77–85.
2. van Rensburg, P. 2016. Overcoming global water reuse barriers: The Windhoek experience. *International Journal of Water Resources Development* **32**(4): p. 622–636.
3. Radcliffe, J.C. 2015. Water recycling in Australia – during and after the drought. *Environmental Science: Water Research & Technology* **1**(5): p. 554–562.
4. Law, I.B. 2008. The future direction for potable reuse. *Water* **35**(8): p. 58–63.
5. Radcliffe, J.C. 2010. Evolution of water recycling in Australian cities since 2003. *Water Science and Technology* **62**(4): p. 792–802.

6. Lefebvre, O. 2018. Beyond NEWater: An insight into Singapore's water reuse prospects. *Current Opinion in Environmental Science and Health* **2**: p. 26–31.
7. Gregory, A. 2009. Beyond the millennium drought. *Water* **36**(8): p. 6.
8. Bixio, D., T. Wintgens, A. Ravazzini, C. Thoeye, A. Aharoni, J. De Koning, and T. Melin. 2007. *Water Reuse, in Wastewater Quality Monitoring and Treatment* p. 329–349. Wiley.
9. Dimitriadis, S., Issues encountered in advancing Australia's water recycling schemes/Sophia Dimitriadis. 2005. *Research Brief* (Australia. Dept. of Parliamentary Services. Parliamentary Library); 2005–06, no. 2., ed. L. Australia. Dept. of Parliamentary Services. Parliamentary, [Canberra]: Dept. of the Parliamentary Services.
10. Chapman, H. 2006. WRAMS, sustainable water recycling. *Desalination* **188**(1–3): p. 105–111.
11. Dolnicar, S., A. Hurlimann, and L.D. Nghiem. 2010. The effect of information on public acceptance – The case of water from alternative sources. *Journal of Environmental Management* **91**(6): p. 1288–1293.
12. McFallan, S. and I. Logan. 2008. *Barriers and drivers of new public-private infrastructure: Sewer mining, in Report No. 6 [CIBE-2007–032A]*, Brisbane, Australia.

7 Biorefinery System for Producing Energy and Biofuels Based on Black Liquor Gasification Technology

Guijin Wang, Hongyou Yuan, Lijun Jin, and Haoquan Hu

CONTENTS

7.1 INTRODUCTION

7.1.1 Properties of Black Liquor

Black liquor, named for its dark brown color, is the spent cooking liquor from the chemical pulping process in the pulp and paper industry. Weak black liquor from the liquid stream at the end of cooking and washing processes contains 10–15 wt.% dry solids. The content of organic and inorganic matters in black liquor is at the level of 60–70 and 30–40 wt.% on dry basis, respectively (Sricharoenchaikul et al. 2003). In particular, organic species in black liquor mainly involve dissolved alkaline lignin, monosaccharides decomposition fragments, phenols and organic (fatty and resin) acids, etc.; inorganic components mainly consist of the spent pulping chemicals, such as Na_2CO_3, NaOH and NaCl (Azadi et al. 2013).

At present, chemical cooking technologies adopted in pulp mills mainly include the kraft (sulfate) method and the alkaline (soda) method. Raw materials and cooking chemicals used in these two pulping processes are different from each other. For the kraft process, the cooking chemicals

DOI: 10.1201/9781003335740-7

TABLE 7.1
Composition of Different Types of Black Liquor

Element/Component	Wood Kraft Black Liquor (Andersson et al. 2015)	Straw Alkaline Black Liquor (Wang et al. 2015)
Ultimate analysis (wt.% on dry basis)		
C	32.5	35.1
H	3.4	5.3
N	0.1	0.6
S	5.5	0.9
Cl	0.1	4.3
Na	20.2	17.9
K	2.2	2.3
Si	–	1.3
O*	36.0	32.3
Proximate analysis (wt.% on dry basis)		
Ash	35.6	47.5
Volatile matter	55.6	43.3
Fixed carbon	8.8	9.2
Lower heating value (LHV, MJ/kg)	12.1	10.5

* Calculated by difference.

are NaOH and Na_2S, and the cooking materials are wood fibers (pine wood, etc.), while the alkaline process employs NaOH as the main cooking chemical, along with nonwood fibers (wheat straw, etc.) as the raw materials. Due to the distinct cooking manners and raw biomass materials, composition and characteristics of the produced black liquors also show variations. The composition of different types of black liquor is shown in Table 7.1. It can be seen that the kraft black liquor is more enriched in sulfur (S), while the content of silicon (Si) in alkaline black liquor is higher. These two elements will seriously affect the subsequent recovery process. As to the wood kraft black liquor, the existing forms of sulfur mainly include sulfur-containing organic substances and inorganic salts, such as Na_2S, Na_2SO_4 and $Na_2S_2O_3$. Since chemical reactions involving the release of sulfur will occur in the recovery process, purification and recovery of H_2S and other sulfur-containing species must be considered. As to the straw alkaline black liquor, the high content of silicon, mainly in forms of silicon dioxide (SiO_2) and sodium silicate (Na_2SiO_3), leads to high viscosity and then makes the subsequent recovery more difficult (Yuan et al. 2016). Therefore, "silicon interference" is one of the key technical problems to be solved in efficient recovery of straw alkaline black liquor.

According to the statistical data from the Food and Agriculture Organization (FAO), the global pulp production was about 161 million tons in 2019. In general, about 1.7–1.8 tons of black liquor on a dry basis is produced incidentally per air-dried ton of pulp (Larson et al. 2000). Therefore, it can be estimated that more than 273 million tons of black liquor solids are produced annually worldwide. Calculated on the basis of 12.0 MJ/kg as the typical lower heating value (LHV) of black liquor solid, the energy contained in black liquor is equivalent to about 110 million tons of standard coal (LHV ≈29.3MJ/kg). In other words, the full energy exploitation of black liquor has the potential to greatly reduce the consumption of conventional fossil fuels, which may bring huge environmental benefits.

Chemical pulp production of the top 10 countries in the world and their estimated black liquor production in 2019 are shown in Table 7.2. Among these top 10 countries, the United States is the largest producer of chemical pulp followed by Brazil and China. Except for China and India, wood

TABLE 7.2
Chemical Pulp Production and Estimation of Black Liquor Production of the Top 10 Countries in the World in 2019

		Production (Million Tons, on Dry Basis)			
Rank	**Country**	**Pulp from Wood Fibers**	**Pulp from Nonwood Fibers**	**Pulp (Sum.)**	**Black Liquor**
1	USA	45.26	0.15	45.41	77.20
2	Brazil	19.20	0.06	19.27	32.75
3	China	10.48	5.88	16.35	27.80
4	Canada	9.02	0.04	9.06	15.40
5	Sweden	8.40	0.00	8.40	14.28
6	Indonesia	8.26	0.11	8.37	14.23
7	Finland	8.32	0.00	8.32	14.14
8	Japan	7.86	0.01	7.86	13.37
9	Russia	5.88	0.10	5.98	10.16
10	India	2.43	3.02	5.45	9.27

Source: http://www.fao.org/faostat/en/#data/FO. (Accessed on 10 June 2022.)

pulp production of other countries accounts for more than 98% of their respective total chemical pulp production. It's worth noting that nonwood fiber pulping in China and India accounted for about 36% and 55% of the total chemical pulp production, respectively. As mentioned earlier, the composition characteristics of wood kraft black liquor and straw alkaline black liquor are varied from each other. Therefore, China and India may face more complex technical problems in the subsequent black liquor recovery process.

7.1.2 Traditional Recovery Process

Black liquor, as an industrial waste, is a special kind of biomass feedstock. From the perspective of energy savings and emission reductions, measures should be taken to recycle it (Bhattacharya et al. 1985). At present, the traditional recovery process, i.e., the direct combustion approach, is widely used in black liquor treatment. Figure 7.1 shows the flow diagram of the traditional recovery cycle in the pulp mill. The core equipment of this process is a Tomlinson recovery boiler, invented by G. H. Tomlinson in 1934 (Vakkilainen 2014). A schematic diagram of a typical modern recovery boiler is shown in Figure 7.2 (Hagga and Laitinen 2006). The function of the recovery boiler is twofold: first, to burn the organic matter and recover the energy in black liquor; second, to recover the inorganic salts through a subsequent causticizing process (Connolly 2006).

As shown in Figure 7.1, in the traditional recovery process, weak black liquor (dry solid content of about 15 wt.%) produced in the cooking and washing sections is first evaporated and concentrated to about 65–80 wt.% solid content in the evaporator, and then the heavy black liquor is sprayed into the recovery boiler through a nozzle for combustion at about 1000–1100 °C. The heat generated from the combustion of black liquor can be used to produce electricity and steam, while residual inorganic matters will flow out in the form of molten salts to turn into green liquor (the main component is Na_2CO_3, etc.) after adding water. The chemicals in green liquor are recovered through the subsequent limestone causticizing reactions [Eqs. (7.1) and (7.2)] at about 95 °C (Azadi et al. 2013). After causticization, the green liquor is converted into white liquor (the main component is NaOH, etc.). Thereafter, the white liquor is recycled to the cooking section, while lime mud

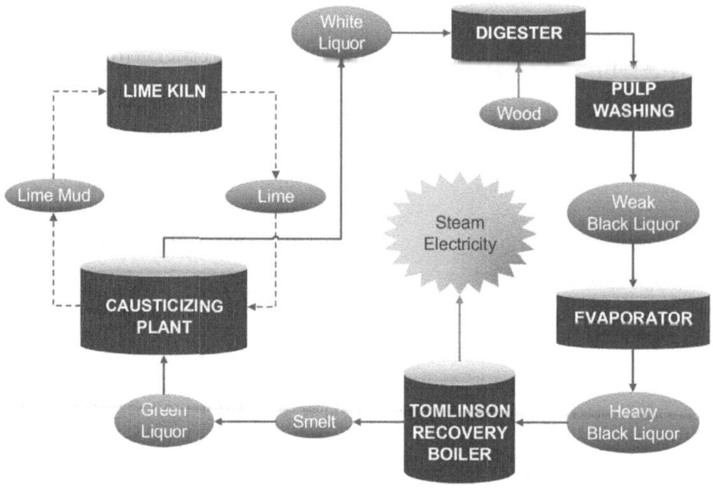

FIGURE 7.1 Flow Diagram of the Traditional Recovery Cycle in the Pulp Mill. The left part of the flow chart is the lime causticization cycle. From clockwise starting from top, the cycle state is "Lime kiln", "Lime", "Causticizing plant" and "Lime mud". The right part of the flow chart is the whole pulping process cycle. From clockwise starting from top left the cycle state is "White liquor", "Digester" with "Wood" input, "Pulp washing", "Weak black liquor", "Evaporator", "Heavy black liquor", "Tomlinson recovery boiler" with "Steam electricity" output, "Smelt", "Green liquor" and "Causticizing plant". The intersection of these two cycles is the "Causticizing plant".

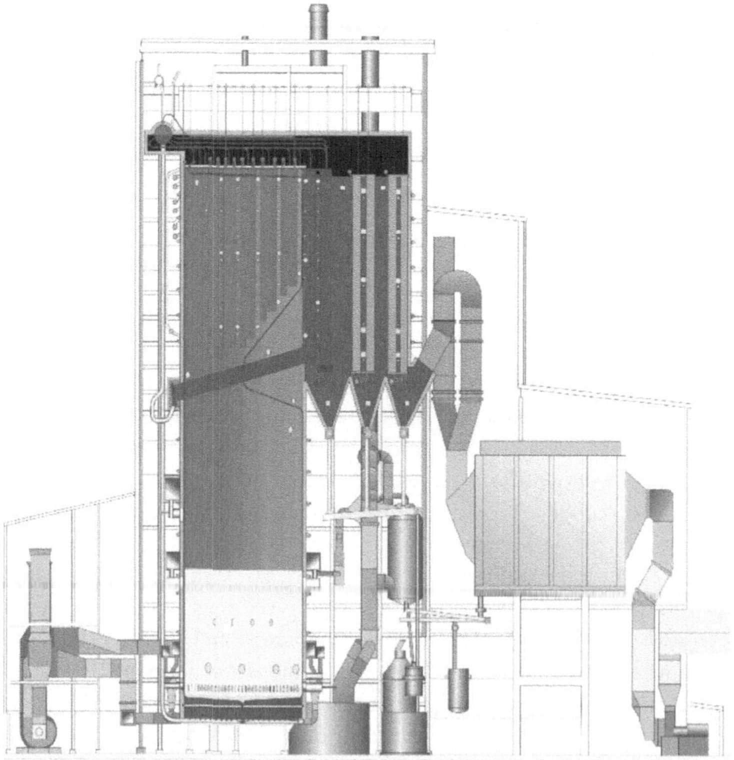

FIGURE 7.2 Schematic Diagram of a Typical Modern Recovery Boiler. Available from (Hagga and Laitinen 2006). The overall outline is similar to that of a building, with the highest in the middle, one ladder on the left and two ladders on the right. There are various regular shapes and lines inside. This image is considered decorative.

($CaCO_3$) generated during the causticizing reactions is calcined in a lime kiln and regenerated into lime (CaO) for reuse [Eq. (7.3)].

$$CaO(s) + H_2O(l) \rightarrow Ca(OH)_2(l) \quad (7.1)$$

$$Ca(OH)_2(l) + Na_2CO_3(l) \rightarrow 2NaOH(l) + CaCO_3(s) \quad (7.2)$$

$$CaCO_3(s) \rightarrow CaO(s) + CO_2(g) \quad (7.3)$$

This traditional recovery process based on a Tomlinson boiler has been running for decades in the kraft pulp industry, and its thermal efficiency is constantly improving with the progress of technology and time. At present, the traditional recovery process has got rid of the disadvantage of low thermal efficiency. Modern recovery boilers have already met about 75–100% of the steam supply for the pulp and paper mills.

Although the traditional Tomlinson recovery process has been gradually improved and upgraded in applications, its inherent drawbacks still exist, primarily the following two points (Gea et al. 2003; Larson et al. 1999; Naqvi et al. 2010a; Whitty 2005):

1. **High investment cost:** The recovery boiler is the most expensive device in the entire pulping process, accounting for about 20% of the investment cost of a pulp mill, and the recovery process module will account for more than 50% of the investment cost. As the investment is huge, it's hard for the ordinary small and medium-sized pulping mills to afford it.
2. **High safety risk:** Although the traditional recovery process has been widely used in the past decades, alkali recovery boiler explosions have occurred from time to time. The reason is primarily the explosion of the mixture of molten salts and water. The unavoidable formation of molten salts in the recovery boiler makes it difficult to eliminate the risk of explosion. In addition, the internal structure components in the recovery boiler are complex, and a large amount of alkali ash deposition will corrode the pipeline surface and bring about the explosion risk. Consequently, pulp and paper mills with traditional recovery processes still need to bear the safety issues and economic risks.

Therefore, it's necessary to develop alternative technologies for the traditional recovery process to meet the demand of various sizes and types of pulp and paper mills. Among them, black liquor gasification technology has emerged, which is considered to be a promising alternative process for traditional recovery and has attracted widespread attention.

7.2 BLACK LIQUOR GASIFICATION (BLG)

7.2.1 Fundamental Principles of BLG

7.2.1.1 Overview of the Conversion Process During BLG

A black liquor droplet undergoes a three-step conversion process in the gasification reactor: drying, pyrolysis (also referred to as devolatilization) and char gasification. The progression of these three stages is shown vividly in Figure 7.3 (Whitty et al. 1998).

Drying: As shown in Figure 7.3, drying is the first step that occurs when the droplet is introduced into a heating zone, and mainly involves evaporation of water from the droplet. The drying rate depends on how quickly heat can be supplied to the droplet since it's mainly heat-transfer controlled. In addition, some degree of swelling occurs during the drying of the black liquor droplet. The volume of the residual solid is about 1.5 times the original volume of the droplet when drying is complete.

Pyrolysis: After the drying stage, organic matter in the black liquor solid begins thermal decomposition at above 200 °C and is complete when the temperature reaches 650–750 °C (Bhattacharya

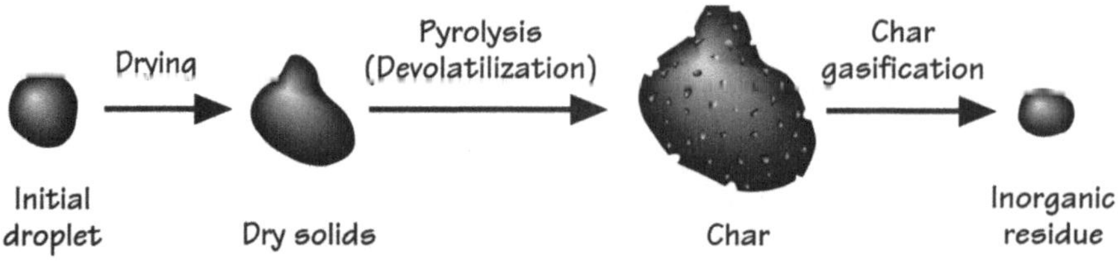

FIGURE 7.3 Conversion Stages During Gasification of a Black Liquor Droplet. Available from (Whitty et al. 1998). From left to right, the "initial droplet" becomes "dry solids" through the "drying" process. Next, the "dry solids" are converted to "char" with significant volume expansion after the "pyrolysis (devolatilization)" process. Then, the char gasification reaction is carried out, with "inorganic residue" remaining.

et al. 1986; Frederick et al. 1994; Kubes et al. 1982). In general, during the devolatilization of black liquor solids, 25–45% of the original mass is converted to noncondensable light gases and other condensable volatiles (Sricharoenchaikul et al. 2002; Sricharoenchaikul et al. 2003; Sricharoenchaikul et al. 2001). Char is formed as the result of completion of the devolatilization process, which mainly contains fixed carbon and various inorganic salts, such as Na_2CO_3 and Na_2SO_4. Remarkably, the ash mass in black liquor char is up to 35–45% of the initial dry solid mass, which is higher than that of coal and other biomass feedstocks (Reis et al. 1995). As to the carbon release during pyrolysis of black liquor solid, it can be found that the carbon content of light gases increases with longer residence time and higher reaction temperature. About 15% of the total carbon in black liquor solid is converted to light gases during devolatilization (Sricharoenchaikul et al. 2002).

In terms of composition of black liquor, the organic matters in black liquor are mainly composed of chemically degraded lignin and carboxylic acids. Dissociation reactions of a variety of functional groups in black liquor, such as decarboxylation, decarbonylation, dealkylation and aromatic ring rupture, occur during devolatilization, with formation of various volatile species (Sricharoenchaikul et al. 2002). These volatiles mainly include CO_2, CO, H_2, CH_4 and other light hydrocarbons (e.g., C_2H_4, C_2H_6); sulfur-containing compounds such as H_2S, COS, CS_2; mercaptans (e.g., CH_3SH, CH_3SCH_3, $CH_3S_2CH_3$, etc.); and condensable liquids (H_2O and tar). Compounds in the primary tar will undergo secondary reactions to produce additional noncondensable light gases (CO, CO_2, CH_4, and other light hydrocarbons), lighter tar species and coke. Additionally, it's worth noting that substantial swelling also occurs during the devolatilization of the black liquor solid. The volume of char after pyrolysis can be more than 30 times that of the original droplet. The swelling behavior, the most important physical change in the pyrolysis process of black liquor, will be described in detail in the following section (7.2.1.2).

Gasification: The gasification reactions of organic constituents (fixed carbon) in black liquor char generally occur in the presence of gasifying agents (e.g., CO_2 or H_2O) at temperatures above 800 °C [Eqs. (7.4) and (7.5)]. Char gasification is an endothermic process and is generally the slowest stage of the whole thermochemical conversion of black liquor. The gasification reactions are very temperature-sensitive, and the reaction rate is improved with the rising of temperature. During char gasification, the water-gas shift reaction [Eq. (7.6)] will occur simultaneously, which plays a significant role in the distribution of carbon between CO and CO_2 as well as H_2 yield in gasification environments at longer residence time and higher reactor temperature (Sricharoenchaikul et al. 2002). Noteworthily, gasification of black liquor char is strongly catalyzed by inherent alkali species in char and proceeds at reaction rates several orders of magnitude higher than those of pure carbon (Li and Vanheiningen 1990a; Li and Vanheiningen 1989). As for the solid physical changes, the char particle has been observed to shrink during the gasification process. When gasification of black liquor char is complete, only inorganic salts remain in the residual solid.

$$C(s)+CO_2(g)\rightleftharpoons 2CO(g) \tag{7.4}$$

$$C(s)+H_2O(g) \rightleftharpoons CO(g)+H_2(g) \tag{7.5}$$

$$CO(g)+H_2O(g) \rightleftharpoons CO_2(g)+H_2(g) \tag{7.6}$$

7.2.1.2 Thermoplastic Properties of Black Liquor: Swelling

One of the most important phenomena in thermochemical conversion of black liquor is swelling, which is closely related to its inherent composition characteristics. It has been clearly indicated that organic compositions of black liquor have significant effects on its swelling behavior (Hupa et al. 1987; Wang et al. 2015). Organic matter in black liquor consists mainly of degraded lignin (27–33%), polysaccharide degradation products (aliphatic carboxylic acids) (31–32%), along with a minor fraction of extracts (4–5%) (Alen 1994). By comparing the swelling behavior of the major separate organic components (lignin and hydroxy carboxylic acids) in black liquor with that of the real black liquor, it can be concluded that the aliphatic acid fraction is the main source of the volatiles contributing to swelling, while the lignin component inhibits the release of volatile fractions within the particle.

Based on the comprehensive consideration of the composition characteristics of black liquor and the devolatilization process, a salt-induced mechanism can be used to explain the swelling behavior of black liquor (Chen et al. 2017). This mechanism is based on the study of pyrolysis of black liquor model compounds, i.e., salt-polymer mixtures. Sodium acetate (CH_3COONa) and sodium hydroxide (NaOH) are selected as representatives for salts and are mixed with the polymers (lignin, lignosulfonates, xylan, and cellulose) to prepare the salt-polymer mixtures as the black liquor model compounds. As shown in Figure 7.4, no obvious swelling has occurred for the polymers without the addition of sodium salts, and the residual chars from pyrolysis are still in a powder state. However, all the chars from pyrolysis of salt-polymer mixtures show a remarkable swelling state. It's revealed that swelling is an interaction between the melting salts and organic polymers existing in the initial black liquor. Furthermore, it can be concluded that the swelling phenomenon occurring in the pyrolysis process of black liquor is accompanied by multiple physical and chemical changes simultaneously, which mainly include:

1. Plastic state generated through the consecutive and concurrent melting of aliphatic carboxylates and NaOH at 230–500 °C.
2. Volatiles released from decomposition of lignin and other organics.
3. Formation of the rigid char surface with nascent carbon materials formed during the swelling process.

FIGURE 7.4 Physical Photos of the Chars Derived from Pyrolysis of Polymers and Salt-Polymer Mixtures (500 °C, 5 min, 1 atm). Available from (Chen et al. 2017). Photos of 12 kinds of char arranged in three rows and four columns. Each photo is marked with the name of the initial reactant, and a metal cylinder with a diameter of 13 mm is placed on the right side of each char as a size reference. From left to right, the first row is lignin char, lignosulfonates char, xylan char and cellulose char; the second row is acetate-lignin char, acetate-lignosulfonates char, acetate-xylan char and acetate-cellulose char; and the third row is sodium hydroxide-lignin char, sodium hydroxide-lignosulfonates char, sodium hydroxide-xylan char and sodium hydroxide-cellulose char.

As mentioned earlier, swelling is a characteristic behavior determined by the composition of black liquor. Meanwhile, it's also affected by various operating parameters (e.g., temperature, pressure and atmosphere) during the pyrolysis process. In general, black liquor begins to swell considerably in the temperature range of 180–270 °C. The maximum swollen volume of the heated solid can reach 40–60 times their initial volumes. As the reactor temperature rises from 300 to 500 °C, the amount as well as release rate of volatile matters will be enhanced, which leads to an increase in the degree of swelling (Gea et al. 2003). Meanwhile, the time required for reaching the maximum swollen volume is shortened with increasing reactor temperature. Under pressurized conditions, the swollen volume of black liquor during pyrolysis decreases in a logarithmic trend over the pressure range from 1 to 20 bar (Whitty et al. 2008). The swollen volume measured at 20 bars is about one-quarter of that measured at 1 bar. In addition, it can be observed that black liquor droplets swell less in oxidizing (O_2) or gasifying (CO_2 and/or H_2O) atmospheres than those in an inert (N_2) atmosphere, which is mainly due to the destructive effects of oxidizing gases (O_2, CO_2 and H_2O) surrounding the rigid carbon surface. In the presence of O_2, CO_2 or H_2O, the maximum swollen volume is the balanced result of competition between release of volatile matter (constructive) and the gasification reactions (destructive) that occur on the solid surface (Gea et al. 2003).

The obvious swelling of black liquor during pyrolysis and gasification will inevitably affect the operation of the reactor. Swelling characteristics of black liquor droplets determine their trajectories in the reactor. These trajectories in turn have an influence on the capacity of the reactor. Therefore, it's necessary to pay special attention to the thermoplastic properties of black liquor in order to design and optimize the performance of reactors.

7.2.1.3 Inorganic Reactions During BLG

In view of the fact that black liquor is rich in inorganic salts, inorganic reactions will occur simultaneously during the devolatilization and gasification of black liquor. These reactions are closely related to the migration of three elements: carbon (C), sodium (Na) and sulfur (S).

Transformation of carbon is the basic route to determine product distribution and energy conversion in the whole thermochemical conversion process of black liquor. There are two types of carbon in black liquor char: organic fixed carbon and inorganic carbonate carbon (Na_2CO_3) (Sricharoenchaikul et al. 2003; Sricharoenchaikul et al. 2001). Formation of carbonate may occur via thermal decomposition of the sodium-containing organic compounds present in black liquor, e.g., sodium carboxylates derived from cellulosic matter and sodium phenolates derived from lignin (Cerfontain et al. 1987). In the presence of oxidizing gases, reactions of CO_2 and H_2O with sulfur-containing inorganic species also lead to the formation of carbonate (Li and Van Heiningen 1994). Some of these reactions are as follows [Eqs. (7.7)–(7.9)]: sulfide with CO_2 in the presence of water vapor, producing carbonate and H_2S; sulfide with CO_2 in the absence of water vapor, producing carbonate and COS; and thiosulfate, a prevalent sulfur-containing ion in black liquor, with CO_2 to produce carbonate, SO_2, and elemental sulfur.

$$Na_2S(s)+H_2O(g)+CO_2(g) \rightleftharpoons Na_2CO_3(s)+H_2S(g) \tag{7.7}$$

$$Na_2S(s)+2CO_2(g) \rightleftharpoons Na_2CO_3(s)+COS(g) \tag{7.8}$$

$$Na_2S_2O_3(s)+CO_2(g) \rightleftharpoons Na_2CO_3(s)+SO_2(g)+S(s) \tag{7.9}$$

Moreover, the thermochemical conversion process of black liquor will also be accompanied by the occurrence of decomposition reactions of Na_2CO_3. Among them, the most typical reactions are the carbothermal reduction reactions between Na_2CO_3 and organic carbon in black liquor char [Eqs. (7.10) and (7.11)]. These decomposition reactions are temperature sensitive and can be enhanced at higher temperatures. In most practical situations, these reactions take place during the char gasification stage. However, they may also occur when black liquor remains in an inert environment with high temperature and prolonged holding times.

$$Na_2CO_3(s) + C(s) \rightarrow 2Na(v) + CO(g) + CO_2(g) \quad (7.10)$$

$$Na_2CO_3(s) + 2C(s) \rightarrow 2Na(v) + 3CO(g) \quad (7.11)$$

As mentioned earlier, at least two types of competing reactions, i.e., formation and decomposition, are involved in the conversion of carbonate (Sricharoenchaikul et al. 2003). Some of the formation reactions are dominant at lower temperatures (700–800 °C), while the decomposition reactions become significant at higher temperatures (900–1000 °C).

Sodium and sulfur are important process elements that need to be recycled in the pulping process (Frederick and Hupa 1991). The kraft black liquor dry solid typically contains 17–20 wt.% sodium and 4–7 wt.% sulfur (Dayton and Frederick 1996; Li and Vanheiningen 1990b). Part of the sodium will be volatilized to form a fume during the gasification of black liquor, which in turn leads to fouling of heat transfer surfaces, plugging of gas passages and emission of particulates. As a result, serious potential problems such as equipment corrosion and environmental pollution will arise in integrated black liquor gasification combined-cycle power generation (Dayton and Frederick 1996). In addition, reactions of alkali metals involving the release of sulfur-containing compounds also occur during the black liquor gasification process [Eqs. (7.7)–(7.9)], which also make the subsequent gas purification and sulfur recovery key issues. Therefore, it will be of great significance to study the transformation pathways of sodium and sulfur.

As for the release of sodium, a little is volatilized as vapor species (e.g., Na, NaCl) during devolatilization at lower temperatures. During char gasification, elemental sodium vapor is produced via the reduction reactions of Na_2CO_3 with carbon [Eqs. (7.10) and (7.11)]. Therefore, sodium vaporization is closely related to the increase or decrease of Na_2CO_3 content. These reactions are extremely temperature dependent (Frederick et al. 1995): 80–90% of sodium in black liquor feed is retained in a residual solid at lower temperatures (700–800 °C) (Sricharoenchaikul et al. 2003), while the sodium loss from carbonate reduction proceeds more rapidly at higher temperatures.

Release of sulfur starts when decomposition of organic fractions in black liquor begins, which is also dependent on the reactor temperature (Frederick et al. 1995). During the char gasification, reactions between Na_2S with CO_2 and water vapor will form H_2S and COS [Eqs. (7.7) and (7.8)]. When water vapor is present, the rate of H_2S formation is much faster than that of COS. Formations of H_2S and COS are both strongly affected by chemical equilibrium. The equilibria for both reactions [Eqs. (7.7) and (7.8)] favor more sulfur release at lower temperatures and less at higher temperatures. Therefore, all of the Na_2S in black liquor char may be converted to H_2S during a low-temperature gasification process. The equilibria favor formation of Na_2S rather than H_2S and COS with the increase of temperature. The H_2S release rate is also influenced by the degree of sulfate reduction reactions [Eqs. (7.12) and (7.13)]. It can be seen that availability of Na_2S, limiting the rate of H_2S production, is controlled by the sulfate reduction reactions.

$$Na_2SO_4(s) + 2C(s) \rightarrow Na_2S(s) + 2CO_2(g) \quad (7.12)$$

$$Na_2SO_4(s) + 4C(s) \rightarrow Na_2S(s) + 4CO(g) \quad (7.13)$$

To summarize, the reaction system is very complex during the black liquor gasification process. It's accompanied by devolatilization of the feed, gasification of the char and inorganic salt reactions. The contribution of each of these categories of reaction to the total conversion of black liquor during gasification is presented schematically in Figure 7.5. In general, pyrolysis (devolatilization) reactions occur first, and then gasification and inorganic decomposition reactions proceed. The individual curves indicate the fraction of each type of reaction attributable to the total conversion.

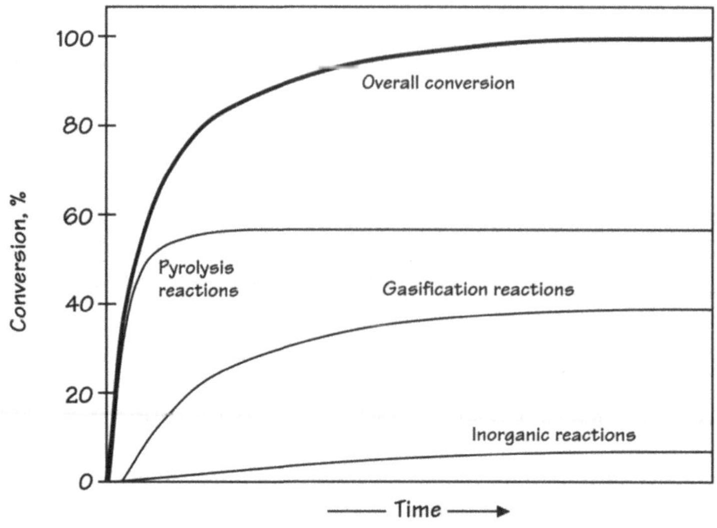

FIGURE 7.5 Schematic Diagram Indicating the Contribution of Various Types of Reactions to Overall Conversion of Black Liquor. Available from (Whitty 1997). A curve graph indicating the contribution of various types of reactions to overall conversion of black liquor. The Y axis states "Conversion, %" running from 0 to 100, and the X axis states "Time" with a horizontal rightwards arrow. Four curves are plotted, and each curve shows an upward trend from left to right. From top to bottom, these four curves represent "Overall conversion", "Pyrolysis reactions", "Gasification reactions" and "Inorganic reactions", respectively.

Source: Adapted from (Whitty 1997)

It can be indicated that both gasification and inorganic decomposition reactions cease at the end of the gasification stage because no more organic carbon is available.

7.2.2 Typical BLG Technologies

Research and development of BLG technology rose in the 1980s (Larson et al. 2007). Compared with the traditional Tomlinson recovery process, BLG technology has many advantages in terms of safety, flexibility, functionality and environmentally friendliness (Naqvi et al. 2010a).

Figure 7.6 shows the comparison between BLG technology and the traditional recovery process. It can be seen that the flexibility of the system based on BLG is greatly improved. The gasifier can be combined with a traditional power boiler to improve the comprehensive utilization rate of various biomass resources in the pulp and paper mill. The gas produced by gasification of black liquor can be used in various ways. On one hand, it can be used as a fuel gas to produce steam and electricity; on the other hand, it can be used as a syngas to synthesize biofuels or other high-value-added chemicals. Therefore, it can be expected that with the application of BLG technology, a pulp and paper mill coupled with a biorefinery system based on BLG has the potential to become an integrated factory for simultaneous production of pulp and biobased fuels and chemicals, just as shown in Figure 7.7.

A series of BLG technologies have emerged in the past decades. However, most of them were abandoned due to technical inferiority and failed to realize commercial operation (Whitty and Verrill 2004). According to the gasification temperature in the reactor, BLG can be divided into low-temperature gasification and high-temperature gasification. The operating temperature of low-temperature gasification is usually at 580–620 °C, which is lower than the melting points of inorganic salts in black liquor, while the operating temperature of the high-temperature gasification process is generally at 900–1000 °C, which will cause the formation of molten salts. Typical representatives for these two types of gasification technologies are the fluidized bed steam reforming

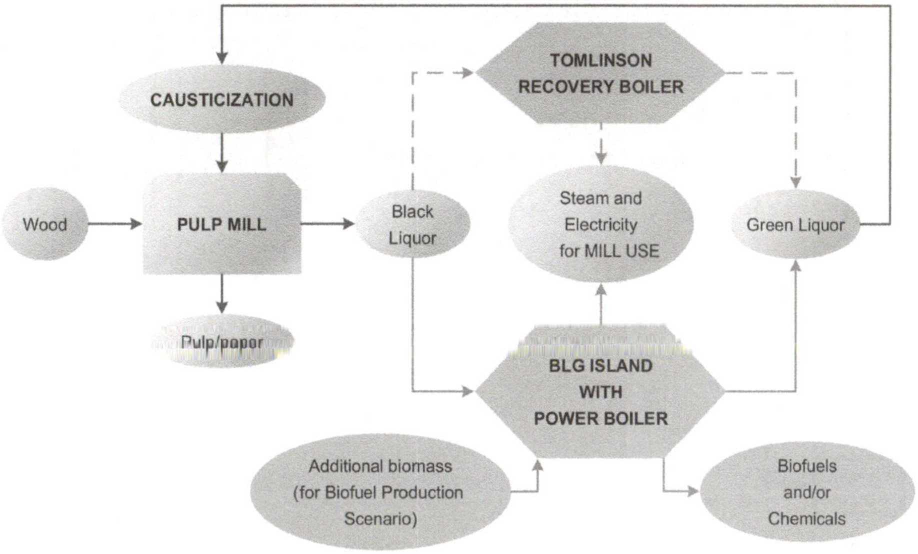

FIGURE 7.6 Comparison between BLG Technology and Traditional Recovery Process. "Wood" is input to the "Pulp mill", and "Pulp and/or paper" products along with "Black liquor" are output. There are two ways to realize the recycling of black liquor. One is to input black liquor into the "Tomlinson recovery boiler", and the other is to transport black liquor to the "BLG island with power boiler". Both paths can produce "Steam and electricity for mill use" and "Green liquor". The green liquor is then recycled to the pulp mill through a "Causticization" unit. "Additional biomass" is supplemented to the "BLG island with power boiler" for production of "Biofuels and/or chemicals".

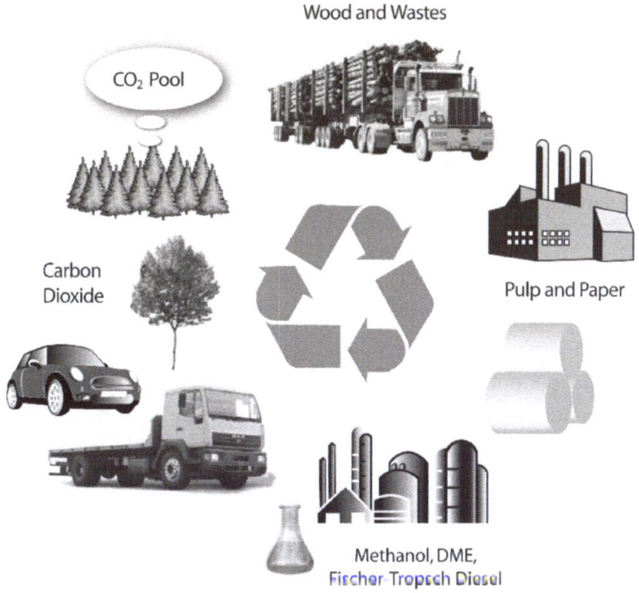

FIGURE 7.7 Schematic Drawing of the Bio-Refinery Cycle in a Pulp and Paper Mill. Available from (IEA Bioenergy ExCo54 Workshop 2007). A triangular cycle composed of clockwise arrows in the middle, surrounded by five types of scenes. From clockwise starting from top left the callouts are as follows: a forest stating "CO_2 Pool", a truck loaded with timber stating "Wood and Wastes", a factory building and three rolls of paper stating "Pulp and Paper", three sets of chemical facilities and a conical bottle containing liquid stating "Methanol, DME, Fischer-Tropsch Diesel", a truck, a car and a tree stating "Carbon Dioxide".

Source: Adapted from (IEA Bioenergy ExCo54 Workshop 2007)

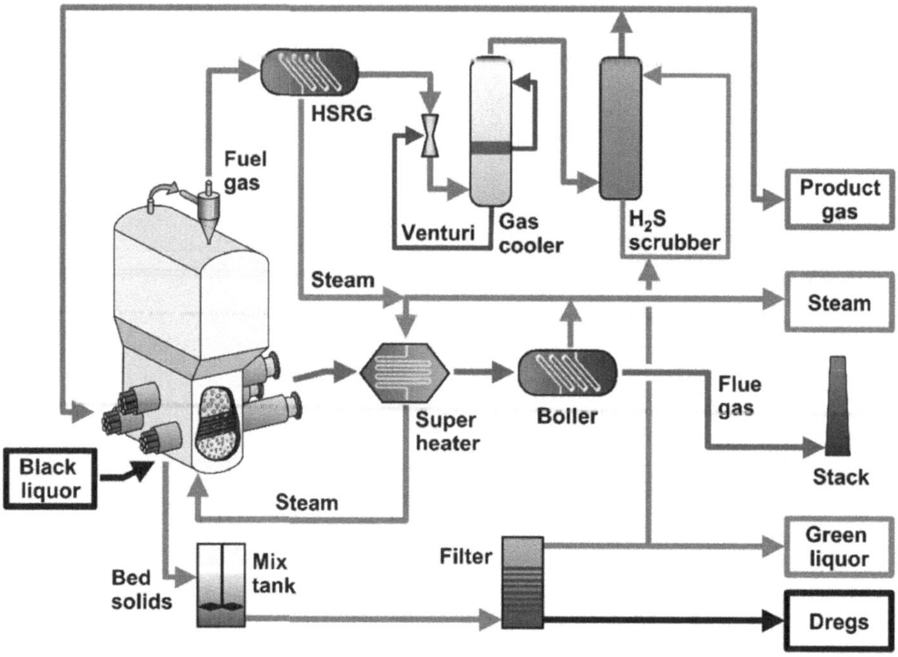

FIGURE 7.8 Conceptual Design of the MTCI Gasification System. Available from (Whitty and Verrill 2004). "Black liquor" is input to a reactor with tube bundles at the bottom. "Fuel gas" is output from the top of the reactor and sent to "HSRG". The gas from HSRG then passes through the "Gas cooler" and "H_2S scrubber" successively and becomes "Product gas" as the output. Part of the product gas is circulated to the reactor. "Steam" from HSRG can be output together with that produced by the "Boiler", and part of the steam returns to the reactor by passing through the "Super heater". The reactor is connected with the boiler via the super heater, and "Flue gas" from the boiler is then discharged from the "Stack". "Bed solids" from the bottom of the reactor enter the "Mix tank" and then are separated into "Green liquor" and "Dregs" through the "Filter". Part of the green liquor is transferred to the H_2S scrubber.

Source: Adapted from (Whitty and Verrill 2004)

process of Manufacturing and Technology Conversion International (MTCI) in the United States and the pressurized entrained flow gasification (PEFG) of Chemrec AB in Sweden. Both of them have been in pilot-scale operation or close to commercialization.

Figure 7.8 shows the conceptual design diagram of the MTCI gasification system. The core device of the system is a pulsed, indirect-heating, bubbling, fluidized bed steam reforming reactor. The bed temperature is maintained at 580–640 °C, which can avoid the melting of inorganic salts, thus getting away from the dangers of a smelt-water explosion. The black liquor is quickly dried and pyrolyzed on a special heating tube bundle to become char. Subsequently, char reacts with water vapor to produce hydrogen-rich gas (~73 vol.%).

This system has been demonstrated in Norampac and Georgia-Pacific pulp and paper mills (United States) (Patrick and Siedel 2003). It has many advantages, such as heating tube bundles with extremely high heat transfer efficiency, without hot moving parts, and higher combustion efficiency (Whitty 2005). However, relatively low carbon conversion is achieved due to the lower bed temperature in this process. The thermal efficiency of this system is about 70%, which has little advantage compared with a traditional recovery boiler (about 65%) (Naqvi et al. 2010a). Currently, the MTCI process has been abandoned for further promotion and application for a long time.

Figure 7.9 shows the schematic drawing of the BLG system developed by Chemrec AB. The core equipment of the system is a pressurized entrained flow reactor integrated with a quench unit. In this

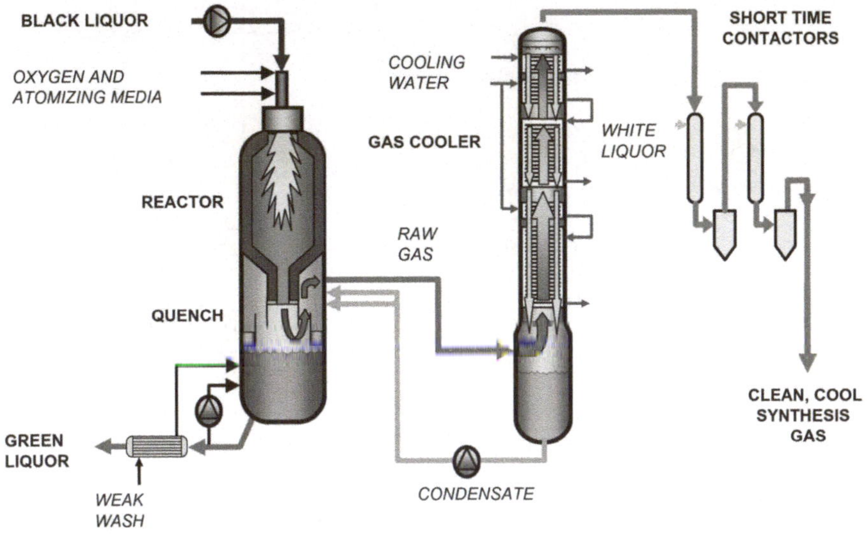

FIGURE 7.9 Schematic Drawing of the BLG System Developed by Chemrec AB. Available from (Wiinikka et al. 2012). "Black liquor" along with "Oxygen and atomizing media" are pumped into the "Reactor" from the top, and a jet combustion flame is generated inside the reactor. A shrinking structure is in the middle of the reactor, and some liquid is at the bottom of the reactor. The liquid at the bottom of the reactor flows out in the form of "Green liquor" after passing through the "Weak wash" unit, and part of the liquid is pumped back into the reactor for "Quench". "Raw gas" from the reactor passes through the "Gas cooler" unit and "Short time contactors" unit continuously, and then "Clean, cool synthesis gas" is output as the product. There are input and output cycles of "Cooling water" and "White liquor" in the upper middle position of the gas cooler. The liquid at the bottom of the gas cooler is pumped into the reactor after passing through a "Condensate".

Source: Adapted from (Wiinikka et al. 2012)

system, heavy black liquor with a solid content of 72.5–80.0 wt.% is pumped and atomized by air or oxygen (as a gasification agent) from the top of the gasification reactor and then injected into the furnace. The atomized droplets will be partially combusted to provide the necessary energy for the entire reactor. Generally, the pressure in the reactor is about 3.0 MPa and the temperature is 950–1000 °C. Residual molten inorganic salts generated from the gasification zone fall into the quench unit to form green liquor. The synthesis gas generated by gasification can be used in the subsequent combined cycle system for power generation or a catalytic conversion system to produce biofuels after condensation and purification.

This high-temperature and high-pressure BLG process has been demonstrated for more than 25,000 h at a 3 MWth pilot plant in Piteå, Sweden, which was commissioned in 2005. Currently, it's owned and run by Luleå University of Technology (LTU) in Sweden (Furusjö et al. 2014). The system has reached a higher industrial standard and process control level. The high operating temperature of this process can make the carbon conversion reach a high level (Jafri et al. 2016). However, the corrosion of equipment materials is inevitable due to the formation of molten salts, which is one of the key issues that needs to be solved urgently.

Comparing the BLG system of Chemrec (LTU) and the traditional recovery system, it can be found that both of them employ limestone causticization to convert green liquor (Na_2CO_3) into white liquor ($NaOH$). The subsequent calcination of lime mud ($CaCO_3$) is an energy-intensive process. In response to this problem, BLG technology with direct causticization using TiO_2 is developed to eliminate the energy consumption of calcination (Le et al. 2000; Naqvi et al. 2012a; Nohlgren et al. 2003; Pels et al. 1997). A system flow chart of this integrated system is shown in Figure 7.10. It can be seen that direct causticization with TiO_2 and the gasification of black liquor can be carried out

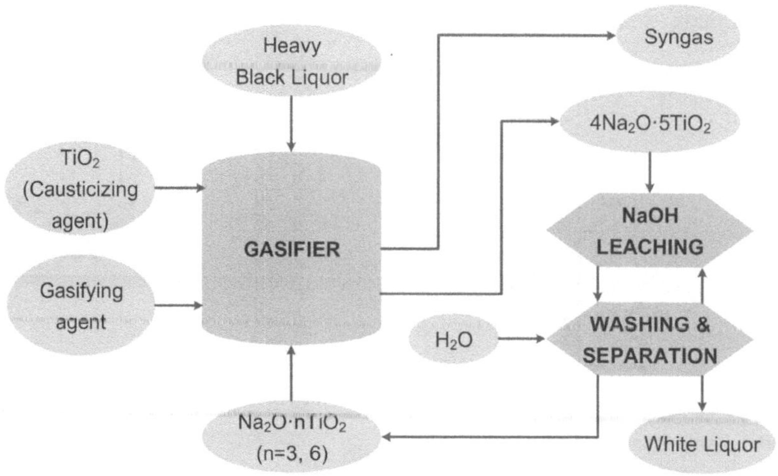

FIGURE 7.10 System Flow Chart of BLG with Direct Causticization Using TiO_2. Three materials, "Heavy black liquor", "Titanium dioxide (Causticizing agent)" and "Gasifying agent", are transported into the "Gasifier". Two streams, "Syngas" and "Sodium penta-titanate", are output from the gasifier'. Through "Sodium hydroxide leaching" along with "Washing and separation" units with the addition of "Water", sodium penta-titanate is then transformed into "White liquor" and "Sodium tri-titanate and/or Sodium hex-titanate". Sodium tri-titanate and/or sodium hex-titanate is transported to the gasifier for cyclic utilization.

in the gasifier simultaneously, which simplifies the process flow of the system obviously. Relevant research results show that the power generation efficiency can be increased to 30–35%, which is 2–3 times that of the traditional recovery process (10–14%) when the limestone causticization section is replaced by direct causticization using TiO_2 (Dahlquist and Jones 2005).

During the direct causticizing process using TiO_2, the reactions are mainly divided into two steps: the decarburization reactions in the gasification reactor and the hydrolysis reactions in the subsequent leaching device. Here, the decarburization reactions can also be directly referred to as direct causticizing reactions [Eqs. (7.14)–(7.16)]. In these reactions, Na_2CO_3 combines with TiO_2 to form $Na_2O{\cdot}nTiO_2$ with the release of CO_2. In the hydrolysis reactions [Eqs. (7.17) and (7.18)], $Na_2O{\cdot}nTiO_2$ is hydrolyzed into NaOH and $Na_2O{\cdot}3TiO_2$. NaOH is reused in the cooking section of pulping, while $Na_2O{\cdot}3TiO_2$ is recycled for the decarbonization reactions of Na_2CO_3. Studies have shown that direct causticizing reactions [Eqs. (7.14)–(7.16)] can be carried out at lower temperatures. The decarburization reactions of Na_2CO_3 can be completed at 680 °C with a residence time of 60 min (Chen and Van Heiningen 2006). Therefore, direct causticization of TiO_2 can be applied to lower-temperature gasification processes. It should be noted that the physical contact (i.e., mixing conditions) between Na_2CO_3 and TiO_2 or $Na_2O{\cdot}3TiO_2$ has significant impacts on the conversion rate and completion degree of direct causticization (Nohlgren et al. 2000). Through the balance calculation, it can be found that the decarburization reaction of Na_2CO_3 can be completed when the molar ratio of TiO_2/Na_2O reaches 0.5. Depending on the molar ratio of TiO_2/Na_2O, different types of $Na_2O{\cdot}nTiO_2$ are formed (Nohlgren and Sinquefield 2004).

Decarburization reactions

$$3TiO_2(s)+Na_2CO_3(l)\rightarrow Na_2O\cdot 3TiO_2(s)+CO_2(g) \quad (7.14)$$

$$Na_2O\cdot 6TiO_2(s)+Na_2CO_3(l)\rightarrow 2(Na_2O\cdot 3TiO_2)(s)+CO_2(g) \quad (7.15)$$

$$5(Na_2O\cdot 3TiO_2)(s)+7Na_2CO_3(l)\rightarrow 3(4Na_2O\cdot 5TiO_2)(s)+7CO_2(g) \quad (7.16)$$

Hydrolysis reactions

$$3\left(4Na_2O\cdot 5TiO_2\right)(s)+7H_2O(l)\rightarrow 14NaOH(aq)+5\left(Na_2O\cdot 3TiO_2\right)(s) \quad (7.17)$$

$$2\left(Na_2O\cdot 3TiO_2\right)(s)+H_2O(l)\rightarrow 2NaOH(aq)+Na_2O\cdot 6TiO_2(s) \quad (7.18)$$

7.2.3 Process Performance of BLG

Based on the pressurized entrained-flow high-temperature black liquor gasification (PEHT-BLG) process described earlier (developed by Chemrec AB and run by LTU), pilot-scale experiments were carried out with a series of operating points (OP) under relatively stable working conditions (Jafri et al. 2016). The process operating variables are listed in Table 7.3, and each OP was run for about 24 h. Composition along with LHV of the gaseous products (syngas) from PEHT-BLG process are shown in Table 7.4. It can be found that the gas production characteristics under different operating conditions (OP–1→OP–5) are at a steady level. The syngas is mainly composed of H_2, CO and CO_2, and the total concentration of these three components in the syngas is more than 95.0 vol.% (96.1 vol.% on average). In addition, there is a small amount of CH_4 (0.5–1.2 vol.%) and C_2–C_3

TABLE 7.3
Values of the Main Variable Process Parameters in the Pilot-Scale BLG Plant

Process Parameters	OP-1	OP-2	OP-3	OP-4	OP-5
Black liquor (BL) feeding rate (kg/h)	1250	1250	1060	1060	1250
O_2/BL (kg/kg)	0.287	0.295	0.284	0.288	0.285
λ[a]	0.491	0.505	0.486	0.493	0.49
Gasifier pressure (bar)	28.7	28.7	27.4	24.1	28.9
Reactor temperature (°C)	1008	1036	984	1001	1004

[a] Ratio of actual O_2 supplied to the stoichiometric requirement for a given mixture.
Source: Available from (Jafri et al. 2016)

TABLE 7.4
Composition and LHV of the Gaseous Products Derived from PEHT-BLG Process

	OP-1	OP-2	OP-3	OP-4	OP-5
Main Gas Components (vol.%)					
H_2	35.3	35.2	35.3	35.4	35.4
CO	25.3	25.7	25.8	25.9	25.0
CO_2	35.3	35.7	34.5	35.2	35.7
Minor Gas Components (vol.%)					
CH_4	1.15	0.47	1.14	0.84	1.10
H_2S	1.62	1.49	1.65	1.56	1.70
N_2	1.27	1.15	1.62	1.06	0.95
Trace Components (ppm)					
COS	68	70	72	64	64
C_2-C_3	8.2	1.3	8.9	5.8	20.7
BTX[a]	125	16	127	92	150
LHV (MJ/Nm³)	7.74	7.55	7.80	7.71	7.73

[a] That is, benzenes, including benzene (C_6H_6), toluene (C_7H_8) and xylene (C_8H_{10}).
Source: Adapted from (Jafri et al. 2016)

and BTX at a ppm level in the produced gas. As the gas composition is stable, LHV of the syngas under various operating conditions also remains at the level of 7.5–7.8 MJ/Nm^3. Moreover, in view of the high sulfur content (6.20 wt.%, d) in the raw sulfate black liquor, a considerable amount of H_2S (1.5–1.7 vol.%) along with trace amounts of COS (64–72 ppm) are also contained in the syngas. Considering the downstream utilization of syngas and the recycling of sulfur in the pulping process, purification of the sulfur-containing gas becomes an indispensable unit.

The energy conversion efficiency of the BLG process can be evaluated by carbon conversion efficiency and cold gas efficiency. The carbon conversion efficiency (η_C) is calculated based on the mass balance of organic carbon in the solid before and after the gasification reactions, and cold gas efficiency (CGE) is defined as the ratio between the chemical energy of syngas produced by gasification and the chemical energy of the raw black liquor material. It can be noted that these two parameters are directly proportional to the energy conversion during the gasification process. Figure 7.11 shows the carbon conversion and CGE under different operating conditions calculated based on the previous definitions. As shown in Figure 7.11, under the working conditions of temperature ~1000 °C, pressure ~28 bar and $\lambda \approx 0.5$, the carbon conversion of the PEHT-BLG process could reach a high level of 98.3–99.2%, and the CGE was also maintained in the range of 52.3–54.5%. As mentioned earlier, this pilot plant could be operated continuously and stably and achieve high energy conversion efficiency while maintaining the output quality of syngas.

7.2.4 Biorefinery System Based on BLG

In the current pulp and paper industry, abundant and diverse biomass resources, such as black liquor, bark and wood residues, are available for centralized comprehensive utilization through gasification technology. Therefore, the existing pulp and paper mills have the potential to be upgraded into a modern integrated biorefinery plant based on BLG for coproducing electricity and/or biofuels in addition to pulp and paper products. According to the output type of products, the integrated biorefinery system based on BLG can be divided into power output mode and biofuel output mode,

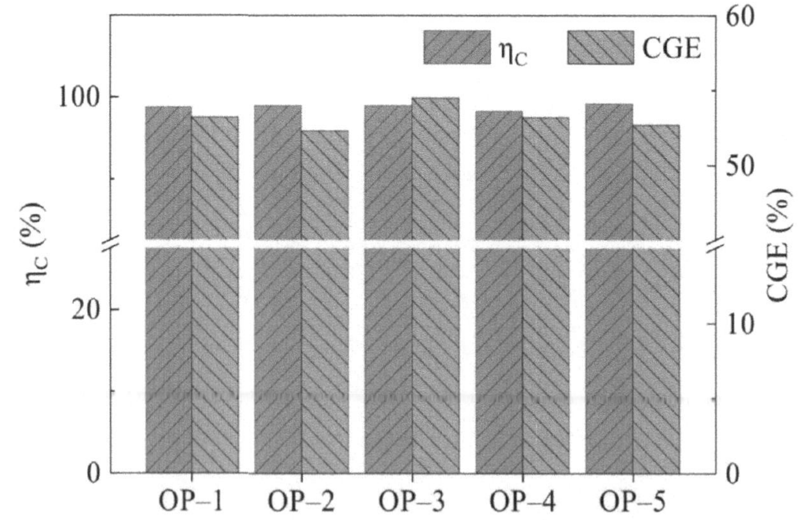

FIGURE 7.11 Carbon Conversion Efficiency of PEHT-BLG Process. Adapted from (Jafri et al. 2016). A column chart showing the carbon conversion (η_C) and cold gas efficiency (CGE) of the PEHT-BLG process. The X axis lists five scenarios at equal intervals, from left to right, OP–1, OP–2, OP–3, OP–4 and OP–5. There are both left and right sides of the Y axis. The left Y axis reads "η_C, %" running from 0 to 100, and the right Y axis reads "CGE" running from 0 to 60. There are two columns for each scenario: the left column reads η_C, while the right column reads CGE.

respectively. For the purpose of power output, a BLG combined cycle (BLGCC) system can be employed to achieve cogeneration of heat and electricity efficiently. If the target products are biofuels—such as hydrogen (H_2), synthetic natural gas (SNG, i.e., CH_4), methanol (MeOH), dimethyl ether (DME) and Fischer-Tropsch synthesis (F-T) fuels—syngas generated by BLG can be used as a raw material to efficiently produce these biofuels through continuous purification, separation and catalytic conversion.

7.2.4.1 BLG Combined Cycle

Figure 7.12 shows a schematic representation of a BLGCC system. The crude gas produced by the gasifier is cleaned by cooling, desulfurization (H_2S) and other purification operations, and the clean gas is transmitted to the gas turbine for combustion. The heat in high-temperature flue gas from the gas turbine can be recovered in a heat recovery steam generator (HRSG), which can synchronously generate high-pressure, medium-pressure and low-pressure steam. Since the high-pressure steam generated from HRSG is relatively limited and insufficient, part of the high-pressure steam still needs to be supplemented by a biomass boiler fired with existing bark and additional biomass. High-pressure steam from both HRSG and the biomass boiler can be first delivered to a back-pressure steam turbine for extra power generation before being transported to the pulping process. The thermal efficiency of the whole process is significantly improved through the multistage gradient utilization of steam. Therefore, BLGCC integrated with a conventional pulp mill has the potential to establish a modern combined heat and power (CHP) plant with higher energy efficiency than the traditional recovery process (Berglin and Berntsson 1998; Eriksson and Harvey 2004).

Table 7.5 lists the mass and energy flow of the traditional recovery process and BLGCC system (Consonni et al. 1998). As shown in Table 7.5, compared with a traditional recovery process based on a Tomlinson boiler, more biomass must be supplemented in these BLGCC systems in order to

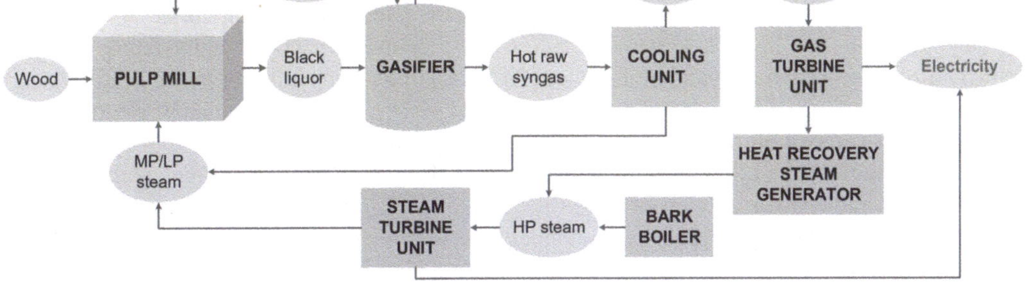

FIGURE 7.12 Simplified Flow Diagram of a BLGCC System. "Wood" is input to the "Pulp mill", and "Black liquor" is produced from the pulp mill. Both black liquor and the "Gasifying agent" flow into the "Gasifier". "Green liquor" and "Hot raw syngas" are output from the gasifier. The hot raw syngas becomes "Cooled syngas" through a "Cooling unit". The cooling unit can also recover "MP and/or LP steam" to the pulp mill. Then, the cooled syngas flows into a "Hydrogen sulfide absorber" and becomes "Clean syngas". The "Sulfur-rich liquid" from the hydrogen sulfide absorber is mixed into the green liquor and then converted to "White liquor" through a "Causticization unit". The white liquor is then recycled to the pulp mill. The clean syngas is introduced into the "Gas turbine unit" with the output of "Electricity". A downstream "Heat recovery steam generator" is connected with the gas turbine unit. It can produce "HP steam" together with a "Bark boiler". HP steam is transferred to a "Steam turbine unit" to generate electricity and becomes MP and/or LP steam, which is recycled by the pulp mill.

TABLE 7.5
Mass and Energy Flow of Traditional Recovery Process and BLGCC Systems (Based on HHV)

	Traditional Recovery	BLGCC		
Variable	Tomlinson recovery boiler	LPLT, air-blown	HPHT, air-blown	HPHT, O_2-blown
Pulp capacity (ADt/day)	1241	1098	1251	1307
Black liquor mass flow (tDS/day)	2159	1911	2177	2274
BL energy flow (MW)	359	318	362	378
Biomass energy flow (MW)	91.21	177.79	123.94	237.50
Total fuel input (MW)	450.21	495.79	485.94	615.50
Total process steam (MW)	234.12	207.15	236.01	246.58
Gross power output (MW)	48.40	136.70	104.70	151.80
Auxiliary consumption (MW)	1.64	22.04	4.19	19.77
Net power output (MW)	46.76	114.66	100.51	132.03
Net excess output (MW)	12.80	84.70	66.30	96.30
Fuel to electricity (%)	10.39	23.13	20.68	21.45
Fuel to steam (%)	52.00	41.78	48.57	40.06
Total efficiency (%)	62.39	64.91	69.25	61.51
Electricity/steam	0.20	0.55	0.43	0.54

Source: Adapted from (Consonni et al. 1998)

meet the steam demand of the pulp mill itself. In all scenarios, including the Tomlinson-based system, the net power output can be realized—that is, electricity production is in excess of the reference pulp mill's own demand. Specifically, BLGCC systems produce more extra power (66.3–96.3 MW) than the Tomlinson-based system (12.8 MW). As for the product output of the systems, it can be found that the power output ratio (electricity/steam) of BLGCC systems is significantly higher than that of the traditional recovery process. As mentioned earlier, BLGCC systems have the ability to output more excess electricity while ensuring the steam demand of the pulp mills.

Comparing the two BLGCC systems based on low-pressure (2 bar), low-temperature (700 °C) (LPLT) air-blown gasification and high-pressure (25 bar), high-temperature (950 °C) (HPHT) air-blown gasification, it can be found that the LPLT-based system provides the highest electrical efficiency (23.13%) among all the systems considered here. This is mainly because the lower gasifier outlet temperature of the LPLT system enables most of the gas output to enter the combined cycle for utilization, rather than only exporting to the Rankine cycle with a larger irreversible loss. In general, with the increase of gasifier outlet temperature, a larger fraction of its output will be recovered in the form of steam, which can only be utilized in the Rankine cycle. In addition, the essential pressurization units of the HPHT system also bring additional power consumption. Therefore, the LPLT system enjoys the advantage of electrical efficiency.

Comparing the two combined cycles using air-blown and O_2-blown gasification under HPHT conditions, it can be seen that the electricity/steam of the O_2-blown system is higher than that of the air-blown system, which means most of the energy input is converted into electricity rather than into steam in the O_2-blown system. This is related to the assistance of the combined cycle, which can make more efficient use of low-temperature heat. Since a necessary oxygen preparation unit is needed in the O_2-blown system, more accessional energy is consumed for the system. It's worth noting that in order to meet the same process steam demand, the O_2-blown system consumes more biomass (up to 237.50 MW) than the air-blown system (123.94 MW). However, the net excess electricity production of the O_2-blown system (96.30 MW) is much higher than that of the air-blown system (66.30 MW).

In summary, the BLGCC system has the potential to output more electricity while meeting the process steam demands for the pulp and paper mills. Based on the coupling and integration of gasification technology and combined cycle, the ratio of electricity to steam of the system can be greatly adjusted, showing higher flexibility of product output. At present, in order to realize the industrialization practice of the BLGCC system, research and development on optimization should be focused on the key process technologies, such as strict cleanup of gaseous products from BLG to meet the quality requirements of the gas turbine, regulation of gas compositions to ensure its stable combustion in gas turbine, heat recovery and utilization in all process streams, gas and residual solids, the cost-effective recovery of chemicals from residual solids and the reduction of heat loss in the whole pulp and paper mills

7.2.4.2 BLG for Biofuel Production

A series of biofuels, such as H_2, SNG, MeOH, DME, and F-T fuels, can be produced from the integrated biorefinery systems based on BLG. Figure 7.13 shows a generic polygeneration of biofuels based on BLG in a pulp mill. As shown in Figure 7.13, before the clean syngas is converted into a variety of biofuels, the corresponding routes need to go through some conventional gas processing units, including CO shift, methanation, catalytic conversion and CO_2 separation, etc. Under the

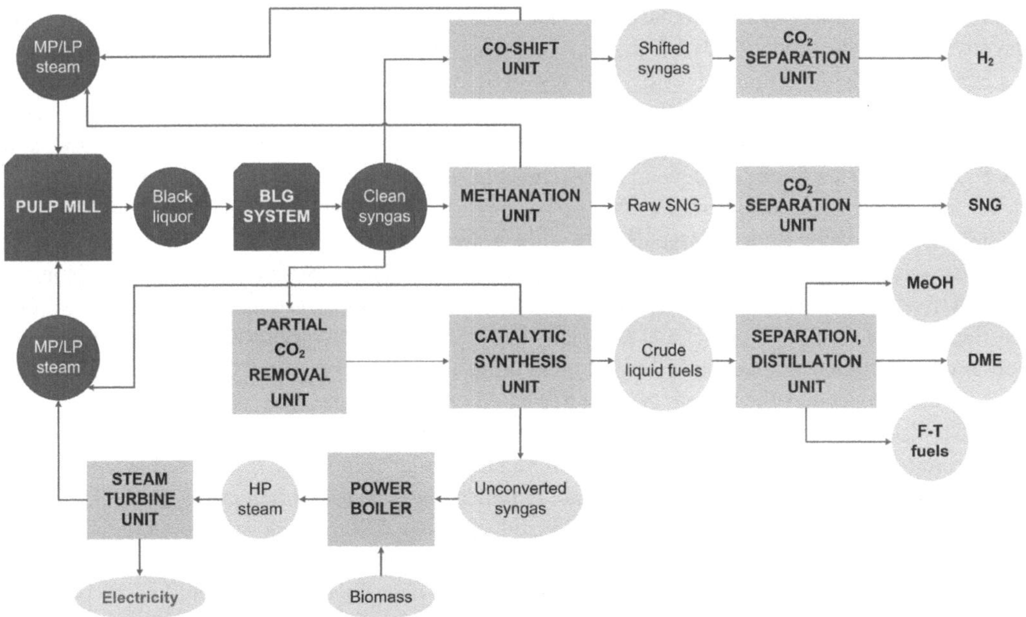

FIGURE 7.13 Potential Polygeneration of Biofuels Based on BLG Process in a Pulp Mill. Adapted from (Naqvi et al. 2010a). "Black liquor" from the "Pulp mill" is transferred to a "BLG system" to produce "Clean syngas". It can be transformed into "Shifted syngas" through a "CO-shift unit", or "Raw SNG" through a "Methanation unit", or "Crude liquid fuels" through the continuous process of a "Partial CO_2 removal unit" and "Catalytic synthesis unit". All these operating units can generate "MP and/or LP steam" synchronously, which can be recycled to the "Pulp mill" for utilization. The "Shifted syngas" and "Raw SNG" can output "Hydrogen" and SNG, respectively, through a "CO_2 separation unit". Crude liquid fuels can produce products such as "Methanol", "Dimethyl ether" and "F-T fuels" via a "Separation, distillation unit". Some "Unconverted syngas" from the catalytic synthesis unit can be transferred to the "Power boiler" with "Biomass" input to produce "HP steam". HP steam is then transferred to a downstream "Steam turbine unit" to generate "Electricity" and becomes "MP and/or LP steam", which is recycled by the pulp mill.

TABLE 7.6
Mass and Energy Flow During Biofuel Production Based on BLG

Product	H_2	SNG	MeOH	DME
Pulp capacity (ADt/day)	1000	1000	1000	1000
BL flow (tDS/day)	1700	1700	1700	1700
BL flow energy content (LHV, MW)	243.5	243.5	243.5	243.5
Available bark flow (tDS/day)	181	181	181	181
Available bark flow energy content (LHV, MW)	40.5	40.5	40.5	40.5
Biomass (MW) import/export (-/+)	-61.8	-36.7	-65.2	-69.0
Electricity (MW) import/export (-/+)	-28.4	-21.8	-28.0	-26.5
Biofuel production (LHV, MW)	130.5	162.2	135.5	131.9
Fuel to product efficiency[a] (FTPE, %)	46.0	56.9	47.4	46.0
Ref.	(Andersson and Harvey 2006)	(Naqvi et al. 2012b)	(Andersson and Harvey 2006)	(Naqvi et al. 2010b)

[a] Fuel to product efficiency (FTPE) = Exportable biofuel output / Total fuel input × 100

condition of biofuel output, compensation measures should be implemented to meet the power and steam demands for the operation of the pulp mill. Here, a power boiler is introduced into the system in order to solve this energy demand issue. Bark and other wood residues, along with part of the unconverted syngas produced by BLG, can be used as supplementary fuels for boiler combustion.

Table 7.6 shows the estimated results of energy and mass flow for different biofuel production systems based on BLG. It can be found that in the scenarios with H_2, SNG, MeOH and DME as output targets, the SNG output system has the highest fuel-to-product efficiency (FTPE) (up to 56.9%), and the FTPE of the other three product output systems is basically maintaining at the level of 46.0–47.5%. In the application scenarios listed in Table 7.6, the syngas produced by BLG is completely used to synthesize biofuels. which leads to large deficits in the power and steam required by the pulping process and the system installations for producing biofuels. Consequently, imports of more additional biomass and electricity are necessary to guarantee the normal operations of the systems. Just as shown in Table 7.6, there are net inputs of biomass and electricity in the production of various biofuels. It's known that consuming additional biomass and electricity will bring about an increase in CO_2 emissions. However, the biofuels produced by the biorefinery systems based on BLG can be used as alternative automotive fuels to replace the traditional gasoline derived from fossil fuels. From this perspective, the CO_2 emissions from traditional gasoline combustion will be partially or completely suppressed by application of biofuels. In addition, in the BLG integrated biorefinery systems for biofuel production, CO_2 can be captured on-site by applying the carbon capture and storage (CCS) technology. Therefore, when considering the pulp and paper mill and the downstream market for biofuels application as a continuous integrated system, so-called "well-to-wheels" (WTW), the biorefinery system plant has the potential to make a substantial reduction of CO_2 emissions on the basis of efficient production of green biofuels.

7.3 SUPERCRITICAL WATER GASIFICATION (SCWG) OF BLACK LIQUOR

7.3.1 Overview of SCWG Process for Black Liquor

Supercritical water gasification (SCWG), an innovative thermochemical method, takes advantage of the special properties of water beyond its critical point (374 °C, 22.1 MPa) and can convert wet biomass and organic wastes into hydrogen-rich gases. It has attracted extensive attention since it was developed in the 1970s, especially for the organic wastewater, because it can obtain hydrogen-rich gases and eliminate the contamination to the environment simultaneously (Cao et al. 2011). The main advantages of SCWG are as follows (Fedyaeva et al. 2019):

1. SCWG conversion is carried out in a closed-loop system. Some hazardous emissions from combustion, such as SO_2, NOx and particulates, will be eliminated in SCWG, even when the feedstock contains sulfur and nitrogen elements, which shows strong environmental friendliness.
2. Most of the organics and gases can be dissolved in supercritical water, so the mass transfer resistance between heterogeneous phases is decreased sharply, which improves the reaction rate and process efficiency.

It has been confirmed that SCWG becomes more efficient than the gasification process when the feedstock moisture is more than 40 wt.% (Yoshida et al. 2003). Since the water content of weak black liquor derived from the cooking and washing sections of the pulping process is generally higher than 80 wt.%, SCWG technology is considered as a promising treatment method for black liquor (De Blasio et al. 2016; Guo et al. 2010; Sricharoenchaikul 2009). As is known to all, evaporation of weak black liquor is indispensable for the traditional recovery process as well as BLG technology. However, this step is very energy-intensive and consumes about one-third of the steam produced from the boiler (Magdeldin and Järvinen 2020). From this point of view, SCWG is more suitable for black liquor recovery in that it has the potential to reduce the process energy consumption. It's supposed that the water in weak black liquor can become a reaction medium and an active chemical reagent in the SCWG process (Akiya and Savage 2002; Brunner 2009; Hawangchu et al. 2017). Moreover, SCWG can take advantage of the inherent alkali in black liquor to carry out an in situ catalytic process to promote the water-gas shift (WGS) reaction [Eq. (6)] in supercritical water, which can improve H_2 production and decrease CO fraction in the gaseous products (Muangrat et al. 2010; Watanabe et al. 2003).

However, some drawbacks such as clogging and corrosion of the reactor are present in the SCWG process of black liquor. As there is a large amount of alkali metals in black liquor, salts become insoluble precipitates in the supercritical water inside the reactor. Figure 7.14 shows the practical effects of salt formation in the reactor inlet during the SCWG of black liquor at 600 °C. It can be found that the blockage is very serious during the continuous operation. Furthermore, alkali salts in black liquor also provoke corrosion under SCWG conditions.

(a) (b) (c)

FIGURE 7.14 Salts Formation (a) in the Reactor Inlet and the Clogging Problem in the Channel Inside the Bolt (b–c) during SCWG of Black Liquor at 600 °C. Available from (De Blasio et al. 2019). Three physical photos labeled (a), (b) and (c) are arranged from left to right. There is a glass beaker containing black massive solid residue in photo (a), and photo (b) is the axial observation photo of a stainless reactor, and blocked solid residues exist in the pipe. Photo (c) shows a stainless reactor placed obliquely, with bulging blocked solid residues at its outlet.

Source: Adapted from (De Blasio et al. 2019)

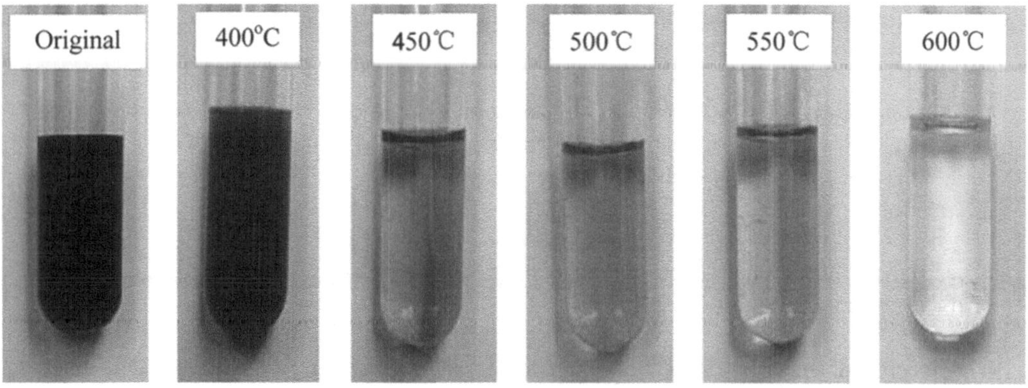

FIGURE 7.15 The Photos of the Treated Black Liquor at Different Temperatures (25 MPa). Available from (Cao et al. 2011). Six photos are arranged from left to right on the horizontal line. In each photo, there is a round bottom sample tube of the same size, which contains an approximate amount of liquid. From left to right, the photos are marked "Original", "400 °C", "450 °C", "500 °C", "550 °C" and "600 °C", and the color of the liquid gradually changes from black to light, and finally becomes transparent.

7.3.2 Process Performance of SCWG for Black Liquor

The main process parameters are temperature, residence time and concentration of black liquor during the SCWG process in a continuous flow system. It has been found that gasification reactions will be enhanced with an increase of temperature (400–600 °C) and residence time (4.94–13.71 s) and a decrease of feeding concentration (Cao et al. 2011). The main product gases are H_2, CH_4 and CO_2 with a small amount of CO, C_2H_4 and C_2H_6. Yields of H_2, CO_2 as well as the total gas yield will be improved with increasing temperature. H_2 fraction is always at a high level (40.26–61.02 vol.%), while CO fraction is maintained at a low level (<1.51 vol.%) even at higher temperatures, which is probably because the alkali in black liquor plays a catalytic role in the WGS reaction [Eq. (6)]. The gasification efficiency can be improved from 55.67% to 68.66%, and H_2 gasification efficiency increases from 84.90% to 200.23% when the black liquor is diluted from 9.5 to 3 wt.%.

The removal efficiency of chemical oxygen demand (COD) will be achieved to 88.69% at 600 °C during the SCWG process of black liquor. As shown in Figure 7.15, the colors of the initial black liquor and the liquids after reaction at different temperatures are compared. It can be seen that the color of the liquids after SCWG treatment gradually change from dark to light with increasing reaction temperature, and the liquid treated at 600 °C is almost as clear as pure water. Inorganic alkali salts (mainly NaOH and Na_2CO_3) in black liquor are precipitated when the fluid temperature in the reactor is above 360 °C at the pressure of 25 MPa for their terribly low solubility in supercritical water. Under these reaction conditions, the liquid from the reactor contains only a small amount of sodium (Na) with the pH in the neutral region (6.4–8.0).

As mentioned earlier, SCWG is an effective route for producing hydrogen-rich gases from black liquor, as well as eliminating environmental and health risks associated with black liquor concurrently. However, it should be noted that alkali precipitation is inevitable during the SCWG process of black liquor with the extension of operation time. As a result, the reactor will be up against the risk of blockage, and its operation stability is still challenged.

7.4 CONCLUSIONS AND PERSPECTIVES

BLG is a promising alternative technology for energy and chemical recovery of black liquor in the pulp and paper industry. The gas produced by BLG can be used downstream as a fuel gas to establish a CHP plant, or used as a syngas to synthesize various types of biofuels, such as H_2,

SNG, methanol, DME and F-T fuels. Through integration of BLG and various biofuel production routes, the current pulp and paper mills can be upgraded to sustainable modern biorefinery plants, which has the potential to bring higher economic and environmental benefits. Simultaneously, CO_2 resource utilization as well as CCS are necessary to be introduced to the biorefinery system to achieve the carbon neutralization from WTW.

In addition, as black liquor is a special biomass resource with high water content, advantages of the SCWG process should be fully exploited. SCWG of black liquor is an effective way to produce hydrogen-rich gas, with elimination of the energy-intensive evaporation section. However, clogging and corrosion of the reactor caused by deposition of inorganic salts in black liquor are key technical bottlenecks limiting the continuous operation of SCWG. Therefore, extensive efforts on reactor structural design, materials with high temperature and corrosion resistance and optimization of operating mode are prerequisites for the future large-scale application of SCWG for black liquor.

In future work, technical-economic evaluations for the biorefinery systems based on BLG and SCWG should be carried out continuously to provide corresponding theoretical references for policy formulations and commercial applications.

ACKNOWLEDGMENTS

This work was performed with the support of the Fundamental Research Funds for the Central Universities (DUT21TD103), Liaoning Revitalization Talent Program (XLYC1908033) and Dalian Innovation Team Support Plan in Key Areas (2019RT10).

REFERENCES

Akiya, N., Savage, P.E. 2002. Roles of water for chemical reactions in high-temperature water. *Chemical Reviews*, **102**(8), 2725–2750.

Alen, R. 1994. Swelling behaviour of kraft black liquor and its organic constituents. *Bioresource Technology*, **49**(2), 99–103.

Andersson, E., Harvey, S. 2006. System analysis of hydrogen production from gasified black liquor. *Energy*, **31**(15), 3426–3434.

Andersson, J., Lundgren, J., Furusjo, E., Landalv, I. 2015. Co-gasification of pyrolysis oil and black liquor for methanol production. *Fuel*, **158**, 451–459.

Azadi, P., Inderwildi, O.R., Farnood, R., King, D.A. 2013. Liquid fuels, hydrogen and chemicals from lignin: A critical review. *Renewable and Sustainable Energy Reviews*, **21**, 506–523.

Berglin, N., Berntsson, T. 1998. CHP in the pulp industry using black liquor gasification: Thermodynamic analysis. *Applied Thermal Engineering*, **18**(11), 947–961.

Bhattacharya, P.K., Parthiban, V., Kunzru, D. 1986. Pyrolysis of black liquor solids. *Industrial & Engineering Chemistry Process Design and Development*, **25**(2), 420–426.

Bhattacharya, P.K., Shrinath, A.S., Kunzru, D. 1985. Pyrolysis of black liquor. *Journal of Chemical Technology and Biotechnology. Chemical Technology*, **35 A**(5), 223–233.

Brunner, G. 2009. Near critical and supercritical water. Part I. Hydrolytic and hydrothermal processes. *The Journal of Supercritical Fluids*, **47**(3), 373–381.

Cao, C.Q., Guo, L.J., Chen, Y.A., Guo, S.M., Lu, Y.J. 2011. Hydrogen production from supercritical water gasification of alkaline wheat straw pulping black liquor in continuous flow system. *International Journal of Hydrogen Energy*, **36**(21), 13528–13535.

Cerfontain, M.B., Meijer, R., Kapteijn, F., Moulijn, J.A. 1987. Alkali-catalyzed carbon gasification in CO/CO_2 mixtures: An extended model for the oxygen exchange and gasification reaction. *Journal of Catalysis*, **107**(1), 173–180.

Chen, C., Alén, R., Lehtimäki, E., Louhelainen, J. 2017. A salt-induced mechanism for the swelling of black liquor droplet during devolatilization. *Fuel*, **202**, 338–344.

Chen, X., Van Heiningen, A.R.P. 2006. Kinetics of the direct causticizing reaction between sodium carbonate and titanium dioxide or sodium trititanate. *Journal of Pulp and Paper Science*, **32**(4), 245–251.

Connolly, T.S. 2006. *CO_2 pyrolysis and gasification of kraft black liquor char.* Orono, ME: The University of Maine.

Consonni, S., Larson, E.D., Kreutz, T.G., Berglin, N. 1998. Black liquor gasifier gas turbine cogeneration. *Journal of Engineering for Gas Turbines and Power–Transactions of the Asme*, **120**(3), 442–449.

Dahlquist, E., Jones, A. 2005. Presentation of a dry black liquor gasification process with direct causticization. *Tappi Journal*, **4**(6), 15–19.

Dayton, D.C., Frederick, W.J. 1996. Direct observation of alkali vapor release during biomass combustion and gasification .2. Black liquor combustion at 1100 degrees C. *Energy & Fuels*, **10**(2), 284–292.

De Blasio, C., De Gisi, S., Molino, A., Simonetti, M., Santarelli, M., Bjorklund-Sankiaho, M. 2019. Concerning operational aspects in supercritical water gasification of kraft black liquor. *Renewable Energy*, **130**(Jan.), 891–901.

De Blasio, C., Lucca, G., Özdenkci, K., Mulas, M., Lundqvist, K., Koskinen, J., Santarelli, M., Westerlund, T., Järvinen, M. 2016. A study on supercritical water gasification of black liquor conducted in stainless steel and nickel-chromium-molybdenum reactors. *Journal of Chemical Technology & Biotechnology*, **91**(10), 2664–2678.

Eriksson, H., Harvey, S. 2004. Black liquor gasification–consequences for both industry and society. *Energy*, **29**(4), 581–612.

Fedyaeva, O.N., Vostrikov, A.A., Shishkin, A.V., Dubov, D.Y. 2019. Conjugated processes of black liquor mineral and organic components conversion in supercritical water. *The Journal of Supercritical Fluids*, **143**, 191–197.

Frederick, W., Hupa, M. 1991. Evidence of sodium fuming during pyrolysis of black liquor. *Tappi Journal*, **74**(11), 192–194.

Frederick, W., Hupa, M., Uusikartano, T. 1994. Volatiles and char carbon yields during black liquor pyrolysis. *Bioresource Technology*, **48**(1), 59–64.

Frederick, W.J., Iisa, K., Wag, K., Reis, V., Boonsongsup, L., Forssen, M., Hupa, M. 1995. *Sodium and sulfur release and recapture during black liquor burning.* Corvallis, OR: Oregon State University.

Furusjö, E., Stare, R., Landälv, I., Löwnertz, P. 2014. Pilot scale gasification of spent cooking liquor from sodium sulfite based delignification. *Energy & Fuels*, **28**(12), 7517–7526.

Gea, G., Murillo, M.B., Arauzo, J., Frederick, W.J. 2003. Swelling behavior of black liquor from soda pulping of wheat straw. *Energy & Fuels*, **17**(1), 46–53.

Guo, L., Cao, C., Lu, Y. 2010. Supercritical water gasification of biomass and organic wastes. *Biomass*, 165.

Hagga, K., Laitinen, M. 2006. Experiences of the world's biggest recovery boiler at Jinhai Pulp & Paper Co. in Yang Pu, Hainan, China. *The 7th international colloquium on black liquor combustion and gasification*, Jyvaskyla, Finland, July 31–Aug. 2, 2006.

Hawangchu, Y., Atong, D., Sricharoenchaikul, V. 2017. The effect of alkali on the product distribution from black liquor conversion under supercritical water. *Environmental Technology*, **38**(13–14), 1742–1750.

Hupa, M., Solin, P., Hyoty, P. 1987. Combustion behaviour of black liquor droplets. *Journal of Pulp and Paper Science*, **13**(2), 67–72.

IEA Bioenergy ExCo54 Workshop. 2007. *Black liquor gasification: Summary and conclusions.* http://www.ieabioenergy.com/publications/exco54-workshop-black-liquor-gasification-summary-and-conclusions/.

Jafri, Y., Furusjö, E., Kirtania, K., Gebart, R. 2016. Performance of a pilot-scale entrained–flow black liquor gasifier. *Energy & Fuels*, **30**(4), 3175–3185.

Kubes, G., Fleming, B., MacLeod, J., Bolker, H. 1982. Thermal analysis of spent pulping liquor: activation energies. *Journal of Wood Chemistry and Technology*, **2**(3), 279–295.

Larson, E.D., Consonni, S., Katofsky, R.E., Iisa, K., Frederick, W. 2007. *A cost-benefit assessment of gasification–based biorefining in the kraft pulp and paper industry.* https://www.osti.gov/servlets/purl/912520.

Larson, E.D., Consonni, S., Kreutz, T.G. 2000. Preliminary economics of black liquor gasifier/gas turbine cogeneration at pulp and paper mills. *Journal of Engineering for Gas Turbines and Power–Transactions of the Asme*, **122**(2), 255–261.

Larson, E.D., Kreutz, T.G., Consonni, S. 1999. Combined biomass and black liquor gasifier/gas turbine cogeneration at pulp and paper mills. *Journal of Engineering for Gas Turbines and Power–Transactions of the Asme*, **121**(3), 394–400.

Le, Z., Pels, J.R., Van Heiningen, A.R. 2000. Direct causticization of kraft black liquor solids with TiO_2 in a fluidized bed. *Tappi Journal*, **83**(12).

Li, J., Vanheiningen, A.R.P. 1989. Reaction–kinetics of gasification of black liquor char. *Canadian Journal of Chemical Engineering*, **67**(4), 693–697.

Li, J., Vanheiningen, A.R.P. 1990a. Kinetics of CO_2 gasification of fast pyrolysis black liquor char. *Industrial & Engineering Chemistry Research*, **29**(9), 1776–1785.

Li, J., Vanheiningen, A.R.P. 1990b. Sodium emission during pyrolysis and gasification of black liquor char. *Tappi Journal*, **73**(12), 213–219.

Li, J., Vanheiningen, A.R.P. 1994. The rate process of H_2S emission during steam gasification of black liquor char. *Chemical Engineering Science*, **49**(24), 4143–4151.

Magdeldin, M., Järvinen, M. 2020. Supercritical water gasification of Kraft black liquor: Process design, analysis, pulp mill integration and economic evaluation. *Applied Energy*, **262**, 114558.

Muangrat, R., Onwudili, J.A., Williams, P.T. 2010. Influence of alkali catalysts on the production of hydrogen–rich gas from the hydrothermal gasification of food processing waste. *Applied Catalysis B: Environmental*, **100**(3–4), 440–449.

Naqvi, M., Yan, J., Dahlquist, E. 2010a. Black liquor gasification integrated in pulp and paper mills: A critical review. *Bioresource Technology*, **101**(21), 8001–8015.

Naqvi, M., Yan, J., Dahlquist, E. 2012a. Energy conversion performance of black liquor gasification to hydrogen production using direct causticization with CO_2 capture. *Bioresource Technology*, **110**, 637–644.

Naqvi, M., Yan, J., Dahlquist, E. 2012b. Synthetic gas production from dry black liquor gasification process using direct causticization with CO_2 capture. *Applied Energy*, **97**, 49–55.

Naqvi, M., Yan, J., Froling, M. 2010b. Bio–refinery system of DME or CH_4 production from black liquor gasification in pulp mills. *Bioresource Technology*, **101**(3), 937–944.

Nohlgren, I., Sinquefield, S.A. 2004. Black liquor gasification with direct causticization using titanates: Equilibrium calculations. *Industrial & Engineering Chemistry Research*, **43**(19), 5996–6000.

Nohlgren, I., Sricharoenchaikul, V., Sinquefield, S., Theliander, H., Frederick, W.J. 2003. Black liquor gasification with direct causticization using titanates in a pressurized entrained–flow reactor. part II: Carbon and carbon species transitions. *Journal of Pulp and Paper Science*, **29**(10), 348–355.

Nohlgren, I., Theliander, H., Zhuang, Q., Van Heiningen, A.R. 2000. Model study of the direct causticization reaction between sodium trititanate and sodium carbonate. *The Canadian Journal of Chemical Engineering*, **78**(3), 529–539.

Patrick, K., Siedel, B. 2003. Gasification edges closer to commercial reality with three New NA mill startups. *Paper Age*, **10**, 30–33.

Pels, J.R., Zeng, L., Van Heiningen, A.R.P. 1997. Direct causticization of kraft black liquor with TiO_2 in a fluidized bed – Identification and analysis of sodium titanates. *Journal of Pulp and Paper Science*, **23**(12), 549–554.

Reis, V.V., Frederick, W.J., Wag, K.J., Iisa, K., Sinquefield, S.A. 1995. Effects of temperature and oxygen concentration on potassium and chloride enrichment during black–liquor combustion. *Tappi Journal*, **78**(12), 67–76.

Sricharoenchaikul, V. 2009. Assessment of black liquor gasification in supercritical water. *Bioresource Technology*, **100**(2), 638–643.

Sricharoenchaikul, V., Frederick, W.J., Agrawal, P. 2002. Black liquor gasification characteristics. 1. Formation and conversion of carbon–containing product gases. *Industrial & Engineering Chemistry Research*, **41**(23), 5640–5649.

Sricharoenchaikul, V., Frederick, W.J., Agrawal, P. 2003. Carbon distribution in char residue from gasification of kraft black liquor. *Biomass & Bioenergy*, **25**(2), 209–220.

Sricharoenchaikul, V., Hicks, A.L., Frederick, W.J. 2001. Carbon and char residue yields from rapid pyrolysis of kraft black liquor. *Bioresource Technology*, **77**(2), 131–138.

Vakkilainen, E. 2014. *Recovery boilers-history and future*. Continuous development of recovery boiler technology-50 years of cooperation in Finland. Proceedings: International Recovery Boiler Conference, Tampere, Finland, 19–31.

Wang, G., Yuan, H., Guo, D., Zhou, Z., Yin, X., Wu, C. 2015. Effects of titanium dioxide as direct causticization agent on the pyrolysis and steam gasification characteristics of straw black liquor solid. *Tappi Journal*, **14**(6), 361–369.

Watanabe, M., Inomata, H., Osada, M., Sato, T., Adschiri, T., Arai, K. 2003. Catalytic effects of NaOH and ZrO_2 for partial oxidative gasification of n–hexadecane and lignin in supercritical water. *Fuel*, **82**(5), 545–552.

Whitty, K. 1997. *Pyrolysis and gasification behavior of black liquor under pressurized conditions*. Department of Chemical Engineering, Åbo Akademi University, Finland.

Whitty, K. 2005. Black liquor gasification: development and commercialization update. *ACERC annual conference*, Provo, Utah.

Whitty, K., Backman, R., Hupa, M. 1998. Influence of char formation conditions on pressurized black liquor gasification rates. *Carbon*, **36**(11), 1683–1692.

Whitty, K., Backman, R., Hupa, M. 2008. Influence of pressure on pyrolysis of black liquor: 1. Swelling. *Bioresource Technology*, **99**(3), 663–670.

Whitty, K., Verrill, C. 2004. A historical look at the development of alternative black liquor recovery technologies and the evolution of black liquor gasifier designs. *Proceeding of 2004 international chemical recovery conference*, South Carolina.

Wiinikka, H., Carlsson, P., Marklund, M., Gronberg, C., Pettersson, E., Lidman, M., Gebart, R. 2012. Experimental investigation of an industrial scale black liquor gasifier. Part 2: Influence of quench operation on product gas composition. *Fuel*, **93**(1), 117–129.

Yoshida, Y., Dowaki, K., Matsumura, Y., Matsuhashi, R., Li, D., Ishitani, H., Komiyama, H. 2003. Comprehensive comparison of efficiency and CO_2 emissions between biomass energy conversion technologies—position of supercritical water gasification in biomass technologies. *Biomass and Bioenergy*, **25**(3), 257–272.

Yuan, H., Wu, S., Yin, X., Guo, D., Wang, G., Zhou, Z., Wu, C. 2016. Effects of silica on chemical recovery in the direct causticization of wheat straw black liquor. *Tappi Journal*, **15**(8), 557–566.

8 Functionalized Bacterial Cellulose-Based Biopolymers for Biomedical Applications

Current Research Trends and Challenges

Baishali Dey, Bunushree Behera, Karthika Parvathy K. R., Sivaraman Jayaraman, and Balasubramanian Paramasivan

CONTENTS

8.1 INTRODUCTION

Over the years much emphasis has been placed on the development of biomaterials from renewable resources, driving the paradigm of research towards more sustainable and greener technologies to develop a circular bioeconomy. Cellulose is said to be the most abundant biopolymer found on Earth with a myriad of biological sources like plants, tunicates, fungi, algae, and several nonpathogenic, aerobic bacteria of the genera *Sarcina, Rhizobium, Agrobacterium* and *Gluconaacetobacter* (formerly *Acetobacter*) (**Ullah *et al.*, 2016**). The formation of bacterial cellulose pellicle floating in an acetic acid medium was first noticed and reported by Brown in the year 1886 (**Brown, 1886**). It was produced by *Bacterium xylinum*, which was later retitled as *Acetobacter xylinum*, then given the name of *Gluconacetobacter xylinus*, and since the last two years again renamed as *Komagataeibacter medellinensis* (**Römling & Galperin, 2015**). These bacteria are often found in rotten fruits and vegetables, fermented foods (for example, kombucha, nata de coco, vinegar) and have the capacity to produce acetic acid by oxidizing alcohols,

DOI: 10.1201/9781003335740-8

aldehydes, and sugar in the presence of oxygen. *Gluconoacetobacter* and *Komagateibacter* genera extracellularly release the cellulose into the fermentation medium for their protection against desiccation and ultraviolet (UV) light damage, also known as bacterial cellulose (BC) (**Raghavendran *et al.*, 2020**). The highly porous network of BC helps in the exchange of nutrients and oxygen necessary for the survival of bacteria. The unique range of properties presented by these bacteria-derived cellulose has been exploited for several applications in both native and composite forms, as discussed in the subsequent sections.

Komagataeibacter (formerly *Gluconacetobacter*) *xylinus* is the most commonly exploited, gram-negative obligate aerobic bacteria used for the commercial production of BC, as it can utilize a wide variety of carbon and nitrogen substrates (**Saxena & Brown, 2016**). The traditional synthetic medium for growing acetic acid bacteria is the Hestrin and Schramm (HS) medium, which consists of glucose, peptone, and yeast extract. The yield of BC can be increased by adding methanol in the HS medium and small amounts of endoglucanase in the production culture. The production costs can be reduced by using cheaper substrates such as agricultural wastes like rotten fruits and vegetables, industrial by-products like tea processing waste, corn-steep liquor, food-agro residues, molasses and date syrup, and petrochemical and wood processing wastes. BC can be produced through static, agitated, or stirred fermentation operations, with each producing varied forms of cellulose. For the production of BC, under static conditions, aerosol, horizontal lift bioreactors, and rotary biofilm contractors have been developed to provide a large surface area, as the BC is synthesized at the air-liquid interface. Fibrous suspensions, spheres, and pellets are formed under agitated and stirred fermenter conditions. BC produced under static conditions has superior mechanical strength, higher yield, and less branching as compared to BC synthesized under stirred and agitated conditions (**Gregory *et al.*, 2021**). Thus, based on the targeted application and required properties, the most suitable operation mode can be chosen to alter the BC microstructure as per requirements. BC produced via static operation was used for wound dressing and scaffold material in biomedical applications, whereas the pellicles synthesized under agitated operation were used in the adsorption of heavy metal ions, oils, and organic solvents along with immobilization of enzymes. The downstream processing of BC involves harvesting the culture medium, either manually or by filtering, followed by mild alkali treatment to remove cell debris and other contaminants.

In 2019, the global BC market was approximately USD 299 million and is expected to reach around 777 million by 2027. Although most of the research is still being conducted at the lab-scale level using the biopolymer, only some of the industries like Xylos Co. (USA) have been able to commercialize BC under the name Prima Cel for wound dressing along with Biofill and Gengiplex developed for tissue regeneration. A German company, Fzmb GmbH, has established itself as the largest producer of BC for cosmetic and biomedical applications (**Ullah *et al.*, 2016**). The present chapter emphasizes the physicochemical properties of the versatile BC, which makes it suitable for use in a variety of biomedical applications. Further, out of the pool of diversified applications researched so far, a few have been discussed in detail. Lastly, the development of BC composites to overcome the limitations and an overview of the techno-economic challenges faced have also been provided.

8.2 PHYSIOCHEMICAL PROPERTIES CONTEXTUALIZATION OF BACTERIAL CELLULOSE

Researchers have already established novel ways of producing microbial or bacterial cellulose in both lab and pilot scales. For widening the utilization of the sustainable BC-based material, its intrinsic properties need to be enhanced for its valorization in the paper, textile, food, pharmaceutical, materials and packaging, biomedical, and cosmeceutical industries, among others. It is a natural, homogenous polymer with a linear arrangement of D-glucopyranose units connected with β linkages and has an average degree of polymerization of about 9000–10,000 (**Gallegos *et al.*, 2016**). The operon which regulates the multistep BC synthesis called *bcs*ABCD and its related genes (first identified in *K. xylinus*) helps in encoding the required proteins and enzymes, further aids in developing the 3D network of cellulose through linear polymerization of glucose units (**Portela *et al.*, 2019**). Also, the numerous reactive hydroxyl groups available help in establishing interactions with over 90% of water molecules by forming plentiful intermolecular and intramolecular hydrogen bonds, which results in high water absorption and retention, chirality, and biodegradability and leaves tremendous scope for functional modification, allowing its use in a plethora of applications (**Gregory *et al.*, 2021**). Thus, BC offers the possibility of developing custom-modified cellular matrices to introduce new or improve existing properties according to the requirements. It has a water-holding capacity of 70–800 times its dry weight, depending on the culture conditions. The structure of BC is almost the same as that of plant cellulose, with the advantage of being devoid of lignin, hemicellulose, pectin, and other biogenic products commonly present in plants (**Rahman & Netravali, 2016**). It is 100 times thinner, presenting a dense 3D network, thus, making it highly pure with special physicochemical characteristics due to its super molecular network with simpler harvesting resulting in a minimal requirement for downstream processing.

Due to the diverse nature of BC compared to plant cellulose, its applicability is not restricted only to biological sciences and chemistry but also extends to engineering and physics (**Hu *et al.*, 2014**). It possesses fiber-binding ability, transparency, biodegradability, biocompatibility, and moldability, among others (**Shezad *et al.*, 2010**), making it a versatile biopolymer of commercial interest. Due to its high porosity, it allows the free exchange of fluids, and it has a very high water absorption capacity, which helps in maintaining a moist environment. It stands out to be a potential candidate for several biomedical applications due to its ribbon-shaped macro-fibrils and micro-fibrils forming an ultrafine network, providing a higher surface area, elasticity, resistance, and flexibility along with biocompatibility, nontoxicity, and noncarcinogenicity (**Ullah *et al.*, 2016**) (Figure 8.1). Despite 99% water content, the synthesized cellulose pellicle has excellent stress-strain properties resembling a soft tissue with stress break at 2×10^6 Pa (**Svensson *et al.*, 2005**). The tensile strength is comparable to Kevlar and steel. It also possesses exceptional mechanical properties even in dried form, making it suitable for applications requiring high mechanical strength (**Picheth *et al.*, 2017**). It mimics the

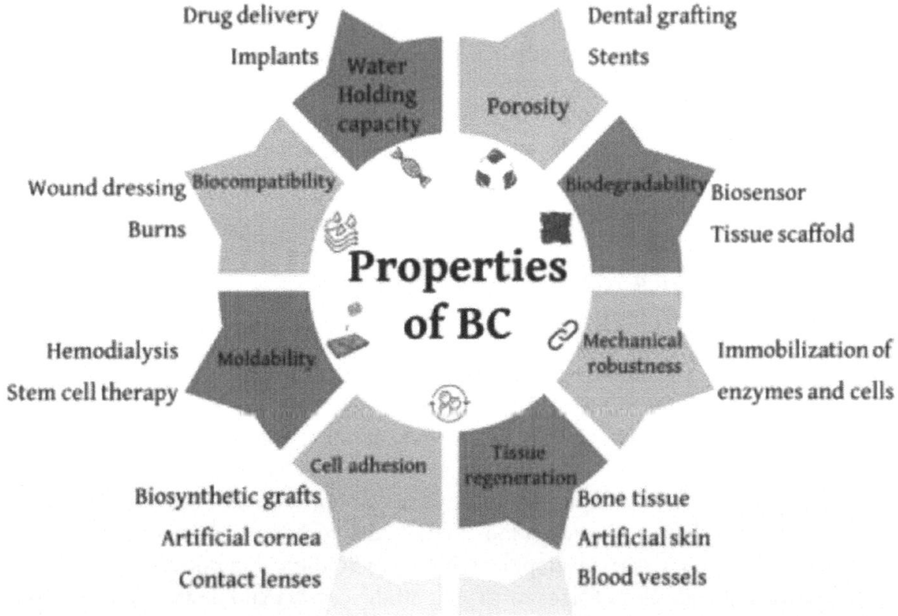

FIGURE 8.1 Properties of BC that make it suitable for biomedical and pharmaceutical applications.

extracellular matrix, helping in the growth and proliferation of cells (**Dutta *et al.*, 2019**). For *in vivo* applications, it can be made susceptible to degradation through modification, and glucose being its degradation product makes it suitable for bioresorbable purposes. For incorporating additional properties, such as antibacterial and anti-inflammatory activity, good biocompatibility, enhanced mechanical strength, improved cell adhesion and proliferation, biomimetic capabilities, and a more even distribution of cells, BC-based composites have been fabricated, which consist of a matrix and reinforcement materials for acquiring the intended property.

Fermentation conditions, along with the culture medium composition, greatly impact the arrangement, chemical structure, and viscosity of the microbial polysaccharides, including BC, which in turn may also affect its properties (**Shezad *et al.*, 2010**). The experiments performed by the previously mentioned authors further revealed that batch cultivation yielded smaller-sized sheets compared to fed-batch cultivation, which gave highly extended and uniform sheets. The cultivation mode has a greater effect on the mass fraction of cellulose $I\alpha$ and $I\beta$, the size of crystallite, and the crystallinity index. However, the crystallite size and crystallinity index were also influenced by the culture medium. The intrinsic properties of the BC also depend on the source, i.e., the synthesizing microorganism. During the biosynthesis of BC, gluconic acid is released, which decreases the medium pH and thus causes a decrease in the synthesis of the polymer. To overcome the issue, **Wulf *et al.* (1996)** created a mutant strain of *Gluconacetobacter xylinus* with a decreased ability to produce gluconic acid, which resulted in a higher yield than wild strains. Yield can also be doubled by restricting the production of glutamate dehydrogenase enzyme in mutant strains (**Shigematsu *et al.*, 2005**). Moreover, the cellulose (cellulose II) produced by the mutant strains has an antiparallel arrangement of the glucan chains instead of a parallel arrangement, as in cellulose I, produced by the wild type. The change in the arrangement of glucan chains in cellulose II is due to the average length of the fibers being about ten times the width, resulting in the folding of the former. Cellulose synthesized by *G. europaeus* showed different properties compared to the one formed by *G. xylinus* in terms of porosity (higher) and film thickness (lesser). After drying, it showed better absorption capacity (110 times its dry weight) and robust mechanical properties with higher elasticity irrespective of the drying method compared to the cellulose synthesized by *G. xylinus*. It also showed lower hardness, higher penetration depth, and lower Young's modulus in the case of room-temperature drying and freeze-drying, contrary to the results of supercritical drying (**Zeng *et al.*, 2014**). *Asaia bogorensis* produced relatively thinner (5–10 nm) films suggesting a difference in organization of synthesis sites in different strains (**Ullah *et al.*, 2016**). Further, genetically engineered strains of *G. xylinus* with genes from *Candida albicans* produced cellulose with enhanced biodegradability. The presence of N-acetyl glucosamine (NAcG) disturbs the highly ordered crystalline structure of BC, making it vulnerable to the action of lysozyme, thus reducing its crystallinity and making it more susceptible to biodegradation (**Yadav *et al.*, 2010a**).

8.3 APPLICATIONS OF BACTERIAL CELLULOSE IN BIOMEDICAL INDUSTRIES

In recent years, rapid progress has been made in the field of biomedical devices utilizing both natural and composite polymers due to the increasing need for medical engineering products for wound care, diagnosis of diseases, vascular grafts, regeneration of organs, and drug transportation (Figure 8.2).

BC has potential applications individually or as a composite across several biomedicine sectors and permits the development of novel biomaterials due to its ideal structure, biocompatibility, positive cellular interaction, tissue development, biodegradability, high porosity, sustainability, and moldability into various shapes and sizes satisfying all the criteria laid down by Davis (2003) to be considered as an ideal biomaterial. The following section further discusses the diversified applicability of BC in different fields of biomedical science.

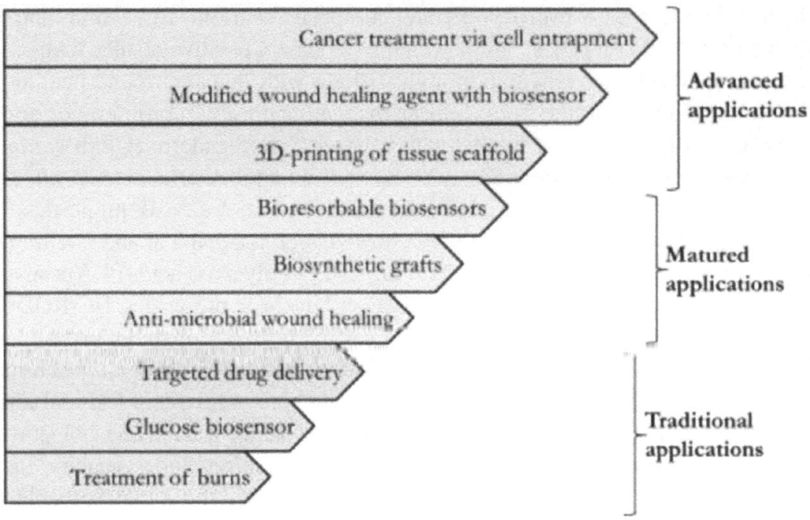

FIGURE 8.2 Technological development in BC-based research leading to more diversified cellulose-based products.

8.3.1 Wound Healing Agent

In health-related applications, natural or biobased materials pose several advantages over synthetic ones, as their basic molecular structures are already present in the structure of living tissues. As discussed previously, BC has several unique properties making it suitable for biomedical applications. Wound healing is a complex biological process with the aim of restoring the skin barrier function of preventing bacterial infection and maintaining hydration. BC has been used as a wound healing agent since the 1980s due to its high clinical potential, specifically for wound and burn care. The mechanical robustness of BC is due to its unique 3D structure consisting of an aggregation of long fibrils with 1.5 nm width resulting in high surface area, elastic property, and resistance, besides being nontoxic, noncarcinogenic, and biocompatible (**Portela *et al.*, 2019**). Its ability to maintain a moist environment; absorb wound exudate; reduce pain; attenuate thrombogenicity; and allow for fluid exchange, cell adhesion, and proliferation makes it a potential candidate for wound healing and dressing applications. Compared to a dry environment, the wound heals faster in a moist environment maintained by BC hydrogel due to faster re-epithelization and lesser scar formation, especially helpful in cases of chronic dry wounds with necrotic tissues. The transparency of BC membranes helps in monitoring the rate and extent of wound healing. Various BC-based wound dressing agents are already available in the market, like XCell, which is recommended for treating pressure sores, biopsy sites, ischemic and diabetic wounds, second-degree burns, skin tears, skin graft donor sites, venous stasis, traumatic abrasions, and lacerations (**Rajwade *et al.*, 2015**). Nanocell is used for treating burns; along with Biofill it can be used for treating ulcers. Gengiflex is used to recover periodontal tissue. Thus, it can also be utilized in the field of dentistry, especially after tooth removal for wound healing, drug delivery for reducing pain and bacterial infections, and also for fabricating bioabsorbable membranes via periodate oxidation.

One major disadvantage of BC is that it doesn't inherently have antimicrobial properties, which can be overcome by loading it with silver nanoparticles (**Yang *et al.*, 2014**), which have both bacteriostatic and bactericidal effects. BC, along with superoxide dismutase and poviargol, are used for treating thermal burns from acute radiation (**Legeza *et al.*, 2004**), and with benzalkonium chloride solution for treating wounds from trauma (**Wei *et al.*, 2011**). Various composites have also been developed using BC for potential wound healing like BC/chitosan, BC/magnetite,

BC/poly(3-hydroxybutyrate-co-4-hydroxybutyrate) composite scaffold, BC/kaolin, and BC/poly(2-hydroxyethyl methacrylate) (PHEMA), all of which have shown positive results. Chitosan and alginate imposed BC showed better rehydration ability along with quicker wound healing due to the release of N-acetyl-β-D-glucosamine from chitosan instigating fibroblast proliferation and controlled collagen deposition earlier than those treated with native BC or Tegaderm (**Ullah *et al.*, 2016**). BC impregnated with silver sulfurdiazine (SSD) not only showed antibacterial activity but also remarkable wound healing ability (**Wen *et al.*, 2015**). Agarose alone shows weak properties for use as a wound healing agent but in combination with BC showed good mechanical and swelling properties, thermal stability, and biodegradability (**Portela *et al.*, 2019**). Polylactic acid (PLA) can also be used as a coating material for enhancing its wound healing and intrinsic properties. BC-RGDC (arginine, glycine, aspartic acid, cysteine peptide)-gentamicin can be utilized for multipurpose applications, such as wound healing coupled with drug delivery (**Rouabhia *et al.*, 2014**). Collagen-incorporated BC is a novel material that has been developed, possessing both antioxidant and anti-inflammatory properties (**Wiegand *et al.*, 2006**). Addition of plasticizers like glycerol and polyethylene glycol aided in increasing its porosity, toughness, and water absorption/retention capacity. Incorporating cellulose-degrading enzymes into the BC composites helped in making it bioresorbable, and as it is made of simple glucose units, the degradation products are not toxic (**Hu & Catchmark, 2011a**). However, a challenge that is often faced in the practical applicability of the product is that due to variation in the pH of the wound or tissue, cellulose might lose up to 90% of its activity. In order to overcome the issue and maintain optimal pH conditions, **Hu & Catchmark (2011b)** impregnated the BC with buffer ingredients. They achieved positive results with a 67% increase in glucose release as compared to bufferless BC, and due to the proper pH microenvironment, the activity of the enzyme was also retained. By genetically engineering bacterial strains **Yadav *et al.* (2010)** biosynthetically incorporated NAcG into the BC matrix, making it biodegradable *in vivo*. The degradation product of BC being harmless glucose units, these composites may be ideal for many wound care and tissue engineering applications with the purpose of bioresorbability.

The outstanding qualities of BC can also be used in treating burn injuries which require expensive treatments, longer hospital stays, and are difficult to manage. Traditional treatment procedures involved the removal of the necrotic tissue, similar to the case of cancer lesions such as basal cell carcinoma or skin lesions related to chemotherapy and radiation therapy. All the previously mentioned skin conditions require a perfect dressing that will maintain a moist environment, reduce pain and healing time, enhance tissue regeneration, and prevent infection, all of which are met by BC. Despite the intrinsic features of BC and the scope for development, not many products have made it to the market. Most of the commercial producers are mainly concentrated in the United States, Brazil, and Poland (**Portela *et al.*, 2019**).

8.3.2 Drug Delivery Materials

The major hurdle faced by topical drug formulation is its loss due to contact with garments or surfaces (**Almeida *et al.*, 2014**). Various polymers, including BC, have been studied extensively in drug delivery applications. The drug release property of BC has been enhanced by fabricating various nanocomposites. Due to its porous 3D network, BC membrane hydrogels can work as excellent carriers for precise drug delivery applications without loss due to the superior retention capacity (**Rajwade *et al.*, 2015**). It can be modified for the sustained release of the drugs and modulated for bioavailability of the drug during percutaneous administration. It has the additional advantage of absorbing wound exudates and sticking to the uneven skin surface. It can be used for both topical and transdermal drug delivery systems verified through 24-h patch tests (**Almeida *et al.*, 2014**). Usually, the simplest method of incorporating drugs in BC membranes is via immersion in the drug solution followed by lyophilization for maximum absorption of the drug. Anti-inflammatory drugs, such as ibuprofen and diclofenac, and antimicrobial drugs, such as gentamycin and amoxicillin, are most commonly loaded into the BC membranes (**Swingler *et al.*, 2021**).

BC hydrogels can also be used for oral drug delivery systems by modifying them to become pH- and thermo-responsive. They were synthesized using lyophilized BC and acrylic acid (AA) and showed a lower drug release rate at reduced pH levels, which can be exploited in a gastric environment for sustained drug release. The addition of AA improved its swelling and increased its water transmission rate (**Mohamad *et al.*, 2014**). Monolayer and multilayer biodegradable films of BC incorporated with sorbic acid (SA) were designed for antibacterial activity against *Escherichia coli* K12-MG1655. The release of SA can be controlled either by fabricating into the multilayer BC or by altering the concentration of SA, as at higher concentrations, the dissolution of crystals was less (**Jipa *et al.*, 2012**). BC modified with hydroxypropyl methylcellulose through in situ fermentation improved its rehydration and capacity to absorb small molecules, making it efficient for smaller drug delivery applications (**Rajwade *et al.*, 2015**). Post-synthesis loading of bovine serum albumin (BSA) by *in vitro* dissolution or ex vivo permeation showed potential for oral delivery with sustained drug release for 13–14 h directly proportional to irradiation dose (**Amin *et al.*, 2012**). Another study suggested the use of BC hydrogels facilitating caffeine release for the purpose of cellulite attenuation (**Silva *et al.*, 2014**). Due to the ability to control the drug release of BC, it can also be used for cancer treatment by targeting tumor sites specifically and reducing the exposure of harsh chemicals to healthy tissues.

For transdermal delivery applications, drugs are loaded into partially dehydrated BC membranes along with glycerol, which gives a plasticizing effect and facilitates swelling and rehydration, making it more suitable. Drug release from BC membrane was evaluated and compared with ibuprofen and lidocaine hydrochloride gel and aqueous and PEG400 solution (**Trovatti *et al.*, 2012**). It was observed that the permeation rate was higher for ibuprofen (lipophilic drug) as compared to other formulations. BC-modified membranes with diclofenac sodium were also explored for *in vitro* studies, which showed comparable permeation rates to commercial patches but lower than commercial gel (**Silva *et al.*, 2014**). Enantioselective delivery of racemic propranolol was studied using BC composites with molecularly imprinted polymeric (MIP) matrices showing successful selective delivery of the drug due to the presence of selective markers (**Bodhibukkana *et al.*, 2006**).

Currently, the use of BC in native form or as a conjugate remains restricted to oral and topical applications (**Yang *et al.*, 2017**). Emulsions using drug-loaded BC nanocrystals can be explored for oral and dermal drug delivery. Further research needs to be conducted to widen its approach for developing simpler drug loading methods and comparable drug release profiles for more specific, tailor-made applications in the previously mentioned area.

8.3.3 Tissue Scaffolds

Exogenous 3D extracellular matrices (ECMs) are commonly used as scaffolds in tissue engineering techniques to adapt cells in tissues, regulate the tissue framework, and maintain the cell phenotype (**Liu *et al.*, 2021**). Soft tissues encompass articular cartilage, nerves, heart valves, arteries, lungs, liver, muscles, skin, and gut, which have a distinct function in the internal organs. Soft tissue engineering is a cutting-edge technique for repairing soft tissue injury by mitigating the limitations of current treatments. Scaffold biomaterials serve a vital role in cell restructuring, angiogenesis, cell migration, mechanical support, and other aspects of soft tissue engineering. BC has stronger mechanical properties than plant-derived cellulose due to its high crystallinity, making it a preferable alternative in the biomedical area, especially when it comes to soft tissue regeneration (**Mbituyimana *et al.*, 2021**).

BC, collagen, different proteoglycans, alginate-based substrates, and chitosan are all biological materials that have been employed to make scaffolds for tissue engineering. Natural polymers, unlike synthetic polymer-based scaffolds, are biologically dynamic and often maintain good cell adhesion and growth. An *in vitro* evaluation of bone regeneration nanocomposite produced from BC, collagen Icol), hydroxyapatite (HA), and osteogenic growth peptide (OGP) or its C-terminal pentapeptide has revealed that Col-based scaffolds have better tissue regeneration **(Bharadwaz & Jayasuriya, 2020)**. Additionally, through sophisticated analysis, BC/Col composites were reported to be composed of needle-like or plate-like apatite crystallites, which mimicked the osteo-calcium phosphate phase, thereby providing better bone repair. Furthermore, the presence of peptides in addition to Col may be linked to the nucleation and growth of plate-like apatite crystallite formation **(Saska *et al.*, 2017)**.

Wang *et al.* (2018) suggested that bacterial cellulose/gelatin (B/G) scaffolds with immobilized heparin can function as a vascular endothelial growth factor (VEGF) delivery matrix and stimulate angiogenesis throughout tissue regeneration. It was confirmed by using the V-B/G/H0.5 scaffold, where it evidenced greater cell proliferation and migration than scaffolds that lack VEGF. Also, the use of B/G scaffolds as releasing systems could be applied to other heparin-binding growth factors. A bacterial cellulose/polyvinyl alcohol (BC/PVA) composite hydrogel was utilized to construct a tissue-engineered matrix for corneal repair, where its characteristics of BC/PVA were better suited for use than BC hydrogels. Therefore, human corneal stromal cells (hCSCs) were used to test the cytotoxicity of the materials, and BC/PVA exhibited tremendous biocompatibility with the cells **(Han *et al.*, 2020)**.

An exopolymer BC film generated from the by-product of kombucha fermentation might be applied as a suitable biomaterial for 3D printing and bioprinting. Further, the material ought to possess good printable and mechanical attributes with exceptional biological properties for tissue regeneration **(Pillai *et al.*, 2021)**. Some findings mentioned that 3D microporous regenerated bacterial cellulose/gelatin (3DMP rBC/G) scaffolds could be exploited as suitable materials for *in vivo* skin regeneration of 94% within two weeks using a porogen-assisted surface modification technique. It showed improved biocompatibility where the scaffolds were regular, with greater porous structure, with an adequate quantity of gelatin during *in vitro* studies **(Khan *et al.*, 2018)**. The inclusion of BC in graphene foam increased the biocompatibility, proliferation, differentiation, and growth of neural networks generated from neural stem cells (NSCs) significantly. Primary cortical neurons were grown on bacterial cellulose/graphene (3D-BC/G) that created a dense neural network with higher activity than those grown on graphene. The 3D-BC/G culture system is not only limited to NSCs and primary cortical neurons—it has the ability to be enhanced for its usage in various fields of regenerative medicine **(Guo *et al.*, 2021)**.

A BC-poly(3-hydroxybutyrate) [P(3HB)] 3D composite scaffold with remarkable dispersion and firm adhesion within BC and [P(3HB)] with sucrose as a porogen was developed with superior load-bearing capabilities, a high surface area–to–volume ratio, consistent pore diameter distribution, and geometrical similarities to ECM. Because of the scaffold's high porosity, mouse chondrogenic ATDC5 cells were able to infiltrate and migrate deep into the material. Furthermore, improved cell attachment and proliferation, as well as the preservation of the chondrocyte phenotype, were discovered **(Akaraonye *et al.*, 2016; Carvalho *et al.*, 2019).** In a standardized bovine cartilage punch model, researchers evaluated the regeneration capacity of BC implants laden with and without cells. The 3D perforation enabled chondrocytes to develop and move across the implant network, which improvised the implant's long-term function. By speeding up cell colonization, the cell loading increased its efficiency. The progressive rise in cell seeding, matrix deposition, and chondrogenic differentiation confirmed the progression of cartilage regeneration **(Horbert *et al.*, 2019)**.

8.3.4 Biosensing Applications

According to the International Union of Pure and Applied Chemistry (IUPAC), a biosensor is defined as an integrated receptor-transducer device that provides semi-quantifiable or quantifiable data using biologically derived sensory components such as proteins, antibodies, enzymes, and whole cells **(Torres *et al.*, 2020)**. In the last decade, BC-based biosensors have gained a lot of attention due to their low cost, ease of application, noninvasiveness, biodegradability, portability, scope for miniaturization and automation, light weight, and faster results compared to conventional testing methods. Since BC can be easily reformed and functionalized with carbon nanotubes, conductive materials, nanoparticles, metal oxides, and biomolecules, it makes BC a versatile material for biosensor applications that can transduce signals of optical, mechanical, or electrochemical types (**Torres *et al.*, 2020**). They have the additional advantage of being used *in vivo* which helps in recording the real-time action of biological signals like proteins and antibodies in response to injury, tissue inflammation, cardiac infarction, infection, and muscle dystrophy, thus helping in the early detection and diagnosis of diseases and alerting health officials regarding any future complications. Furthermore, BC provides the required matrix for immobilization of the bioreceptor or transducer component in a biosensor. The modified matrix for BC-based biosensors may be prepared either by modifying the culture medium or by adding the material once it is synthesized. After synthesis, the material can be incorporated either by dipping, which preserves its native structure, or by allowing it to act as a template for fixation of other materials, or by breaking down the structure into a homogenous suspension with the material to develop a nanocomposite. Based on the type of response, BC sensors can be classified into electrochemical, resonant, and optical detection. Electrochemical biosensors can be further subdivided into potentiometric, amperometric, and conductimetric.

The use of biosensors in glucose determination has been the most profound since its invention. **Wang *et al.* (2011)** reported the fabrication of BC-Au composites for amperometric detection of glucose by immobilizing glucose oxidase and horseradish peroxidase. When BC was used as an outer membrane in a glucose biosensor, it showed better stability as compared to commercially available Cuprophan for both diluted and undiluted blood samples (**Moniri *et al.*, 2017**). BC-Ag nanocomposites have been utilized for both analysis of amino acids (l-histidine, l-glutamine, and l-phenylalanine) and also as a substrate for surface-enhanced raman scattering (SERS) (**Ullah *et al.*, 2016**). The sensitivity of these novel biosensors has been found to be comparable with that of commercial ones. A biosensor with immobilized horseradish peroxide was fabricated to detect H_2O_2 and achieved a sensitivity of less than 1 μM. In another study, polyaniline was used with BC to enhance its electrical conductivity with considerable improvement from 10^{-8} to 10^{-2} S cm^{-1}, working as an electro-conductive hydrogel (**Shi *et al.*, 2012**). *In vivo* detection of electrolytes using PVC-based membranes is often riddled with thrombogenic concerns. In order to overcome the issue, **Badr *et al.* (2015)** sandwiched the PVC-based sensors in between native BC or heparin-modified BC membranes, which showed a comparable response in terms of selectivity and limits of detection to PVC-based membrane electrodes with better biocompatibility. Detection of esterase is performed using fluorogenic sensors, where fluorescein-tetraethylene glycol-azide (FTA) is bound to BC using covalent bonds and its diffusion prevented, while its *in vivo* use prevents its toxic aftereffects (**Derikvand *et al.*, 2016**). BC-Prussian blue nanocubes immobilized with lactate oxidase have been developed for detecting lactate (**Aditya *et al.*, 2022**). Further, inorganic-organic hybrid nanopaper has been developed by functionalizing BC using nanoparticles of oxides of vanadium, titanium, or both (**Gutierrez *et al.*, 2013**). **Li *et al.* (2016)** developed a biosensor for detecting dopamine levels by breaking down BC pellicles using ultrasound and loading the biosensor with palladium nanoparticles. It was further modified using Nafion and laccase, showcasing high specificity and sensitivity. **Fontenot *et al.* (2017)** designed a smart wound dressing substitute based on nanocellulose composites combined with a protease sensor for chronic wounds. Another interesting approach was made by developing a chemically modified wound dressing that could detect pathogens or endogenous enzymes like esterase or lipase, which may slow down wound healing. It works on the principle that when these enzymes cleave BC sheets conjugated with 5(6)-carboxyfluorescein-tetraethylene glycol-azide (FTA), it emits a fluorescence that can be detected by a benchtop illuminator (**Derikvand *et al.*, 2016**). Thus these kinds of successful attempts open the window for its applicability in detecting a wide range of biomarkers for direct *in vivo* diagnosis. **Yao *et al.* (2018)** developed an optical biosensor using cadmium telluride (CdTe) quantum dots and BC for measuring glucose.

Due to the high porosity, interconnectivity, slow diffusion, and acceptable mechanical properties, BC serves as a suitable matrix for the immobilization of biosensor components (**Torres *et al.*, 2020**). It can be further developed to hold various biomarkers for the real-time monitoring of health parameters and to overcome current limitations like reusability and to maintain the activity of the enzymes and proteins under suboptimal conditions.

8.4 FUNCTIONALIZATION OF BACTERIAL CELLULOSE FOR BIOMEDICAL APPLICATIONS

Due to its unique structure, BC can be functionally tailor-made to suit a variety of target-specific applications through chemical and biological functionalization, as shown in Table 8.1. These modifications can be done either during the biosynthesis process by adding it into the growth medium, called the in situ method, or after the formation of the pellicle, known as the ex situ method. In situ alterations can be performed by introducing additives into the culture medium or by changing the

TABLE 8.1
Bacterial Cellulose (BC) and Its Composites with Enhanced Properties for Diverse Biomedical Applications

BC/Composites	Functionalization Method	Enhanced Property	Area of Application	References
Bacterial cellulose	Biological	• Mechanical stability • Permeability to gases and liquids	Biomedical application	**Meftahi *et al.* (2010); Gatenholm & Klemm (2010)**
Bacterial cellulose	Biological	• High liquid loading capacity • High tensile strength • Flexibility	Biomedical application	**Luo *et al.* (2020); Dayal *et al.* (2016)**
Bacterial cellulose	Biological	• Better water retention • Strong hydrophilicity	Biomedical application	**Bashari *et al.* (2018); Portela *et al.* (2019); Nuutila *et al.* (2021)**
Bacterial cellulose	Biological	• Biodegradability • Biocompatibility	Biomedical application	**Mohamad *et al.* (2017); Rühs *et al.* (2018); Torgbo *et al.* (2020)**
Chemically modified 6-carboxyfluorescein (Ch-6CF)/BC	Chemical	• Durability • Environmentally friendly	Drug delivery	**Gao *et al.* (2019)**
6CF-BC	Chemical			
Poly(fluorophenol)/BC	Chemical	• Hydrophobicity • Durability	Textile industry	**Song *et al.* (2019)**
Bio-TENG (triboelectric nanogenerator)/BC	Chemical	• Cellulose-based biocompatible TENG	Self-powered implantable and wearable electronics	**Boonchai *et al.* (2021)**
Silver (Ag)/BC nanocomposite	Chemical	• Antimicrobial activity	Wound healing	**Pal *et al.* (2017)**
Aminated nanocrystalline bacterial cellulose (NCC-BC)	Chemical	• Enhances protein loading	Targeted therapeutic drug delivery	**Cook (2013)**
ε-Poly-L-Lysine/BC	Chemical	• Wound dressing	Biomedical application	**Fürsatz *et al.* (2018)**
Silver nanoparticles (*Ag40A and Ag20CN1.1A)/BC	Chemical	• Burn dressing • Good porosity • Antimicrobial activity • Cell viability	Biomedical application	**Teixeira *et al.* (2019)**
Buckminsterfullerene (C_{60})/BC (BCC_{60})	Chemical	• Wound dressing • Cancer therapy	Biomedical application	**Chu *et al.* (2018)**

*Ag40A: Autoclaved silver nanoparticles (40 ppm) treated BC.
Ag20CN1.1A: Autoclaved silver nanoparticles (20 ppm) + 1.1% cerium nitrate treated BC.

carbon source to achieve distinct mechanical and physical properties. On the other hand, in the ex situ method, the synthesized pellicle can be either immersed into or cast with the active agents and remolded as membrane/films to be used as tissue engineering scaffolds or wound dressing agents, or a tubular form to be used as artificial veins, catheters, urethras, etc. The micro/nanoporous morphological structure of BC allows the activated compounds to easily penetrate and provide the desired functionality. The properties are easily functionalized because of the cross-linking characteristics facilitate BC's use as a composite. In situ modification of BC membrane with sodium fluoride by **Sun *et al.* (2020)** demonstrated higher Young's modulus and tensile strength resulting in bulkier and stiffer fibers. The inference drawn was attributed to the formation of hydrogen fluoride (HF) that interferes with the hydrogen bonding in BC membranes. **Gao *et al.* (2019)** demonstrated that in situ modification of BC via the addition of glucose modified with carboxyfluorescein would provide fluorescence during UV spectroscopy and confocal microscopy. In situ modifications are often done with specific chemicals targeted for tissue engineering applications and mostly facilitate extracellular membrane recapitulations **(Gorgieva & Trček, 2019)**. For instance, **Romanov *et al.* (2014)** reported the in situ addition of hydroxyapatite (HAp) and tricalcium phosphate during the culturing of BC for use in bone tissue regeneration. **Wang *et al.* (2012)** prepared a BC composite with chitosan and heparin as anticoagulation agents for vascular tissue engineering. Apart from biomedical properties, the porosity enhancement of the BC membrane could be achieved via the addition of potato starch **(Lv *et al.*, 2016)** and a foaming procedure of the medium utilizing mannitol followed by stabilization with surfactant and xanthan to increase the viscosity **(Rühs *et al.*, 2018)**. The disadvantage of in situ modification strategies is mostly related to the inhibition of bacteria owing to the antibacterial activity of chemical additives, their insolubility in a medium, high surface tension, and lack of structural stability of BC membranes **(Gorgieva & Trček, 2019)**.

The ex situ chemical modification process following post-treatment via chemical reactions involving carboxy-methylation, sulfo-ethylation, phosphorylation, esterification, methylation, oxidation with 2,2,6,6-tetramethyl-1-piperidinyloxy, and cationization is summarized by **Gregory *et al.* (2021)**. Ex situ functionalization with silver (Ag) or zinc (Zn) nanoparticles or oxides is often done to enhance the antimicrobial properties, promoting its utilization as a wound healing agent **(Katepetch *et al.*, 2013; Pal *et al.*, 2017)**. Immersion of BC in polyvinyl alcohol (PVA) **(Millon *et al.*, 2006)** and silk fibroin **(Oliveira Barud *et al.*, 2015)** improves the mechanical strength, noncytotoxicity, and nongenotoxicity of BC membranes. **Zhu *et al.* (2014)** reported the external molding of Kombucha Bacterial Cellulose (KBC) membranes into interwoven ribbons (20–100 nm width) to be utilized as nerve conduits, showing its *in vivo* application to be biocompatible without any toxicological or hematological effects.

Another aspect to be considered for use as a tissue engineering scaffold for biomedical applications is biodegradability. Though BC is degradable, the lack of a cellulase enzyme in the human body, which facilitates the breakdown of cellulose's main chain and branches, thereby disturbing its crystallinity, hydrophobic-hydrophilic balance, and morphology, often hinders its degradation *in vivo*. Oxidation is an essential approach for stimulating BC degradation, as suggested in the study by **Yang *et al.* (2016)**. To functionalize and enhance the degradability (>90%), the BC membrane was modified with HAp and gelatin for utilization as a bone scaffold in the previously mentioned study. Oxidation is accompanied by modification (phosphorylation) of the BC membrane, which further rearranges the bonds and alters the physical/mechanical structure **(Oksman *et al.*, 2016)**. Similar to the earlier study, **(Gorgieva *et al.*, 2017)** also reported that periodate oxidation and the gelatin biopolymer showed improved biodegradation. Nevertheless, these technological approaches help in producing BC of desirable quality for biomedical applications.

8.5 TECHNO-ECONOMIC CHALLENGES OF FUNCTIONALIZED BACTERIAL CELLULOSE IN BIOMEDICAL APPLICATIONS

The field of tissue engineering and regenerative medicine is currently evolving in the domain of material science, where cellulosic biopolymers have currently gained attention because of their ease in use, structural compatibility, and economic feasibility. The technical and economic feasibility is largely dependent on the process conditions of BC formation. Several researchers have optimized the operational parameters concerning BC production during kombucha fermentation (**Treviño-Garza *et al.*, 2020; Villarreal-Soto *et al.*, 2021**). However, the detailed techno-economic feasibility analysis is limited. **Dourado *et al.* (2018)** reported that the economic feasibility is influenced by the low yield and high production costs linked with the bacterial nanocellulose commercialization. **Ul-Islam *et al.* (2020)** also have reported the majority of costs for BC are linked with the direct plant investments. With a similar conclusion as that of the previously mentioned research, a recent study by **Behera *et al.* (2022)** projected that increasing the production volume as well as exploring multiple products in a biorefinery is expected to improve the overall technical feasibility of the BC production. Bacterial nanocellulose production kinetics must be addressed for low-cost scale-up and potential biomedical applications. Future research trends must focus on identifying green manufacturing options for biomedical applications from BC to establish easier market penetration (**Mishra *et al.*, 2022**). Commercial wound healing agents like Nanoderm™ functionalized with Ag nanoparticles are economical and also have a prolonged efficacy and changing time because of their antibacterial properties (**Zhong, 2020**). These wound dressing agents are also economical compared to conventional wound healing agents (***Axcelon Dermacare Inc*, 2022**).

8.6 CONCLUSION

Bacterial cellulose (BC) is a highly versatile biopolymer with huge potential for fabricating novel, smart, and green materials for varied uses in the field of biomedicine. In the quest for green technologies, BC provides the solution as a potential raw material, but it still has to overcome many limitations like low yield and high production costs that currently limiting its commercialization. Owing to its superior properties in pure form, BC has drawn a lot of attention in the last decade in becoming the biomaterial of choice compared to plant cellulose. BC has intrinsic properties like high mechanical strength, water absorptivity, biocompatibility, porosity, and biodegradability that makes it a pioneer for use as a raw material in the field of biomedicine. Efforts should be made for the large-scale production of BC and its nanocomposites under optimized conditions in a biorefinery approach, which will help in enhancing its widespread applications. Future research should focus on overcoming these limitations so that the potential of BC can be utilized to the full extent.

ACKNOWLEDGEMENT

The authors are grateful to the Department of Biotechnology and Medical Engineering, NIT Rourkela, for providing their research facilities. The research was funded by the Biotechnology Ignition Grant (BIG), Biotechnology Industry Research Assistance Council (BIRAC), Department of Biotechnology, Government of India (BIRAC/KIIT0471/BIG-13/18). The authors greatly acknowledge the Ministry of Human Resources Development (MHRD) of the Government of India (GoI) for sponsoring the PhD program of the first author.

REFERENCES

Aditya, T., Allain, J. P., Jaramillo, C., & Restrepo, A. M. (2022). Surface modification of bacterial cellulose for biomedical applications. *International Journal of Molecular Sciences*, 23(2), 610.

Akaraonye, E., Filip, J., Safarikova, M., Salih, V., Keshavarz, T., Knowles, J. C., & Roy, I. (2016). Composite scaffolds for cartilage tissue engineering based on natural polymers of bacterial origin, thermoplastic poly(3-hydroxybutyrate) and micro-fibrillated bacterial cellulose. *Polymer International*, 65(7), 780–791.

Almeida, I. F., Pereira, T., Silva, N. H. C. S., Gomes, F. P., Silvestre, A. J. D., Freire, C. S. R., Sousa Lobo, J. M., & Costa, P. C. (2014). Bacterial cellulose membranes as drug delivery systems: An in vivo skin compatibility study. *European Journal of Pharmaceutics and Biopharmaceutics*, 86(3), 332–336.

Amin, M. C. I. M., Ahmad, N., Halib, N., & Ahmad, I. (2012). Synthesis and characterization of thermo- and pH-responsive bacterial cellulose/acrylic acid hydrogels for drug delivery. *Carbohydrate Polymers*, 88(2), 465–473.

Axcelon Dermacare Inc | Nanoderm | Ontario. (2022). Retrieved June 15, 2022, from https://www.axcelondc.com/

Badr, I. H. A., Abdel-Sattar, R., & Keshk, S. M. A. S. (2015). Enhancing biocompatibility of some cation selective electrodes using heparin modified bacterial cellulose. *Carbohydrate Polymers*, 134, 687–694.

Bashari, A., Rouhani Shirvan, A., & Shakeri, M. (2018). Cellulose-based hydrogels for personal care products. *Polymers for Advanced Technologies*, 29(12), 2853–2867.

Behera, B., Laavanya, D., & Balasubramanian, P. (2022). Techno-economic feasibility assessment of bacterial cellulose biofilm production during the Kombucha fermentation process. *Bioresource Technology*, 346, 126659.

Bharadwaz, A., & Jayasuriya, A. C. (2020). Recent trends in the application of widely used natural and synthetic polymer nanocomposites in bone tissue regeneration. *Materials Science and Engineering: C*, 110, 110698.

Bodhibukkana, C., Srichana, T., Kaewnopparat, S., Tangthong, N., Bouking, P., Martin, G. P., & Suedee, R. (2006). Composite membrane of bacterially-derived cellulose and molecularly imprinted polymer for use as a transdermal enantioselective controlled-release system of racemic propranolol. *Journal of Controlled Release*, 113(1), 43–56.

Boonchai, N., Mongkolthanaruk, W., Harnchana, V., Faungnawakij, K., & Pinitsoontorn, S. (2021). Chemical functionalization of bacterial cellulose film for enhancing output performance of bio-triboelectric nanogenerator. *Biointerface Research in Applied Chemistry*, 12(2), 1587–1600.

Brown, A. J. (1886). XIX.—The chemical action of pure cultivations of bacterium aceti. *Journal of the Chemical Society, Transactions*, 49(0), 172–187.

Carvalho, T., Guedes, G., Sousa, F. L., Freire, C. S. R., & Santos, H. A. (2019). Latest advances on bacterial cellulose-based materials for wound healing, delivery systems, and tissue engineering. *Biotechnology Journal*, 14(12), 1900059.

Chu, M., Gao, H., Liu, S., Wang, L., Jia, Y., Gao, M., Wan, M., Xue, C., & Ren, L. (2018). Functionalization of composite bacterial cellulose with C 60 nanoparticles for wound dressing and cancer therapy. *RSC Advances*, 8(33), 18197–18203.

Cook, J. (2013). Amine functionalization of bacterial cellulose for targeted delivery applications. *Electronic Thesis and Dissertation Repository*, 1442. https://ir.lib.uwo.ca/etd/1442.

Davis, J. (2003). *Materials for medical devices*. ASM Handbook Series. ASM International, Ohio, USA.

Dayal, M., & Polymers, J. C.-C. (2016). *Mechanical and structural property analysis of bacterial cellulose composites*. Carbohydrate Polymers, 144, 447–453.

Derikvand, F., Yin, D. L. T., Barrett, R., & Brumer, H. (2016). Cellulose-based biosensors for esterase detection. *Analytical Chemistry*, 88(6), 2989–2993.

De Wulf, P., Joris, K., & Vandamme, E. J. (1996). Improved cellulose formation by an Acetobacter xyhzum Mutant Limited in (Ket6)Gluconate synthesis. *Journal of Chemical Technology & Biotechnology*, 67, 376–380.

Dourado, F., Fontão, A. I., Leal, M., Rodrigues, A. C., & Gama, M. (2018). Process modelling and techno-economic evaluation of an industrial airlift bacterial cellulose fermentation process. *Nanocellulose and Sustainability*, 1–16.

Dutta, S. D., Patel, D. K., & Lim, K. T. (2019). Functional cellulose-based hydrogels as extracellular matrices for tissue engineering. *Journal of Biological Engineering*, 13(1), 1–19.

Fontenot, K. R., Edwards, J. V., Haldane, D., Pircher, N., Liebner, F., Condon, B. D., Qureshi, H., & Yager, D. (2017). Designing cellulosic and nanocellulosic sensors for interface with a protease sequestrant wound-dressing prototype: Implications of material selection for dressing and protease sensor design. *Journal of Biomaterials Applications*, 32(5), 622–637.

Fürsatz, M., Skog, M., Sivlér, P., Palm, E., Aronsson, C., Skallberg, A., Greczynski, G., Khalaf, H., Bengtsson, T., & Aili, D. (2018). Functionalization of bacterial cellulose wound dressings with the antimicrobial peptide ϵ-poly-L-Lysine. *Biomedical Materials (Bristol)*, 13(2), 025014.

Gallegos, A. M. A., Carrera, S. H., Parra, R., Keshavarz, T., & Iqbal, H. M. N. (2016). Bacterial cellulose: A sustainable source to develop value-added products – a review. *BioResources*, 11(2), 5641–5655.

Gao, M., Li, J., Bao, Z., Hu, M., Nian, R., Feng, D., An, D., Li, X., Xian, M., & Zhang, H. (2019). A natural in situ fabrication method of functional bacterial cellulose using a microorganism. *Nature Communications*, 10(1), 1–10.

Gatenholm, P., & Klemm, D. (2010). Bacterial nanocellulose as a renewable material for biomedical applications. *MRS Bulletin*, 35(3), 208–213.

Gorgieva, S., Girandon, L., & Kokol, V. (2017). Mineralization potential of cellulose-nanofibrils reinforced gelatine scaffolds for promoted calcium deposition by mesenchymal stem cells. *Materials Science and Engineering: C*, 73, 478–489.

Gorgieva, S., & Trček, J. (2019). Bacterial cellulose: Production, modification and perspectives in biomedical applications. *Nanomaterials*, 9(10), 1352.

Gregory, D. A., Tripathi, L., Fricker, A. T. R., Asare, E., Orlando, I., Raghavendran, V., & Roy, I. (2021). Bacterial cellulose: A smart biomaterial with diverse applications. *Materials Science and Engineering: R: Reports*, 145, 100623.

Guo, R., Li, J., Chen, C., Xiao, M., Liao, M., Hu, Y., Liu, Y., Li, D., Zou, J., Sun, D., Torre, V., Zhang, Q., Chai, R., & Tang, M. (2021). Biomimetic 3D bacterial cellulose-graphene foam hybrid scaffold regulates neural stem cell proliferation and differentiation. *Colloids and Surfaces B: Biointerfaces*, 200, 111590.

Gutierrez, J., Fernandes, S. C. M., Mondragon, I., & Tercjak, A. (2013). Multifunctional hybrid nanopapers based on bacterial cellulose and sol-gel synthesized titanium/vanadium oxide nanoparticles. *Cellulose*, 20(3), 1301–1311.

Han, Y., Li, C., Cai, Q., Bao, X., Tang, L., Ao, H., Liu, J., Jin, M., Zhou, Y., Wan, Y., & Liu, Z. (2020). Studies on bacterial cellulose/poly(vinyl alcohol) hydrogel composites as tissue-engineered corneal stroma. *Biomedical Materials*, 15(3), 035022.

Horbert, V., Boettcher, J., Foehr, P., Kramer, F., Udhardt, U., Bungartz, M., Brinkmann, O., Burgkart, R. H., Klemm, D. O., & Kinne, R. W. (2019). Laser perforation and cell seeding improve bacterial nanocellulose as a potential cartilage implant in the in vitro cartilage punch model. *Cellulose*, 26(1), 647–664.

Hu, W., Chen, S., Yang, J., Li, Z., & Wang, H. (2014). Functionalized bacterial cellulose derivatives and nanocomposites. *Carbohydrate Polymers*, 101(1), 1043–1060.

Hu, Y., & Catchmark, J. M. (2011a). Integration of cellulases into bacterial cellulose: Toward bioabsorbable cellulose composites. *Journal of Biomedical Materials Research Part B: Applied Biomaterials*, 97B(1), 114–123.

Hu, Y., & Catchmark, J. M. (2011b). In vitro biodegradability and mechanical properties of bioabsorbable bacterial cellulose incorporating cellulases. *Acta Biomaterialia*, 7(7), 2835–2845.

Jipa, I., Stoica-Guzun, A., & Stroescu, M. (2012). *Controlled release of sorbic acid from bacterial cellulose based mono and multilayer antimicrobial films.* LWT, 47(2), 400–406.

Katepetch, C., Rujiravanit, R., & Tamura, H. (2013). Formation of nanocrystalline ZnO particles into bacterial cellulose pellicle by ultrasonic-assisted in situ synthesis. *Cellulose*, 20(3), 1275–1292.

Khan, S., Ul-Islam, M., Ikram, M., Islam, S. U., Ullah, M. W., Israr, M., Jang, J. H., Yoon, S., & Park, J. K. (2018). Preparation and structural characterization of surface modified microporous bacterial cellulose scaffolds: A potential material for skin regeneration applications in vitro and in vivo. *International Journal of Biological Macromolecules*, 117, 1200–1210.

Legeza, V. I., Galenko-Yaroshevskii, V. P., Zinov'ev, E. V., Paramonov, B. A., Kreichman, G. S., Turkovskii, I. I., Gumenyuk, E. S., Karnovich, A. G., & Khripunov, A. K. (2004). Effects of new wound dressings on healing of thermal burns of the skin in acute radiation disease. *Bulletin of Experimental Biology and Medicine*, 138(3), 311–315.

Li, D., Ao, K., Wang, Q., Lv, P., & Wei, Q. (2016). Preparation of Pd/Bacterial Cellulose Hybrid Nanofibers for dopamine detection. *Molecules*, 21(5), 618.

Liu, W., Du, H., Zheng, T., & Si, C. (2021). Biomedical applications of bacterial cellulose based composite hydrogels. *Current Medicinal Chemistry*, 28(40), 8319–8332.

Luo, Z., Liu, J., Lin, H., Ren, X., Tian, H., Liang, Y., Wang, W., Wang, Y., Yin, M., Huang, Y., & Zhang, J. (2020). In situ fabrication of nano zno/bcm biocomposite based on ma modified bacterial cellulose membrane for antibacterial and wound healing. *International Journal of Nanomedicine*, 15, 1–15.

Lv, X., Yang, J., Feng, C., Li, Z., Chen, S., Xie, M., Huang, J., Li, H., Wang, H., & Xu, Y. (2016). Bacterial cellulose-based biomimetic nanofibrous scaffold with muscle cells for hollow organ tissue engineering. *ACS Biomaterials Science and Engineering*, 2(1), 19–29.

Mbituyimana, B., Liu, L., Ye, W., Ode Boni, B. O., Zhang, K., Chen, J., Thomas, S., Vasilievich, R. V., Shi, Z., & Yang, G. (2021). Bacterial cellulose-based composites for biomedical and cosmetic applications: Research progress and existing products. *Carbohydrate Polymers*, 273, 118565.

Meftahi, A., Khajavi, R., Rashidi, A., Sattari, M., Yazdanshenas, M. E., & Torabi, M. (2010). The effects of cotton gauze coating with microbial cellulose. *Cellulose*, 17(1), 199–204.

Millon, L., et al. (2006). The polyvinyl alcohol–bacterial cellulose system as a new nanocomposite for biomedical applications. *Journal of Biomedical Materials Research Part B: Applied Biomaterials*, 79(2), 245–253.

Mishra, S., Singh, P. K., Pattnaik, R., Kumar, S., Ojha, S. K., Srichandan, H., Parhi, P. K., Jyothi, R. K., & Sarangi, P. K. (2022). Biochemistry, synthesis, and applications of bacterial cellulose: A review. *Frontiers in Bioengineering and Biotechnology*, 10, 780409.

Mohamad, N., Buang, F., Mat Lazim, A., Ahmad, N., Martin, C., & Mohd Amin, M. C. I. (2017). Characterization and biocompatibility evaluation of bacterial cellulose-based wound dressing hydrogel: Effect of electron beam irradiation doses and concentration of acrylic acid. *Journal of Biomedical Materials Research - Part B Applied Biomaterials*, 105(8), 2553–2564.

Mohamad, N., Mohd Amin, M. C. I., Pandey, M., Ahmad, N., & Rajab, N. F. (2014). Bacterial cellulose/acrylic acid hydrogel synthesized via electron beam irradiation: Accelerated burn wound healing in an animal model. *Carbohydrate Polymers*, 114, 312–320.

Moniri, M., Moghaddam, A. B., Azizi, S., Rahim, R. A., Ariff, A. Bin, Saad, W. Z., Navaderi, M., & Mohamad, R. (2017). Production and status of bacterial cellulose in biomedical engineering. *Nanomaterials*, 7(9), 257.

Nuutila, K., Care, E. E.-A. W. (2021). Moist wound healing with commonly available dressings. *Liebert Publishers Com*, 10(12), 685–698.

Oksman, K., Aitomäki, Y., Mathew, A. P., Siqueira, G., Zhou, Q., Butylina, S., Tanpichai, S., Zhou, X., & Hooshmand, S. (2016). Review of the recent developments in cellulose nanocomposite processing. *Composites Part A: Applied Science and Manufacturing*, 83, 2–18.

Oliveira Barud, H. G., Barud, H. D. S., Cavicchioli, M., Do Amaral, T. S., De Oliveira Junior, O. B., Santos, D. M., De Oliveira Almeida Petersen, A. L., Celes, F., Borges, V. M., De Oliveira, C. I., De Oliveira, P. F., Furtado, R. A., Tavares, D. C., & Ribeiro, S. J. L. (2015). Preparation and characterization of a bacterial cellulose/silk fibroin sponge scaffold for tissue regeneration. *Carbohydrate Polymers*, 128, 41–51.

Pal, S., Nisi, R., Stoppa, M., & Licciulli, A. (2017). Silver-functionalized bacterial cellulose as antibacterial membrane for wound-healing applications. *ACS Omega*, 2(7), 3632–3639.

Picheth, G. F., Pirich, C. L., Sierakowski, M. R., Woehl, M. A., Sakakibara, C. N., de Souza, C. F., Martin, A. A., da Silva, R., & de Freitas, R. A. (2017). Bacterial cellulose in biomedical applications: A review. *International Journal of Biological Macromolecules*, 104, 97–106.

Pillai, M. M., Tran, H. N., Sathishkumar, G., Manimekalai, K., Yoon, J. H., Lim, D. Y., Noh, I., & Bhattacharyya, A. (2021). Symbiotic culture of nanocellulose pellicle: A potential matrix for 3D bioprinting. *Materials Science and Engineering: C*, 119, 111552.

Portela, R., Leal, C. R., Almeida, P. L., & Sobral, R. G. (2019). Bacterial cellulose: A versatile biopolymer for wound dressing applications. *Microbial Biotechnology*, 12(4), 586–610.

Raghavendran, V., Asare, E., & Roy, I. (2020). Bacterial cellulose: Biosynthesis, production, and applications. *Advances in Microbial Physiology*, 77, 89–138.

Rahman, M. M., & Netravali, A. N. (2016). Aligned bacterial cellulose arrays as "green" nanofibers for composite materials. *ACS Macro Letters*, 5(9), 1070–1074.

Rajwade, J. M., Paknikar, K. M., & Kumbhar, J. V. (2015). Applications of bacterial cellulose and its composites in biomedicine. *Applied Microbiology and Biotechnology*, 99(6), 2491–2511.

Romanov, D. P., Khripunov, A. K., Baklagina, Y. G., Severin, A. V., Lukasheva, N. V., Tolmachev, D. A., Lavrent'Ev, V. K., Tkachenko, A. A., Arkharova, N. A., & Klechkovskaya, V. V. (2014). Nanotextures of composites based on the interaction between hydroxyapatite and cellulose Gluconacetobacter xylinus. *Glass Physics and Chemistry*, 40(3), 367–374.

Römling, U., & Galperin, M. Y. (2015). Bacterial cellulose biosynthesis: Diversity of operons, subunits, products, and functions. *Trends in Microbiology*, 23(9), 545–557.

Rouabhia, M., Asselin, J., Tazi, N., Messaddeq, Y., Levinson, D., & Zhang, Z. (2014). Production of biocompatible and antimicrobial bacterial cellulose polymers functionalized by RGDC grafting groups and gentamicin. *ACS Applied Materials and Interfaces*, 6(3), 1439–1446.

Rühs, P. A., Storz, F., López Gómez, Y. A., Haug, M., & Fischer, P. (2018). 3D bacterial cellulose biofilms formed by foam templating. *NPJ Biofilms and Microbiomes*, 4(1), 1–6.

Saska, S., Teixeira, L. N., de Castro Raucci, L. M. S., Scarel-Caminaga, R. M., Franchi, L. P., dos Santos, R. A., Santagneli, S. H., Capela, M. V., de Oliveira, P. T., Takahashi, C. S., Gaspar, A. M. M., Messaddeq, Y., Ribeiro, S. J. L., & Marchetto, R. (2017). Nanocellulose-collagen-apatite composite associated with osteogenic growth peptide for bone regeneration. *International Journal of Biological Macromolecules*, 103, 467–476.

Saxena, I. M., & Brown, R. M. (2012). Biosynthesis of bacterial cellulose. In: *Bacterial NanoCellulose: A Sophisticated Multifunctional Material*, Edited by Gama, M., Gatenholm, P and Klemm, D. CRC Press. 1–18.

Shezad, O., Khan, S., Khan, T., & Park, J. K. (2010). Physicochemical and mechanical characterization of bacterial cellulose produced with an excellent productivity in static conditions using a simple fed-batch cultivation strategy. *Carbohydrate Polymers*, 82(1), 173–180.

Shi, Z., Zang, S., Jiang, F., Huang, L., Lu, D., Ma, Y., & Yang, G. (2012). In situ nano-assembly of bacterial cellulose–polyaniline composites. *RSC Advances*, 2(3), 1040–1046.

Shigematsu, T., Takamine, K., Kitazato, M., Morita, T., Naritomi, T., Morimura, S., & Kida, K. (2005). Cellulose production from glucose using a glucose dehydrogenase gene (gdh)-deficient mutant of Gluconacetobacter xylinus and its use for bioconversion of sweet potato pulp. *Journal of Bioscience and Bioengineering*, 99(4), 415–422.

Silva, N. H. C. S., Rodrigues, A. F., Almeida, I. F., Costa, P. C., Rosado, C., Neto, C. P., Silvestre, A. J. D., & Freire, C. S. R. (2014). Bacterial cellulose membranes as transdermal delivery systems for diclofenac: In vitro dissolution and permeation studies. *Carbohydrate Polymers*, 106(1), 264–269.

Song, J. E., Silva, C., Cavaco-Paulo, A. M., & Kim, H. R. (2019). Functionalization of bacterial cellulose non-woven by poly(Fluorophenol) to improve its hydrophobicity and durability. *Frontiers in Bioengineering and Biotechnology*, 7, 332.

Sun, B., Zhang, L., Wei, F., AL-Ammari, A., Xu, X., Li, W., Chen, C., Lin, J., Zhang, H., & Sun, D. (2020). In situ structural modification of bacterial cellulose by sodium fluoride. *Carbohydrate Polymers*, 231, 115765.

Svensson, A., Nicklasson, E., Harrah, T., Panilaitis, B., Kaplan, D. L., Brittberg, M., & Gatenholm, P. (2005). Bacterial cellulose as a potential scaffold for tissue engineering of cartilage. *Biomaterials*, 26(4), 419–431.

Swingler, S., Gupta, A., Gibson, H., Kowalczuk, M., Heaselgrave, W., & Radecka, I. (2021). Recent advances and applications of bacterial cellulose in Biomedicine. *Polymers*, 13(3), 412

Teixeira, S. R. Z., Dos Reis, E. M., Apati, G. P., Meier, M. M., Nogueira, A. L., Formolo Garcia, M. C., Dos Santos Schneider, A. L., Testa Pezzin, A. P., & Porto, L. M. (2019). Biosynthesis and functionalization of bacterial cellulose membranes with cerium nitrate and silver nanoparticles. *Materials Research*, 22.

Torgbo, S., & Sukyai, P. (2020). *Biodegradation and thermal stability of bacterial cellulose as biomaterial: The relevance in biomedical applications.* Polymer Degradation and Stability, 179, 109232.

Torres, F. G., Troncoso, O. P., Gonzales, K. N., Sari, R. M., & Gea, S. (2020). Bacterial cellulose-based biosensors. *Medical Devices & Sensors*, 3(5), e10102.

Treviño-Garza, M. Z., Guerrero-Medina, A. S., González-Sánchez, R. A., García-Gómez, C., Guzmán-Velasco, A., Báez-González, J. G., & Márquez-Reyes, J. M. (2020). Production of microbial cellulose films from green tea (Camellia sinensis) kombucha with various carbon sources. *Coatings*, 10(11), 1–14.

Trovatti, E., Freire, C. S., Pinto, P. C., Almeida, I. F., Costa, P., Silvestre, A. J., Neto, C.P. & Rosado, C. (2012). Bacterial cellulose membranes applied in topical and transdermal delivery of lidocaine hydrochloride and ibuprofen: In vitro diffusion studies. *International Journal of Pharmaceutics*, 435(1), 83–87.

Ul-Islam, M., Ullah, M. W., Khan, S., & Park, J. K. (2020). Production of bacterial cellulose from alternative cheap and waste resources: A step for cost reduction with positive environmental aspects. *Korean Journal of Chemical Engineering*, 37(6), 925–937.

Ullah, H., Wahid, F., Santos, H. A., & Khan, T. (2016). Advances in biomedical and pharmaceutical applications of functional bacterial cellulose-based nanocomposites. *Carbohydrate Polymers*, 150, 330–352.

Villarreal-Soto, S. A., Bouajila, J., Beaufort, S., Bonneaud, D., Souchard, J. P., & Taillandier, P. (2021). Physicochemical properties of bacterial cellulose obtained from different Kombucha fermentation conditions. *Journal of Vinyl and Additive Technology*, 27(1), 183–190.

Wang, B., Lv, X., Chen, S., Li, Z., Yao, J., Peng, X., Feng, C., Xu, Y., & Wang, H. (2018). Use of heparinized bacterial cellulose based scaffold for improving angiogenesis in tissue regeneration. *Carbohydrate Polymers*, 181, 948–956.

Wang, J., Wan, Y., & Huang, Y. (2012). Immobilisation of heparin on bacterial cellulose-chitosan nano-fibres surfaces via the cross-linking technique. *IET Nanobiotechnology*, 6(2), 52–57.

Wang, W., Zhang, T. J., Zhang, D. W., Li, H. Y., Ma, Y. R., Qi, L. M., Zhou, Y. L., & Zhang, X. X. (2011). Amperometric hydrogen peroxide biosensor based on the immobilization of heme proteins on gold nanoparticles–bacteria cellulose nanofibers nanocomposite. *Talanta*, 84(1), 71–77.

Wei, B., Yang, G., & Hong, F. (2011). Preparation and evaluation of a kind of bacterial cellulose dry films with antibacterial properties. *Carbohydrate Polymers*, 84(1), 533–538.

Wen, X., Zheng, Y., Wu, J., Yue, L., Wang, C., Luan, J., Wu, Z., & Wang, K. (2015). In vitro and in vivo investigation of bacterial cellulose dressing containing uniform silver sulfadiazine nanoparticles for burn wound healing. *Progress in Natural Science: Materials International*, 25(3), 197–203.

Wiegand, C., Elsner, P., Hipler, U., Cellulose & Klemm, D. (2006). Protease and ROS activities influenced by a composite of bacterial cellulose and collagen type I in vitro. *Cellulose*, 13(6), 689–696.

Yadav, V., Paniliatis, B. J., Shi, H., Lee, K., Cebe, P., & Kaplan, D. L. (2010). Novel in vivo-degradable cellulose-chitin copolymer from metabolically engineered gluconacetobacter xylinus. *Applied and Environmental Microbiology*, 76(18), 6257–6265.

Yang, J., Lv, X., Chen, S., Li, Z., Feng, C., Wang, H., & Xu, Y. (2014). In situ fabrication of a microporous bacterial cellulose/potato starch composite scaffold with enhanced cell compatibility. *Cellulose*, 21(3), 1823–1835.

Yang, M., Zhen, W., Chen, H., & Shan, Z. (2016). Biomimetic design of oxidized bacterial cellulose-gelatin-hydroxyapatite nanocomposites. *Journal of Bionic Engineering*, 13(4), 631–640.

Yang, R., Wei, T., Goldberg, H., Wang, W., Cullion, K., & Kohane, D. S. (2017). Getting drugs across biological barriers. *Advanced Materials*, 29(37), 1606596.

Yao, J., Ji, P., Wang, B., Wang, H., & Chen, S. (2018). Color-tunable luminescent macrofibers based on CdTe QDs-loaded bacterial cellulose nanofibers for pH and glucose sensing. *Sensors and Actuators, B: Chemical*, 254, 110–119.

Zeng, M., Laromaine, A., & Roig, A. (2014). Bacterial cellulose films: Influence of bacterial strain and drying route on film properties. *Cellulose*, 21(6), 4455–4469.

Zhong, C. (2020). Industrial-scale production and applications of bacterial cellulose. *Frontiers in Bioengineering and Biotechnology*, 8, 605374.

Zhu, C., Li, F., Zhou, X., Lin, L., & Zhang, T. (2014). Kombucha-synthesized bacterial cellulose: Preparation, characterization, and biocompatibility evaluation. Journal of Biomedical Materials Research: Part A 102(5), 1548–1557.

9 Nanomaterials in Lignocellulosic Biodiesel Production by Oleaginous Microorganisms

Pritam Bardhan, Kuldeep Gupta, Manabendra Mandal, and Rupam Kataki

CONTENTS

9.1 INTRODUCTION

The lignocellulosic biomass (LCB) is an abundant renewable resource from the agricultural industry, algal biodiesel industry (deoiled algal biomass), paper mills, etc. Primarily, these compounds constitute cellulose, hemicelluloses and lignin. However, the natural recalcitrance provided by the plant cell wall to enzymatic, microbial and other biochemical degradation necessitates a prior pretreatment step to introduce disruptions in the structure (Scordia et al. 2012). Such pretreatment facilitates hydrolysis of hemicellulose, decrystallized cellulose and displaced lignin. Chemical pretreatment methods include the use of corrosive acid and alkali reagents that generate inhibitory compounds such as furfural, hydroxy methyl furfural and acetic acid from the degradation of celluloses and hemicelluloses, respectively, whereas lignin degradation results in the formation of syringaldehyde, vanillin, acetosyringone, etc. (Liu et al. 2020b). Such compounds are considered important platform chemicals for the catalytic conversion into liquid fuels and fine chemicals (Bhanja and Bhaumick 2016). Additionally, lignocellulosic hydrolysate contains fermentable sugars (C5 and C6) that can be converted into microbial lipids (triacylglycerides) using oleaginous microorganisms under certain stress conditions (Park et al. 2017; Poontawee et al. 2018). Some of the most widely discussed oleaginous

DOI: 10.1201/9781003335740-9

microorganisms are microalgae (*Chlorella vulgaris*, *Scenedesmus obliquus*, *Nanochloropsis oculate*, etc.), yeast (*Rhodotorula mucilaginosa*, *Lipomyces starkeyi*, *Yarrowia lipolytica*, etc.), filamentous fungi (*Mortierella isabellina*, *Mucor circinelloides*, etc.) and bacteria (*Rhodococcus opacus*) that are able to accumulate excess lipids as intracellular lipid droplets (Bardhan et al. 2019b). However, the economic feasibility of single cell oil (SCO) production using an LCBlargely depends on the lipid content (% lipid/g dry cell weight) of the producer microorganism (Santek et al. 2018). Besides, the downstream processing that involves harvesting, cell disruption and extraction of microbial lipids contributes to the overall production cost and remains one of the major hindrances towards full economic efficiency of microbial lipids (Ochsenreither et al. 2016). The SCO can be converted into biodiesel by the catalytic transesterification reaction. In this context, a renewable heterogeneous base catalyst and immobilized lipase enzyme are preferred for the catalytic conversion of plant-derived vegetable oils, nonedible oils, waste cooking oils and microbial lipids into biodiesel (Ingle et al. 2020b). Furthermore, the biodiesel production using oleaginous microorganisms will depend on the economic utilization of value-added products such as natural pigments (carotenoids), polyunsaturated fatty acids and other volatile organic acids, renewable oleochemicals and deoiled biomass, which is a rich source of proteins and polysaccharides (Bardhan et al. 2020). Considering the need for new technological interventions for economic production of biomass-derived biodiesel using oleaginous microorganisms, nanomaterials offer solutions by altering the characteristics of feed additives and biocatalysts used for biodiesel production (Rai et al. 2016).

Nanomaterials are materials in the nanoscale dimension, which means that the external dimension or internal structure or surface structure is in the range of 1 to 100 nanometers. These materials are generally classified on the basis of shape (spherical, oval and rod-shaped), composition (single metallic or bimetallic, multiple elements, carbon-based, metal oxides, nanocomposites) and dimension (1D nanoparticles, 2D nanotubes, 3D nanofilms) (Buzea and Pacheco 2017). The physicochemical properties of nanoform materials vary considerably from the bulk material and are largely dependent on its shape, size, composition and functionalization (Gatoo et al. 2014). Nanomaterials find wide applications in various steps of the LCB-derived microbial biodiesel production process due to their biocompatibility, large specific surface area and resistance to shear stress. Nanomaterials could be used at specific process steps in lignocellulosic biodiesel production using oleaginous microbes (Figure 9.1). The role of nanomaterials in the biorefinery approach for

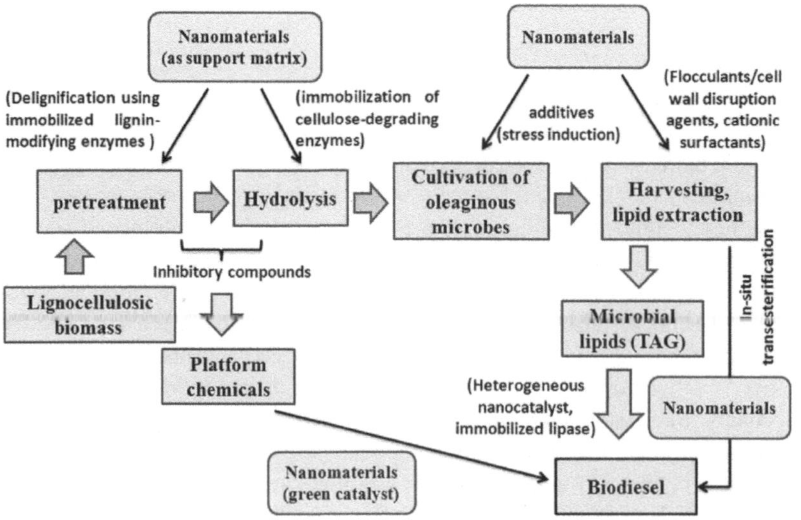

FIGURE 9.1 A schematic representation ofthe process flow in lignocellulosic biodiesel production using oleaginous microbes, highlighting the focus areas for application of nanomaterials.

economic feasibility of microbial biodiesel production has been highlighted as a green catalyst for the synthesis of platform chemicals or catalytic conversion of inhibitory by-products into liquid fuels or fine chemicals. The chapter also discusses the flexibility of using nanomaterials for various added-value products from oleaginous microorganisms.

9.2 NANOBIOMATERIALS IN BIOCONVERSION OF THE LIGNOCELLULOSIC BIOMASS

LCB is the most abundant renewable feedstock for the production of liquid transport fuels and is relatively cheaper than crude oil (Ge et al. 2018). However, the natural recalcitrance of such materials presents a major bottleneck towards economical LCB-derived biofuel production (Phitsuwan et al. 2013). In this context, lignin-modifying enzymes (laccases, lignin peroxidases and manganese peroxidases) and cellulose-degrading enzymes (endoglucanases, xylanases and β-glucosidases) play an important role in the enzymatic pretreatment and hydrolysis of the biomass (Kainthola et al. 2019). Despite the availability of lignocellulolytic enzymes, the industrial-scale application and stability of such enzymes in biomass hydrolysis arelargely influenced by their properties such as catalytic activity, thermal stability, resistance to inhibitory products, nonspecific adsorption, reusability and inactivation due to shear stress (Periyasamy et al. 2018). Nanomaterials such as functionalized magnetic nanoparticles (NPs), carbon-based nanostructures (graphene oxide, nanosheets, nanoclays, etc.) and polymeric supports for enzyme immobilization have created a new domain for stabilization of lignocellulolytic enzymes (Gupta et al. 2018). Moreover, the inhibitory products such as levulenic acid, acetic acid, vanillin, furfural and 5-hydroxymethyl furfural (HMF), generated during the process of biomass pretreatment and hydrolysis can be converted into liquid biofuels. In this regard, porous nanobiomaterials such as micro/mesoporous carbon, zeolites, metal oxides, organic polymers, silica and resin are widely used as a green catalyst for the catalytic conversion of such platform chemicals into biodiesel (Bhanja and Bhaumick 2016).

9.2.1 Nanobiomaterials for Pretreatment and Enzymatic Hydrolysis

Pretreatment is one of the most important steps in the production of biodiesel from LCB. However, the conventional methods of biomass pretreatment involve the use of corrosive reagents, generation of toxic inhibitors and extensive washing to remove the acid or alkali completely. In this regard, acid-functionalized magnetic NPs (MNPs) can be employed as an alternative for pretreatment of LCB. For instance, alkylsulfonic acid, butylcarboxylic acid and sulphonic acid functionalized Fe_3O_4MNPs showed efficient catalytic activity in the pretreatment of sugarcane bagasse as compared to acid pretreatment (Ingle et al. 2020a). Moreover, such MNPs offer easy recovery and reusability and also lower the cost of the production process. For efficient biomass hydrolysis, polymer-enzyme nanobioconjugates, such as cellulases immobilized onto hydrogel reinforced with graphene oxide nanosheets, offer better enzyme stability, hydrolytic activity, storage and enhanced saccharification yield as compared to a free enzyme and enzyme-hydrogel conjugate (Ariaeenejad et al. 2020). Similarly, polymer-coated MNPs such as chitosan-coated Fe_3O_4@SiO_2 NPs or magnetic chitosan cross-linked with genipin are a stable supporting matrix for the immobilization of lignocellulolytic enzymes like *Trichoderma asperellum* laccase and *Thermomyces lanuginosus* xylanase, respectively. Such immobilized enzymes exhibit improved pH and thermal stability, reusability and higher delignification and hydrolysis potential (Gracida et al. 2019; Shanmugam et al. 2020). Furthermore, *Aspergillus niger* cellulases immobilized onto nanobiomaterials such as iron oxide NPs or polymer-based MNPs like β-cyclodextrin-conjugated magnetite NPs was used in combination with different pretreatment methods like alkali or ionic liquid to enhance hydrolysis of pretreated rice straw (Huang et al. 2015; Kumar et al. 2017)

9.2.2 Green Nanocatalyst for Synthesis and Conversion of Platform Chemicals

Pretreatment of LCB is often accompanied by the formation of inhibitory by-products such as phenolic components (vanillin, syringaldehyde, acetosyringone, etc.) originated from lignin degradation, C5 and C6 sugar derivatives (furfural and 5-HMF) and weak acids like acetic acid, formic acids, etc. (Kumar et al. 2015). These intermediate platform chemicals can be converted into liquid fuels using a biorefinery approach. In this context, several nanobiomaterials having high surface acidity and porous nanostructures play a crucial role as heterogeneous solid acid catalysts for the synthesis of platform chemicals from biomass-derived carbohydrates (Bhanja and Bhaumick 2016). For example, acid-base (-SO_3H/-NH_2) bifunctionalized mesoporous silica NPs were used for the one-pot synthesis of HMF from cellulose (Peng et al. 2012). Similarly, several doped metal oxide nanocatalysts such as Zn/CuO NPs, Ru/TiO_2 NPs and polymer-coated magnetic NPs (chitosan cross-linked with magnetite) have been used for the catalytic conversion of cellulosic biomass into platform chemicals and potential fuels like furfural, γ-valerolactone and oligomerization of itaconic anhydride (Hosseini et al. 2019; Mishra et al. 2019; Xu et al. 2018). Biomass-derived platform chemicals such as furan derivatives (2-methyl furan, 5-methy furfural) can be upgraded to biodiesel precursors containing a specific carbon range (C11 to C23) (Balakrishnan et al. 2014). Similarly, biowaste-derived levulinic acid and γ-valerolactone are precursor molecules for valeric biofuels (Galletti et al. 2013). In this context, a hydrotalcite-supported bimetallic (Ni-Cu) nanocatalyst was used as a stable, reusable solid base catalyst for the synthesis of hydrogenated biofuels (2,5-dimethylfuran, 2,5-dimethyltetrahydrofuran) from biomass-derived platform chemicals (Gupta et al. 2020).

9.3 Lipid Production by Oleaginous Microorganisms Using Nanobiomaterials

Conventional biodiesel production relied predominantly on plant-based edible (vegetable oils) and nonedible feedstock (oils derived from seeds of energy crops). However, food security issues pertaining to the use of edible oils and land area requirements for cultivation of energy crops encourage researchers and industrialists to explore sustainable alternatives for biodiesel production. Microbial lipids have garnered immense research interest as an alternative to these conventional sources. Oleaginous microorganisms include microalgae, filamentous fungi and yeasts that are capable of utilizing a widerange of renewable substrates and convert monomeric sugars into excess fats (stored as intracellular lipid droplets) under certain stress conditions (Bardhan et al. 2019b). Nutrient limitations such as depletion of nitrogen, phosphorous, NaCl or certain trace elements in the growth medium have been reported to induce lipogenesis or synthesis and accumulation of long-chain fatty acids in oleaginous microorganisms. However, the economic feasibility of microbial biodiesel largely depends on production and extraction of substantial quantities of lipids (Meng et al. 2009). Therefore, strategies to improve the cultivation of oleaginous microorganisms with higher lipid production parameters (biomass concentration, lipid concentration and lipid content) are of immense need at the present time. In this regard, nanobiomaterials such as metal and metal oxide NPs and carbon-based nanomaterials (nanotubes, nanosheets) find interesting applications in the cultivation, harvesting of and extraction of lipids from SCO-producing microbes (Rana et al. 2020).

9.3.1 Nanomaterials for Cultivation and Induction of Stress Environments

The trace elements present in the lipid production medium strongly influence not only the lipid productivity (g lipid/L/day) but also the fatty acid composition, which in turn determine the biodiesel properties. Besides, trace metal ions like Cu^{2+}, Zn^{2+}, $Mn^{2+}$and La^{3+} act as cofactors of key enzymes involved in lipid biosynthesis (Jernejc and Legisa 2002). Incorporation of metal and metal oxide NPs and carbonaceous nanomaterials in the culture medium have been reported to trigger growth and lipid production in a host of microalgal species, including *Chlorella vulgaris*, *Chlorella pyrenoidosa*, *Chlorella sorokiniana*, *Scenedesmus obliquus* and *Nannochloropsis oculata* (Nguyen et al. 2019; Rana et al. 2020). Additionally, metal NPs are associated with induction of a stress environment, such as oxidative stress (due to the generation of reactive oxygen species [ROS]) that has been reported to stimulate growth, lipid production and a fatty acid profile in oleaginous microorganisms (Shi et al. 2017). Such stress-induced lipid accumulation is always accompanied with increased antioxidant defenses like ROS-scavenging enzymes (stress-responsive proteins). In fact, exogenous exposure to certain nanomaterials like TiO_2 NPs has been shown to trigger intracellular neutral lipid (unsaturated fatty acids) formation in the yeast *Pichia pastoris* as a response to cope with oxidative stress (Yu et al. 2015). Table 9.1 depicts the nanomaterials that has been used for stress induction to improve lipid accumulation in oleaginous microorganisms.

9.3.2 Functionalized Nanomaterials for Harvesting of Oleaginous Microbes

The downstream processing technology is one of the major determinants in the economic commercialization of the microbial lipid-based biodiesel production process. This includes efficient harvesting of oleaginous microorganisms from the production medium using flocculation (for separation of microalgae from a low-concentration culturebroth), centrifugation (for yeasts) and filtration (for filamentous fungi) (Yellapu et al. 2018). In this regard, metallic NPs and polymer conjugated MNPs like aminoclay-TiO_2, bare-Fe_3O_4, Zn-Mg-ferrite, magnesium aminoclay-Fe_3O_4, cetrimonium bromide-Fe_3O_4 and chitosan-Fe_3O_4 find wide application in microalgal harvesting (Lee et al. 2013; Lee et al. 2014b; Seo et al. 2016; Japar et al. 2017; Kim et al. 2018b). Microalgae separation is based on the temporary agglomeration between negatively charged cells in the culture media and external magnetic force of iron oxide NPs (Yang et. al. 2018). Moreover, functionalized Fe_3O_4 NPs are preferred due to their large surface area, biocompatibility and surface charge characteristics that facilitate electrostatic flocculation (Xu et al. 2011; Lee et al. 2014a). In a study reported by (Zhao et al. 2015), magnetic flocculation using Fe_3O_4 NPs coated with polyaluminium chloride and polyacrylamide could harvest 99% of *C. vulgaris* cells in less than 0.5 min. Similarly, amino acid (arginine)–modified Fe_3O_4 NPs at an optimized dosage of 200 mg/L was used to achieve 95% harvesting efficiency and lower cost (347 USD/t of harvested C. *vulgaris* biomass) (Liu et al. 2020a). MNP-based flocculation is considered cost-effective with a high efficiency of cell recovery and is feasible for large-scale application (Yang et al. 2018). Apart from flocculation, some other techniques for microalgal separation include floatation, chemical coagulation and filtration, but they have several limitations, like unfeasibility for marine microalgae harvesting, expensive chemical flocculants and high operational and maintenance cost, respectively (Kurniawati et al. 2014; Barros et al. 2015; Zhao et al. 2017)

Although, filtration is preferred for the separation of filamentous fungi, biomass harvesting for oleaginous yeast and bacteria from the lipid production medium has been predominantly done by using batch centrifugation (Dassey and Theegala 2013; Bardhan et al. 2019a). A huge knowledge gap has been observed in the literature on the applications of nanobiomaterials as an intensive and cost-effective tool for biomass harvesting of yeasts and fungi. However, metal nanomaterials like silver NPs have been reported to act as elicitors for cell growth, lipid and fatty acids in co-cultures of *C.vulgaris and Dictyochloropsis splendida* (Shanab et al. 2019). Similar effects of metal NPs on lipid production could be investigated for co-cultivation of microalgae—yeast or fungi.

9.3.3 Functionalized Nanomaterials for Cell Disruption and Lipid Extraction

Cell lysis is a prerequisite for intracellular lipid extraction from oleaginous microbes. It is used to disrupt the tough outer cell wall, which is made of polysaccharides like chitin (yeast and fungi), cellulose, hemicellulose and small amounts of lignin-like compounds in the case of green microalgae (Domozych et al. 2012; Welker et al. 2015). The economic feasibility of the extraction process

TABLE 9.1
Nanomaterials in Stress Induction for Improved Lipid Accumulation in Oleaginous Microorganisms

Oleaginous Microorganisms	Nanobiomaterials	Size (nm)	Optimum Dosage (mg/L)	Stress	Improved Parameters	Ref.
Yeast Strains						
Candida sp., *Rhodotorula glutinis*, *Trichosporon cutaneum*, and *Yarrowia lipolytica*	Zero-valent iron nanoparticles	50	9–13	Oxidative stress (ROS)	Biomass, lipid and PUFA (linoleic acid)	(Pádrová et al. 2016)
-Do-	Low concentration of lanthanides or monazite	-	0.03–3	Ca^{2+} ion substitutes (inhibition of Ca^{2+}-dependent enzymes)	Increased cell mass/lipids, higher concentration of palmitoleic acid, ω6-linoleic acid	(Kolouchova et al. 2016)
Microalgae						
Chlorella pyrenoidosa	Nitrogen-doped carbon and graphene nanosheets	-	1–5 mg/50 mL	Growth supporting factor/ oxidative stress	2- and 6-fold increase in lipid content for N-doped carbon and graphene, respectively	(Khanra et al. 2018)
Scenedesmus obliquus	α-Fe_2O_3 NPs, MgO NPs and carbon nanotubes (CNT)	α-Fe_2O_3 (<30) MgO (<50) CNT (diameter<2, length<20 μm)	5 for CNT and Fe_2O_3 NPs; 40 for MgO NPs	Oxidative stress	8.9%, 39.6% and 18.9% increase in lipid productivity for α-Fe_2O_3 NPs, MgO NPs and CNT, respectively	(He et al. 2017)
Chlorella pyrenoidosa	α-Fe_2O_3 NPs	< 50	30	Oxidative stress	Biomass (2.01 g/L) and lipid yield (16.89 wt%)	(Rana et al. 2020)
Chlorella sp. KR-1	Magnesium aminoclay NPs	-	10–100	Oxidative stress	Lipid productivity (410 mg FAME/L/d), inhibit contaminating microbes	(Kim et al. 2016)

depends on lysis methods with low-energy inputs and high yield (Probst et al. 2016). Using metal nanoparticles for cell lysis, wherein the microalgal cell wall is bombarded with high-velocity NPs under an electromagnetic field for efficient penetration, is an innovative method that was patented by Raytheon Company (Costas and Eck 2014). Apart from the physical damage, metallic NPs are effective cell wall–disrupting agents due to the generation of ROS that can cause considerable cytotoxicity, resulting in plasmolysis (Lee et al. 2017). Moreover, charged NPs are easy to separate and reuse by applying an external magnetic field. Nano-sized NiO NPs (<50 nm) were employed by (Huang and Kim 2016) for both disruption and harvesting (with 98.8% efficiency) of *C. vulgaris* cells. Further, cationic surfactant (CTAB)–decorated Fe_3O_4 NPs have found applications in integrated harvesting and cell disruption (Seo et al. 2016). In this context, the quaternary ammonium heads of the surfactants not only flocculate microlagal cells but also weaken its thick cell wall. In another study, cell wall–degrading enzymes (cellulase and lysozyme) were coimmobilized onto amino-functionalized nanocarrier (MNPs) for disruption of *Nannpchrolopsis* sp. cell wall (Chen et al. 2018). The enzymes were reported to exhibit thermal stability and wide pH tolerance and could retain up to 60% of residual activity after six times of recycled use.

For cell disruption of wet oleaginous yeast strains like *Lipomyces starkeyi*, *Cutaneotrichosporon curvatus*, *Rhodotorula toruloides* and *Rhodotorula mucilaginosa*, mild acid treatment at an elevated temperature led to the highest lipid recovery yields as compared to other physico-chemical cell lysis techniques like microwave-assisted, bead beating, cryopulverization, high-pressure homogenization and enzymatic treatments (Khot and Ghosh 2017; Kruger et al. 2018). Although charged silver NPs have been demonstrated to show cellular internalization leading to toxicity in non-oleaginous model yeast like *Saccharomyces cerevisiae*, the metallic NPs, to the best of our knowledge, still have not been used for cell disruption of yeast and fungi.

Nanomaterials also find wide applications in enhanced lipid extraction from the oleaginous microbial biomass. For instance, in the presence of silica-coated MNPs (Fe_3O_4), the amount of lipid obtained from *Chlorella pyrenoidosa* was four times thatcompared to the control (Vashist et al. 2020). In another study, it was observed that a low dosage (0.45 mM concentration) of cationic surfactant (hexadecyltrimethylammonium) was associated with enhanced lipid recovery from *Chlorella* biomass using ethyl acetate as a solvent (Lai et al. 2017). In this context, Fe_3O_4 NPs decorated with cationic surfactants like cetrimonium bromide were efficiently used for improved total extracted lipid content. Moreover, cells could be easily detached from microalgae- Fe_3O_4 flocs using an anionic surfactant like sodium dodecyl sulfate (SDS), and the free CTAB-decorated Fe_3O_4 NPs were reused with 80% efficiency (Seo et al. 2016).

9.4 NANOMATERIALS FOR CONVERSION OF MICROBIAL OIL INTO BIODIESEL

Biodiesel fuel, consisting of long-chain fatty acid methyl esters (FAMEs), is obtained by the catalytic transesterification of vegetable oils, waste cooking oil, animal tallow, microbial lipids or triacylglycerides (TAG) with methanol (Figure 9.2). The physicochemical properties of biodiesel such as flash point, lubricity and cetane number have been found to be better as compared to fossil oil (Datta and Mandal 2016). In this context, nanomaterials are employed as either a heterogeneous nanocatalyst or nanosupport matrix for immobilized enzymes to perform the transesterification reaction.

9.4.1 Renewable Heterogeneous Base Nanocatalyst for Transesterification

For commercial purposes, a heterogeneous base catalyst is predominantly utilized for the transesterification process over a homogeneous catalyst such as carbonates, hydroxides or potassium and sodium alkoxide (Xie et al. 2017). The heterogeneous base catalyst offers several advantages like easy separation, reusability and does not generate any waste product post-removal from the reaction

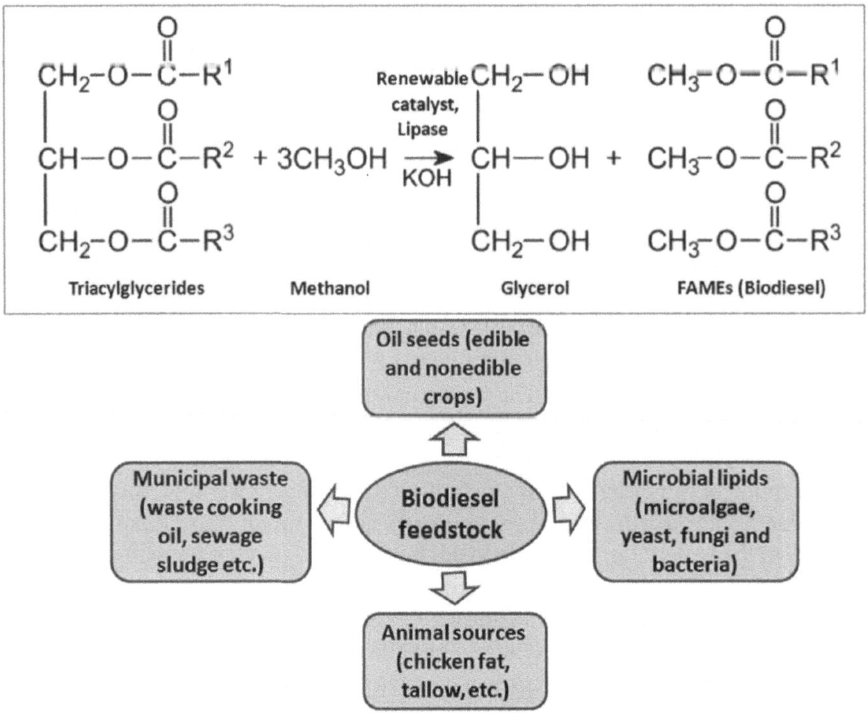

FIGURE 9.2 The catalytic transesterification reaction of triglycerides into fatty acid methyl esters and glycerol. Additionally, the different feedstocks for biodiesel production is depicted in a flow chart below.

mixture. Moreover, it can be derived from a variety of renewable sources like banana peels, yeast deoiled cake, rice husk, etc. (Zhao et al. 2018; Bardhan et al. 2019a, Gohain et al. 2020). The catalytic activity can be further improved by preparing it in a nano-form, thereby increasing its specific surface area. In this regard, a nanocatalyst representing nanomaterials or nanocrystals with a particle size between 1 and 100 nm hasgarnered recent research interest related to biodiesel synthesis (I.B. Banković–Ilić et al. 2017). Several studies have reported the application of heterogeneous nanocatalysts like a polyethylene glycol–encapsulated Mn-ZnO nanocomposite, nano-CaO derived from eggshell waste and magnetic $KOH/Fe_3O_4@Al_2O_3$ in biodiesel production from oils of different oleaginous microalgae such as *Nannochloropsis oculata*, *Scenedesmus armatus* and a mixed culture of microalgae (Kazemifard et al. 2019; Pandit and Fulekar 2019; Raj et al. 2019). The nanocatalysts, due to their unique properties such as high activation energy, specific reactivity, controlled rate of reaction and easy recyclability, have potential applications in biodiesel production from a variety of microbial oil sources, including oleaginous yeast, fungi and bacteria (Ingle et al. 2020b). However, a nanocatalyst has been predominantly used for biodiesel conversion from plant-based edible and nonedible oils, animal fats and waste cooking oil. The various renewable nanocatalysts for biodiesel conversion from a variety of sources are listed in (Table 9.2).

9.4.2 Nanobiomaterials for Immobilized Lipase-Based Biocatalytic Systems

Another approach is the enzymatic transesterification using microbial lipases for the conversion of SCO, waste cooking oil (WCO) and plant-based oils into biodiesel. However, nonedible oil and WCO are rich in free fatty acid and serve as a suitable resource for lipase-mediated transesterification (Kim et al. 2018a). Although lipases are the most promising enzymes for biodiesel conversion, the high cost of enzyme production coupled with the long reaction time remain the major

TABLE 9.2
Renewable Heterogeneous Nanocatalyst in Transesterification of Oil into Biodiesel

Source of Biodiesel	Renewable Heterogeneous Nanocatalyst	Optimum Transesterification Condition (Temperature, Time, Catalyst Loading, Biomass-to-Methanol Ratio)	Biodiesel Yield (%)	Advantages	References
Algal oil					
Neochloris oleoabundans	Fe_2O_3 NPs synthesized from plant extract	65 °C, 6 h, 1 wt%, 1:5	81%	Higher biodiesel yield as compared to HCl/NaOH catalyst, lesser environmental footprints	(Banerjee et al. 2019)
Chlorella vulgaris	CaC/TiO_2 nano-photocatalyst (in supercritical methanol medium)	260 °C, 1 h, 9–10 MPa pressure, 0.3g algal powder: 12 cm^3 methanol	Total product yield (up to 51.6%), FAMEs yield (up to 28.1%)	Generation of valuable by-products (hydrocarbons, oxygenates) No soap formation, higher yield and low reaction time	(Aghilinategh et al. 2019)
Marine brown algae (*Phaeodactylum tricornutum* and *Isochrysis* sp.)	Biochar and iron nanoparticles from plant leaves	9 wt%, 12 h, 100 rpm, 80 °C, 6:1	Biochar (32.8) FeNPs (28.87)	Enhanced recovered FAME mass%	(Anto et al. 2019)
Ulva lactuca seaweed	TiO_2–ZnO nanocomposite	4 wt% catalysts at 4 h under 60 °C	82.8%.	High yield due to increased accessibility on FFA, methanol and triglycerides, low cost, eco-friendly, less energy consuming	(Gurusamy et al. 2019)
Nannochloropsis sp.	Nano- $Ca(OCH_3)_2$ (calcium methoxide)	3 wt.%, 30:1 (methanol to oil), 3 h at 80 °C	99.0%	High yield, excellent catalytic performance, high specific surface area, pore size, methoxy (CH_3O-) species active sites	(Teo et al. 2016)
Plant-based oils, animal fats and waste cooking oil					
Waste cooking oil	Zn doped CaO nanocatalyst (waste eggshell derived)	5 wt%, 65 °C, 4 h, 20:1(methanol to oil)	96.74%	Excellent catalytic activity, reusability	(Borah et al. 2019)
Chicken fat	$AC/CuFe_2O_4$@CaO Activated carbon (AC) derived from lotus leaves	3 wt%, 65 °C, 4 h, 12:1(methanol to oil)	95.6%	Increased biodiesel production efficiency at optimized parameters	(Seffati et al. 2020)
Jatropha curcus	Porous $CaSO_4/Fe_2O_3$-SiO_2 core shell NPs	4 h, 120 °C, 12wt.%, 9:1	94%	Core shell nanostructure, good catalytic activity, easy dispersibility, separation, reusability	(Teo et al. 2019)
Neem oil	Cu-doped ZnO	10%, 55 °C, 1 h, 1:10	97.18%	Porous, rough surface morphology, good stability, reusability	(Gurunathan and Ravi 2015)

bottlenecks for the economic commercialization of the process (Amini et al. 2017; Moazeni et al. 2019). However, sequential integrated processes like oil industry wastewater treatment using lipolytic bacterial strains and transesterification of nonedible plant oils for high-quality biodiesel are an important perspective for overcoming the cost limitation (Rana et al. 2019). Moreover, immobilized lipases are considered highly efficient and environmentally friendly biocatalysts for the production of biodiesel. In this context, nanobiomaterials like MNPs, silica NPs, carbon nanotubes, organic polymers, metal-organic frameworks and nanohybrids are a suitable support matrix for enzyme immobilization due to large surface area, low mass transfer limitation, high enzyme loading capacity, good dispersibility and biocompatibility (Zhong et al. 2020). Lipase immobilized onto functionalized magnetic Fe_3O_4 NPs have been used for biodiesel synthesis from SCO derived from filamentous oleaginous fungi *Aspergillus niger* strain SF2, which was isolated from a tannery effluent–contaminated soil (Jambulingam et al. 2019). Similarly, lipids obtained from microalgal biomass like *C. vulgaris* and *C. salina* were transesterified into biodiesel using immobilized lipase loaded on functionalized super MNPs (Duraiarasan et al. 2016; Nematian et al. 2020). Lipase immobilization strategies on various nanobiomaterials supports include adsorption, cross-linking, covalent binding, entrapment and whole-cell immobilization (Zhong et al. 2020). Further, lipase-based biocatalysts provide high catalytic activity, reusability and mechanical stability that compensate for the high cost of the enzymes and make the biocatalytic process promising for biodiesel production from microbial oil sources. Lipases are primarily obtained from microbial sources like yeast (*Candida rugosa*), filamentous fungi (*Penicillium citrinum*) and bacteria (*Pseudomonas aeruginosa*) (Raoufi and Gargari 2018; Bardhan et al. 2019a; Fernandes et al. 2021). However, lipases derived from plant and animal sources like cocoa pod husk, oil palm fruit and Pacific white shrimps have also been used for biodiesel synthesis (Khanahmadi et al. 2016; Suwanno et al. 2017; Kuepethkaew et al. 2017) Table 9.3 lists some of the lipase-based biocatalysts used for biodiesel production from SCO and other nonmicrobial sources.

9.5 NANOBIOMATERIALS AND ADDED-VALUE PRODUCTS FROM OLEAGINOUS MICROBES

Oleaginous microorganisms are able to produce a variety of high-value products such as pigment molecules (chlorophyll, phycocyanin, astaxanthin and carotenoids), protein- and polysaccharide-rich streams of deoiled biomass, metabolic coproducts (organic acids and polyunsaturated fatty acids) and molecules like free fatty acids and fatty alcohols that are precursors of several renewable oleochemicals like surfactants, plastic and soaps (Figure 9.3) (Bardhan et al. 2020). In this context, microalgal species like *Haematococcus pluvialis*, *Spirulina platensis*, *Chlorella pyrenoidosa*, and *Chlorellasaccharophila* are prolific producers of natural pigment molecules like astaxanthin, phycocyanin, chlorophyll, zeaxanthin and β-carotene (K.W. Chew et al. 2017). Such pigment molecules have antioxidant properties and find wide application as human food ingredients and feed additives for aquaculture. Similarly, several oleaginous yeast and fungi such as *Rhodotorula mucilaginosa*, *Phaffia rhodozyma*, *Sporobolomyces salmonicolor*, and *Penicillium* sp. are suitable for production of lipid-soluble pigments like carotenoids that have anticancer and pro–vitamin A functions (Cheng and Yang 2016). Besides, oleaginous microorganisms also produce a variety of metabolic coproducts, including long-chain omega-fatty acids and volatile fatty acids that are used as nutraceuticals and in the chemical industry, respectively (Athenaki et al. 2018; Filippousi et al. 2019). Moreover, the nutrient-rich deoiled algal biomass has wide applications as a feedstock for biohydrogen, bioethanol and biopolymer (polyhydroxybutyrate) and biodiesel production (Subhash and Mohan 2014; Arora et al. 2016; Kumar et al. 2020). Recently, we demonstrated the feasibility of oleaginous yeast *R. mucilaginosa* deoiled cake as a source of a renewable heterogeneous catalyst (Gohain et al. 2020). Furthermore, deoiled biomass can also be converted into novel biomaterials like nanoparticles and carbon dots, which find immense application in biomedical and bioenergy (Maurya et al. 2016). In this regard, biosynthesis of silver NPs with antioxidant and antimicrobial properties from

TABLE 9.3
Lipase-Based Biocatalysts Used for Biodiesel Production

Source of Lipase	Nanobiomaterial Support	Immobilization Strategy	Source of Biodiesel Extraction	FAME Yield or Biodiesel Conversion (%)	Advantages	References
Pseudomonas sp. ISTPL3	Activated biochar impregnated with calcite (prepared by sol-gel method)	Covalent binding	*Serratia* sp. ISTD04	97.41	Increased enzyme stability and reusability	(Kumar and Thakur 2020)
Aspergillus sp. and *Candida* sp.	-	Whole cell immobilization	*Scenedesmus obliquus*	90.82 ± 1.43	Sustainable, reusable without loss in conversion efficiency	(Guldhe et al. 2016)
Burkholderia sp.	Alkyl-grafted nanocomposites (Fe_3O_4–SiO_2)	Mechanical stirring on magnetic device followed by lyophilization	*Chlorella vulgaris*	(97.3 wt% oil)	Tolerate high methanol to oil molar ratio, recyclability up to six times without loss of activity	(Tran et al. 2012)
Commercial purified lipase	Magnetic NPs (coprecipitation method using $FeSO_4$ and $FeCl_3$)	Covalent binding	*Aspergillus niger* strain SF2	84%	Improved storage stability (60 days), reusability without any change in esterase activity	(Jambulingam et al. 2019)
Candida rugosa	Graphene oxide Fe_3O_4 nanocomposite	Interfacial activation on hydrophobic surfaces	Soyabean oil	92.8%	Easily recovered by application of magnetic field, recyclable	(Xie and Huang 2018)
Candida antarctica	Fe_3O_4 core with silica shell (Fe_3O_4@SiO_2)	Covalent immobilization	Waste cooking oil		Higher thermal stability, methanol tolerance, reusability	(Mehrasbi et al. 2017)
Pseudomonas fluorescens	Single-walled carbon nanotubes derivatized with glycerol diglycidyl ether	Covalent binding	Sunflower oil	99%	High stability, retained 99% catalytic activity after multiple recycling	(Bartha-Vári et al. 2020)
Burkholderia cepacia	Super paramagnetic multiwalled carbon nanotubes (functionalized with polyamidoamine dendrimers)	Covalent binding	Soyabean oil	92.8	Enhanced thermostability and pHresistance, recyclable, easily separated using magnet	(Fan et al. 2017)
Rhizopus arrhizus	Lipase NPs (with methyl-β-cyclodextrin as lyoprotectant)	Co-lyophilization phase separation method	Olive oil	98%	Enhanced activity and increased conversion yield	(Sharma et al. 2018)

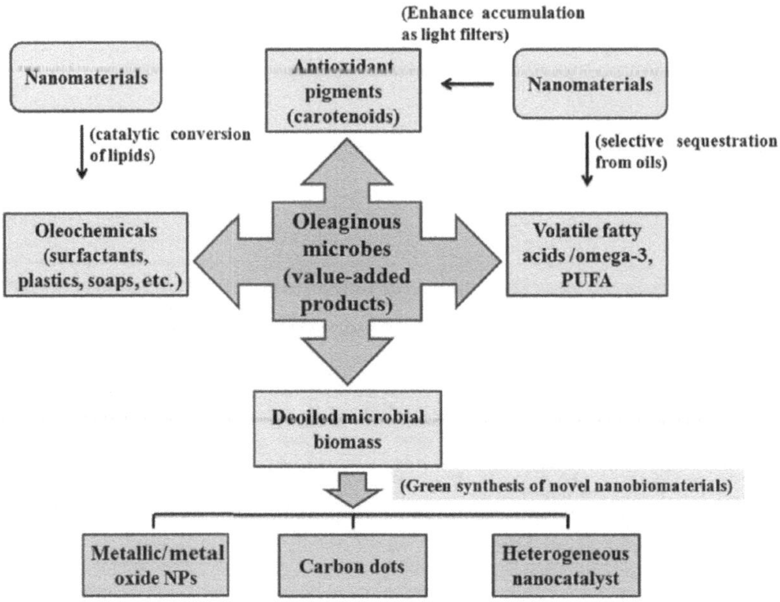

FIGURE 9.3 A schematic representation of the various value-added products that are obtained from oleaginous microbes, highlighting the focus areas for application of nanomaterials.

deoiled biomass of a variety microalgal species such as *Acutodesmus dimorphus*, *Saccharina japonica* and *Trichodesmium erythraeum* has been reported by several researchers (Chokshi et al. 2016; Sivagnanam et al. 2017; Sathishkumar et al. 2019). Biogenic synthesis of metallic and metal oxide NPs from algal waste reduces the cost of biodiesel production and also discourages the use of hazardous chemicals, thereby promoting 'green synthesis' (Sharma et al. 2019). Metal nanoparticles like gold nanorods and silver NPs around microalgal culture flasks have been found to enhance accumulation of natural pigments (chlorophyll, carotenoids) in *Chlorella vulgaris* as a wavelength-specific backscattering light filter (Eroglu et al. 2013). On the contrary, exposure to SiO_2 NPs had a negative effect on chlorophyll and the carotenoid content of *Dunaliella salina* in a concentration-dependent manner, indicating a significant toxic effect (Shariati and Shirazi 2019). Nanomaterials like aminopropyl functionalized mesoporous silica NPs have been used for selective sequestration of omega-3 and other polyunsaturated free fatty acids from microalgal oil (Valenstein et al. 2012). Similarly, hybrid nanostructured materials consisting of Ni NPs and aminopropyl groups immobilized onto silica NPs were used to selectively capture and convert free fatty acids into saturated hydrocarbons (Kandel et al. 2013). In another study, presence of engineered ZnO nanomaterials was found to correlate positively with enhanced accumulation of volatile fatty acids in sludge anaerobic digestion (Zhang et al. 2017). Nanomaterials play an important role in the catalytic conversion of microbial lipids (triacylglycerides) into a variety of renewable oleochemicals. For instance, surfactant-capped tungsten oxide NPs were employed for oxidative cleavage of unsaturated fatty acids (oleic acid) present in lipids and convert them into valuable materials like monocarboxylic and dicarboxylic acids (Kerenkan et al. 2016).

9.6 CURRENT CHALLENGES AND FUTURE PERSPECTIVES

Some of the challenges associated with the use of nanomaterials are as follows:

1. High cost of fabrication techniques for nanomaterials such as molecular self-assembly, electrospinning and nano-patterning. Therefore, research efforts are underway to develop

easy, efficient and low-cost fabrication methods with potential scale up synthesis for application in biotechnology and medicine.

2. Assessment of cytotoxicity and biodegradability of nanomaterials is important for microbial biotechnology applications. Incorporation of nanomaterials in culture media as an additive is associated with induction of oxidative stress mediated through the generation of ROA. Although such stress conditions might enhance lipid accumulation in oleaginous microbes, a higher concentration of nanomaterials is toxic to the producer microorganism. Additionally, biodegradable nanomaterials are considered safe for microbial applications as well as environmentally friendly.

9.7 CONCLUSION

Conventional methods of biodiesel production using edible oils, nonedible oils and animal tallow havegarnered negative criticism owing to huge land area use and compromising food security. LCB-based microbial biodiesel is considered an alternative to sustain the biodiesel industry. Nanomaterials are less energy-demanding, reusable and easy to separate and thus offer great advantage in terms of economic feasibility of the biodiesel production process. Nanobiomaterials are widely used for pretreatment and hydrolysis of LCB as a suitable support matrix for lignocellulolytic enzymes and as acid functionalized nanomaterials. Nanotechnology has the potential to synthesize and convert biomass-derived platform chemicals into liquid fuels; stimulate lipid accumulation during cultivation of oleaginous microorganism; facilitate harvesting, cell disruption, and lipid extraction; and selectively sequester polyunsaturated fatty acids. The use of nanomaterials as heterogeneous nanocatalysts or nanosupports for lipase immobilization offers several advantages over traditional base catalyzed transesterification of oils into biodiesel. Nanotechnology also plays a significant role in the production and transformation of added-value products from oleaginous microbes. Thus, nanotechnology holds the potential to make the biodiesel production process using oleaginous microorganisms sustainable and cost-effective.

9.8 ACKNOWLEDGMENTS

The Department of Biotechnology-Junior research Fellowship (DBT/JRF/BET-16/I/2016/AL/110-501 dated June 29, 2016) to Mr. Pritam Bardhan for his PhD work from Department of Biotechnology, Government of India (GOI), New Delhi, India, is highly acknowledged.

REFERENCES

Aghilinategh, M., Barati, M., and Hamadanian, M. 2019. Supercritical methanol for one put biodiesel production from chlorella vulgaris microalgae in the presence of CaO/TiO2 nano-photocatalyst and subcritical water. *Biomass and Bioenergy* 123: 34–40.

Amini, Z., Ilham, Z., Ong, H. C., Mazaheri, H., and Chen, W. H. 2017. State of the art and prospective of lipase-catalyzed transesterification reaction for biodiesel production. *Energy Conversion and Management* 141: 339–353.

Anto, S., Karpagam, R., Renukadevi, P., Jawaharraj, K., and Varalakshmi, P. 2019. Biomass enhancement and bioconversion of brown marine microalgal lipid using heterogeneous catalysts mediated transesterification from biowaste derived biochar and bionanoparticle. *Fuel* 255: 115789.

Ariaeenejad, S., Motamedi, E., and Salekdeh, G. H. 2020. Stable cellulase immobilized on graphene oxide@ CMC-g-poly (AMPS-co-AAm) hydrogel for enhanced enzymatic hydrolysis of lignocellulosic biomass. *Carbohydrate Polymers*, 230: 115661.

Arora, N., Patel, A., Pruthi, P. A., and Pruthi, V. 2016. Recycled de-oiled algal biomass extract as a feedstock for boosting biodiesel production from *Chlorella minutissima*. *Applied Biochemistry and Biotechnology* 180: 1534–1541.

Athenaki, M., Gardeli, C., Diamantopoulou, P., Tchakouteu, S. S., Sarris, D., Philippoussis, A., and Papanikolaou, S. 2018. Lipids from yeasts and fungi: physiology, production and analytical considerations. *Journal of Applied Microbiology* 124: 336–367.

Balakrishnan, M., Sacia, E. R., and Bell, A. T. 2014. Syntheses of biodiesel precursors: sulfonic acid catalysts for condensation of biomass-derived platform molecules. *ChemSusChem*, 7: 1078–1085.

Banerjee, S., Rout, S., Banerjee, S., Atta, A., and Das, D. 2019. Fe_2O_3 nanocatalyst aided transesterification for biodiesel production from lipid-intact wet microalgal biomass: a biorefinery approach. *Energy Conversion and Management*, 195: 844–853.

Banković–Ilić, I. B., Miladinović, M. R., Stamenković, O. S., and Veljković, V. B. 2017. Application of nano CaO–based catalysts in biodiesel synthesis. *Renewable andSustainable Energy Reviews* 72: 746–760.

Bardhan, P., Gohain, M., Daimary, N., Kishor, S., Chattopadhyay, P., Gupta, K., Chaliha, C., Kalita, E., Deka, D., and Mandal, M. 2019a. Microbial lipids from cellulolytic oleaginous fungus Penicillium citrinum PKB20 as a potential feedstock for biodiesel production. *Annals of Microbiology* 69: 1135–1146.

Bardhan, P., Gupta, K., Kishor, S., Chattopadhyay, P., Chaliha, C., Kalita, E., Goud, V. V., and Mandal, M. 2020. Oleaginous yeasts isolated from traditional fermented foods and beverages of Manipur and Mizoram, India, as a potent source of microbial lipids for biodiesel production. *Annals of Microbiology* 70: 1–14.

Bardhan, P., Gupta, K., and Mandal, M. 2019b. Microbes as bio-resource for sustainable production of biofuels and other bioenergy products. In *New and Future Developments in Microbial Biotechnology and Bioengineering*, ed. J. S. Singh, and D. P. Singh, 205–222. Elsevier.

Barros, A. I., Gonçalves, A. L., Simões, M., and Pires, J. C. 2015. Harvesting techniques applied to microalgae: a review. *Renewable and Sustainable Energy Reviews* 41: 1489–1500.

Bartha-Vári, J. H., Moisă, M. E., Bencze, L. C., Irimie, F. D., Paizs, C., and Toşa, M. I. 2020. Efficient biodiesel production catalyzed by nanobioconjugate of lipase from Pseudomonas fluorescens. *Molecules*, 25: 651.

Bhanja, P., and Bhaumik, A. 2016. Porous nanomaterials as green catalyst for the conversion of biomass to bioenergy. *Fuel* 185: 432–441.

Borah, M. J., Das, A., Das, V., Bhuyan, N., and Deka, D. 2019. Transesterification of waste cooking oil for biodiesel production catalyzed by Zn substituted waste egg shell derived CaO nanocatalyst. *Fuel* 242: 345–354.

Buzea, C., Pacheco, I. 2017. Nanomaterials and their classification. In: *EMR/ESR/EPR Spectroscopy for Characterization of Nanomaterials*, ed. A. K. Shukla, 3–45. Springer.

Chen, Q., Liu, D., Wu, C., Yao, K., Li, Z., Shi, N., Wen, F., and Gates, I. D. 2018. Co-immobilization of cellulase and lysozyme on amino-functionalized magnetic nanoparticles: an activity-tunable biocatalyst for extraction of lipids from microalgae. *Bioresource Technology* 263: 317–324.

Cheng, Y. T., and Yang, C. F. 2016. Using strain *Rhodotorula mucilaginosa* to produce carotenoids using food wastes. *Journal of the Taiwan Institute of Chemical Engineers* 61: 270–275.

Chew, K. W., Yap, J. Y., Show, P. L., Suan, N. H., Juan, J. C., Ling, T. C., Lee, D. J., and Chang, J. S. 2017. Microalgae biorefinery: high value products perspectives. *Bioresource Technology* 229: 53–62.

Chokshi, K., Pancha, I., Ghosh, T., Paliwal, C., Maurya, R., Ghosh, A., and Mishra, S. 2016. Green synthesis, characterization and antioxidant potential of silver nanoparticles biosynthesized from de-oiled biomass of thermotolerant oleaginous microalgae *Acutodesmus dimorphus*. *RSC Advances* 6: 72269–72274.

Costas, C. R., and Eck, C. R. 2014. *U.S. Patent No. 8,722,375.* Washington, DC: U.S. Patent and Trademark Office.

Dassey, A. J., and Theegala, C. S. 2013. Harvesting economics and strategies using centrifugation for cost effective separation of microalgae cells for biodiesel applications. *Bioresource Technology* 128: 241–245.

Datta, A., and Mandal, B. K. 2016. A comprehensive review of biodiesel as an alternative fuel for compression ignition engine. *Renewable and Sustainable Energy Reviews* 57: 799–821.

Domozych, D. S., Ciancia, M., Fangel, J. U., Mikkelsen, M. D., Ulvskov, P., and Willats, W. G. 2012. The cell walls of green algae: a journey through evolution and diversity. *Frontiers in Plant Science* 3: 82.

Duraiarasan, S., Razack, S. A., Manickam, A., Munusamy, A., Syed, M. B., Ali, M. Y., Ahmed, G. M., and Mohiuddin, M. S. 2016. Direct conversion of lipids from marine microalga C. salina to biodiesel with immobilised enzymes using magnetic nanoparticle. *Journal of Environmental Chemical Engineering* 4: 1393–1398.

Eroglu, E., Eggers, P. K., Winslade, M., Smith, S. M., and Raston, C. L. 2013. Enhanced accumulation of microalgal pigments using metal nanoparticle solutions as light filtering devices. *Green Chemistry* 15: 3155–3159.

Fan, Y., Su, F., Li, K., Ke, C., and Yan, Y. 2017. Carbon nanotube filled with magnetic iron oxide and modified with polyamidoamine dendrimers for immobilizing lipase toward application in biodiesel production. *Scientific Reports* 7: 1–13.

Fernandes, K. V., Cavalcanti, E. D., Cipolatti, E. P., Aguieiras, E. C., Pinto, M. C., Tavares, F. A., da Silva, P. R., Fernandez-Lafuente, R., Arana-Pena, S., Pinto, J. C., and Freire, D. M. 2021. Enzymatic synthesis of biolubricants from by-product of soybean oil processing catalyzed by different biocatalysts of Candida rugosa lipase. *Catalysis Today* 362: 122–129.

Filippousi, R., Antoniou, D., Tryfinopoulou, P., Nisiotou, A. A., Nychas, G. J., Koutinas, A. A., and Papanikolaou, S. 2019. Isolation, identification and screening of yeasts towards their ability to assimilate biodiesel-derived crude glycerol: microbial production of polyols, endopolysaccharides and lipid. *Journal of Applied Microbiology* 127: 1080–1100.

Galletti, A. M. R., Antonetti, C., Ribechini, E., Colombini, M. P., Di Nasso, N. N., and Bonari, E. 2013. From giant reed to levulinic acid and gamma-valerolactone: a high yield catalytic route to valeric biofuels. *Applied Energy* 102: 157–162.

Gatoo, M. A., Naseem, S., Arfat, M. Y., Mahmood Dar, A., Qasim, K., and Zubair, S. 2014. Physicochemical properties of nanomaterials: implication in associated toxic manifestations. *BioMed Research International* 2014: 1–8.

Ge, X., Chang, C., Zhang, L., et al. 2018. Conversion of lignocellulosic biomass into platform chemicals for biobased polyurethane application. In *Advances in Bioenergy*, 161–213. Elsevier.

Gohain, M., Bardhan, P., Laskar, K., Sarmah, S., Mandal, M., Bora, U., Kalita, M. C., Goud, V. V., and Deka, D. 2020. *Rhodotorula mucilaginosa*: a source of heterogeneous catalyst for biodiesel production from yeast single cell oil and waste cooking oil. *Renewable Energy* 160: 220–230.

Gracida, J., Arredondo-Ochoa, T., García-Almendárez, B. E., Escamilla-García, M., Shirai, K., Regalado, C., and Amaro-Reyes, A. 2019. Improved thermal and reusability properties of xylanase by genipin cross-linking to magnetic chitosan particles. *Applied Biochemistry and Biotechnology* 188: 395–409.

Guldhe, A., Singh, P., Kumari, S., Rawat, I., Permaul, K., and Bux, F. 2016. Biodiesel synthesis from microalgae using immobilized Aspergillus niger whole cell lipase biocatalyst. *Renewable Energy* 85: 1002–1010.

Gupta, D., Kumar, R., and Pant, K. K. 2020. Hydrotalcite supported bimetallic (Ni-Cu) catalyst: a smart choice for one-pot conversion of biomass-derived platform chemicals to hydrogenated biofuels. *Fuel* 277: 118111.

Gupta, M. N., and Bisaria, V. S. 2018. Stable cellulolytic enzymes and their application in hydrolysis of lignocellulosic biomass. *Biotechnology Journal* 13: 1700633.

Gurunathan, B., and Ravi, A. 2015. Process optimization and kinetics of biodiesel production from neem oil using copper doped zinc oxide heterogeneous nanocatalyst. *Bioresource Technology* 190: 424–428.

Gurusamy, S., Kulanthaisamy, M. R., Hari, D. G., Veleeswaran, A., Thulasinathan, B., Muthuramalingam, J. B., Balasubramani, R., Chang, S. W., Arasu, M. V., Al-Dhabi, N. A., and Alagarsamy, A. 2019. Environmental friendly synthesis of TiO2–ZnO nanocomposite catalyst and silver nanomaterials for the enhanced production of biodiesel from *Ulva lactuca* seaweed and potential antimicrobial properties against the microbial pathogens. *Journal of Photochemistry and Photobiology B: Biology* 193: 118–130.

He, M., Yan, Y., Pei, F., Wu, M., Gebreluel, T., Zou, S., and Wang, C. 2017. Improvement on lipid production by *Scenedesmus obliquus* triggered by low dose exposure to nanoparticles. *Scientific Reports* 7: 1–12.

Hosseini, S. M., Kim, S. M., Sayed, M., Younesi, H., Bahramifar, N., Park, J. H., and Pyo, S. H. 2019. Lipase-immobilized chitosan-crosslinked magnetic nanoparticle as a biocatalyst for ring opening esterification of itaconic anhydride. *Biochemical Engineering Journal* 143: 141–150.

Huang, P. J., Chang, K. L., Hsieh, J. F., and Chen, S. T. 2015. Catalysis of rice straw hydrolysis by the combination of immobilized cellulase from Aspergillus niger on β-cyclodextrin-Fe_3O_4 nanoparticles and ionic liquid. *BioMed Research International* 2015: 1–9.

Huang, W. C., and Kim, J. D. 2016. Nickel oxide nanoparticle-based method for simultaneous harvesting and disruption of microalgal cells. *Bioresource Technology* 218: 1290–1293.

Ingle, A. P., Chandel, A. K., Philippini, R., Martiniano, S. E., and da Silva, S. S. 2020b. Advances in nanocatalysts mediated biodiesel production: a critical appraisal. *Symmetry* 12: 256.

Ingle, A. P., Philippini, R. R., and da Silva, S. S. 2020a. Pretreatment of sugarcane bagasse using two different acid-functionalized magnetic nanoparticles: a novel approach for high sugar recovery. *Renewable Energy* 150: 957–964.

Jambulingam, R., Shalma, M., and Shankar, V. 2019. Biodiesel production using lipase immobilised functionalized magnetic nanocatalyst from oleaginous fungal lipid. *Journal of Cleaner Production* 215: 245–258.

Japar, A. S., Azis, N. M., Takriff, M. S., and Yasin, N. H. M. 2017. Application of different techniques to harvest microalgae. *Trans-Science and Technology* 4: 98–108.

Jernejc, K., and Legiša, M. 2002. The influence of metal ions on malic enzyme activity and lipid synthesis in *Aspergillus niger*. *FEMS Microbiology Letters* 217: 185–190.

Kainthola, J., Kalamdhad, A. S., Goud, V. V., and Goel, R. 2019. Fungal pretreatment and associated kinetics of rice straw hydrolysis to accelerate methane yield from anaerobic digestion. *Bioresource Technology* 286: 121368.

Kandel, K., Frederickson, C., Smith, E. A., Lee, Y. J., and Slowing, I. I. 2013. Bifunctional adsorbent-catalytic nanoparticles for the refining of renewable feedstocks. *ACS Catalysis* 3: 2750–2758.

Kazemifard, S., Nayebzadeh, H., Saghatoleslami, N., and Safakish, E. 2019. Application of magnetic alumina-ferric oxide nanocatalyst supported by KOH for in-situ transesterification of microalgae cultivated in wastewater medium. *Biomass and Bioenergy* 129: 105338.

Kerenkan, A. E., Ello, A. S., Echchahed, B., and Do, T. O. 2016. Synthesis of mesoporous tungsten oxide/γ-alumina and surfactant-capped tungsten oxide nanoparticles and their catalytic activities in oxidative cleavage of oleic acid. *International Journal of Chemical Reactor Engineering* 14: 899–907.

Khanahmadi, S., Yusof, F., Ong, H. C., Amid, A., and Shah, H. 2016. Cocoa pod husk: a new source of CLEA-lipase for preparation of low-cost biodiesel: an optimized process. *Journal of Biotechnology* 231: 95–105.

Khanra, A., Sangam, S., Shakeel, A., Suhag, D., Mistry, S., Rai, M. P., Chakrabarti, S., and Mukherjee, M. 2018. Sustainable growth and lipid production from *Chlorella pyrenoidosa* using N-doped carbon nanosheets: unravelling the role of graphitic nitrogen. *ACS Sustainable Chemistry & Engineering* 6: 774–780.

Khot, M., and Ghosh, D. 2017. Lipids of *Rhodotorula mucilaginosa* IIPL32 with biodiesel potential: oil yield, fatty acid profile, fuel properties. *Journal of Basic Microbiology* 57: 345–352.

Kim, B., Bui, V. K. H., Farooq, W., Jeon, S. G., Oh, Y. K., and Lee, Y. C. 2018a. Magnesium aminoclay-Fe_3O_4 (MgAC-Fe_3O_4) hybrid composites for harvesting of mixed microalgae. *Energies* 11: 1359.

Kim, B., Praveenkumar, R., Lee, J., Nam, B., Kim, D. M., Lee, K., Lee, Y. C., and Oh, Y. K. 2016. Magnesium aminoclay enhances lipid production of mixotrophic *Chlorella* sp. KR-1 while reducing bacterial populations. *Bioresource Technology* 219: 608–613.

Kim, K. H., Lee, O. K., and Lee, E. Y. 2018b. Nano-immobilized biocatalysts for biodiesel production from renewable and sustainable resources. *Catalysts* 8: 68.

Kolouchova, I., Sigler, K., Zimola, M., Rezanka, T., Matatkova, O., and Masak, J. 2016. Influencing fatty acid composition of yeasts by lanthanides. *World Journal of Microbiology and Biotechnology* 32: 1–10.

Kruger, J. S., Cleveland, N. S., Yeap, R. Y., Dong, T., Ramirez, K. J., Nagle, N. J., Lowell, A. C., Beckham, G. T., McMillan, J. D., and Biddy, M. J. 2018. Recovery of fuel-precursor lipids from oleaginous yeast. *ACS Sustainable Chemistry & Engineering* 6: 2921–2931.

Kuepethkaew, S., Sangkharak, K., Benjakul, S., and Klomklao, S. 2017. Optimized synthesis of biodiesel using lipase from Pacific white shrimp (*Litopenaeus vannamei*) hepatopancreas. *Renewable Energy* 104: 139–147.

Kumar, A. N., Chatterjee, S., Hemalatha, M., Althuri, A., Min, B., Kim, S. H., and Mohan, S. V. 2020. Deoiled algal biomass derived renewable sugars for bioethanol and biopolymer production in biorefinery framework. *Bioresource Technology* 296: 122315.

Kumar, A. N., Singh, S., Tiwari, R., Goel, R., and Nain, L. 2017. Immobilization of indigenous holocellulase on iron oxide (Fe_2O_3) nanoparticles enhanced hydrolysis of alkali pretreated paddy straw. *International Journal of Biological Macromolecules* 96: 538–549.

Kumar, G., Bakonyi, P., Periyasamy, S., Kim, S. H., Nemestóthy, N., and Bélafi-Bakó, K. 2015. Lignocellulose biohydrogen: practical challenges and recent progress. *Renewable and Sustainable Energy Reviews* 44: 728–737.

Kumar, V., and Thakur, I. S. 2020. Biodiesel production from transesterification of Serratia sp. ISTD04 lipids using immobilised lipase on biocomposite materials of biomineralized products of carbon dioxide sequestrating bacterium. *Bioresource Technology* 307: 123193.

Kurniawati, H. A., Ismadji, S., and Liu, J. C. 2014. Microalgae harvesting by flotation using natural saponin and chitosan. *Bioresource Technology* 166: 429–434.

Lai, Y. S., Zhou, Y., Martarella, R., Wang, Z., and Rittmann, B. E. 2017. Synergistic integration of C12–C16 cationic surfactants for flocculation and lipid extraction from chlorella biomass. *ACS Sustainable Chemistry & Engineering* 5: 752–757.

Lee, K., Lee, S. Y., Na, J. G., Jeon, S. G., Praveenkumar, R., Kim, D. M., Chang, W. S., and Oh, Y. K. 2013. Magnetophoretic harvesting of oleaginous Chlorella sp. by using biocompatible chitosan/magnetic nanoparticle composites. *Bioresource Technology* 149: 575–578.

Lee, K., Lee, S. Y., Praveenkumar, R., Kim, B., Seo, J. Y., Jeon, S. G., Na, J. G., Park, J. Y., Kim, D. M., and Oh, Y. K. 2014a. Repeated use of stable magnetic flocculant for efficient harvest of oleaginous *Chlorella* sp. *Bioresource Technology* 167: 284–290.

Lee, S. Y., Cho, J. M., Chang, Y. K., and Oh, Y. K. 2017. Cell disruption and lipid extraction for microalgal biorefineries: a review. *Bioresource Technology* 244: 1317–1328.

Lee, Y. C., Lee, H. U., Lee, K., Kim, B., Lee, S. Y., Choi, M. H., Farooq, W., Choi, J. S., Park, J. Y., Lee, J., and Huh, Y. S. 2014b. Aminoclay-conjugated TiO2 synthesis for simultaneous harvesting and wet-disruption of oleaginous *Chlorella* sp. *Chemical Engineering Journal* 245: 143–149.

Liu, P., Wang, T., Yang, Z., Hong, Y., Xie, X., and Hou, Y. 2020a. Effects of Fe_3O_4 nanoparticle fabrication and surface modification on *Chlorella* sp. harvesting efficiency. *Science of The Total Environment* 704: 135286.

Liu, Z., Feist, A. M., Dragone, G., and Mussatto, S. I. 2020b. Lipid and carotenoid production from wheat straw hydrolysates by different oleaginous yeasts. *Journal of Cleaner Production* 249: 119308.

Maurya, R., Paliwal, C., Ghosh, T., Pancha, I., Chokshi, K., Mitra, M., Ghosh, A., and Mishra, S. 2016. Applications of de-oiled microalgal biomass towards development of sustainable biorefinery. *Bioresource Technology* 214: 787–796.

Mehrasbi, M. R., Mohammadi, J., Peyda, M., and Mohammadi, M. 2017. Covalent immobilization of Candida antarctica lipase on core-shell magnetic nanoparticles for production of biodiesel from waste cooking oil. *Renewable Energy* 101: 593–602.

Meng, X., Yang, J., Xu, X., Zhang, L., Nie, Q., and Xian, M. 2009. Biodiesel production from oleaginous microorganisms. *Renewable Energy* 34: 1–5.

Mishra, R. K., Kumar, V. B., Victor, A., Pulidindi, I. N., and Gedanken, A. 2019. Selective production of furfural from the dehydration of xylose using Zn doped CuO catalyst. *Ultrasonics Sonochemistry* 56: 55–62.

Moazeni, F., Chen, Y. C., and Zhang, G. 2019. Enzymatic transesterification for biodiesel production from used cooking oil, a review. *Journal of Cleaner Production* 216: 117–128.

Nematian, T., Salehi, Z., and Shakeri, A. 2020. Conversion of bio-oil extracted from *Chlorella vulgaris* micro algae to biodiesel via modified superparamagnetic nano-biocatalyst. *Renewable Energy* 146: 1796–1804.

Nguyen, M. K., Moon, J. Y., Bui, V. K. H., Oh, Y. K., and Lee, Y. C. 2019. Recent advanced applications of nanomaterials in microalgae biorefinery. *Algal Research* 41: 101522.

Ochsenreither, K., Glück, C., Stressler, T., Fischer, L., and Syldatk, C. 2016. Production strategies and applications of microbial single cell oils. *Frontiers in Microbiology* 7: 1539.

Pádrová, K., Čejková, A., Cajthaml, T., Kolouchová, I., Vítová, M., Sigler, K., and Řezanka, T. 2016. Enhancing the lipid productivity of yeasts with trace concentrations of iron nanoparticles. *Folia microbiologica*, 61: 329–335.

Pandit, P. R., and Fulekar, M. H. 2019. Biodiesel production from *Scenedesmus armatus* using egg shell waste as nanocatalyst. *Materials Today: Proceedings* 10: 75–86.

Park, G. W., Chang, H. N., Jung, K., Seo, C., Kim, Y. C., Choi, J. H., Woo, H. C., and Hwang, I. J., 2017. Production of microbial lipid by *Cryptococcus curvatus* on rice straw hydrolysates. *Process Biochemistry* 56: 147–153.

Peng, W. H., Lee, Y. Y., Wu, C., and Wu, K. C. W. 2012. Acid–base bi-functionalized, large-pored mesoporous silica nanoparticles for cooperative catalysis of one-pot cellulose-to-HMF conversion. *Journal of Materials Chemistry* 22: 23181–23185.

Periyasamy, K., Santhalembi, L., Mortha, G., Aurousseau, M., Boyer, A., and Subramanian, S. 2018. Bioconversion of lignocellulosic biomass to fermentable sugars by immobilized magnetic cellulolytic enzyme cocktails. *Langmuir* 34: 6546–6555.

Phitsuwan, P., Sakka, K., and Ratanakhanokchai, K. 2013. Improvement of lignocellulosic biomass in planta: a review of feedstocks, biomass recalcitrance, and strategic manipulation of ideal plants designed for ethanol production and processability. *Biomass and Bioenergy*, 58: 390–405.

Poontawee, R., Yongmanitchai, W., and Limtong, S. 2018. Lipid production from a mixture of sugarcane top hydrolysate and biodiesel-derived crude glycerol by the oleaginous red yeast, Rhodosporidiobolus fluvialis. *Process Biochemistry* 66: 150–161.

Probst, K. V., Schulte, L. R., Durrett, T. P., Rezac, M. E., and Vadlani, P. V. 2016. Oleaginous yeast: a value-added platform for renewable oils. *Critical Reviews in Biotechnology* 36: 942–955.

Rai, M., dos Santos, J. C., Soler, M. F., Marcelino, P. R. F., Brumano, L. P., Ingle, A. P., Gaikwad, S., Gade, A., and da Silva, S. S. 2016. Strategic role of nanotechnology for production of bioethanol and biodiesel. *Nanotechnology Reviews* 5: 231–250.

Raj, J. V. A., Bharathiraja, B., Vijayakumar, B., Arokiyaraj, S., Iyyappan, J., and Kumar, R. P. 2019. Biodiesel production from microalgae Nannochloropsis oculata using heterogeneous Poly Ethylene Glycol (PEG) encapsulated ZnOMn^{2+} nanocatalyst. *Bioresource Technology* 282: 348–352.

Rana, M. S., Bhushan, S., Sudhakar, D. R., and Prajapati, S. K. 2020. Effect of iron oxide nanoparticles on growth and biofuel potential of Chlorella spp. *Algal Research* 49: 101942.

Rana, Q. U. A., Rehman, M. L. U., Irfan, M., Ahmed, S., Hasan, F., Shah, A. A., Khan, S. and Badshah, M. 2019. Lipolytic bacterial strains mediated transesterification of non-edible plant oils for generation of high quality biodiesel. *Journal of Bioscience and Bioengineering* 127: 609–617.

Raoufi, Z., and Gargari, S. L. M. 2018. Biodiesel production from microalgae oil by lipase from Pseudomonas aeruginosa displayed on yeast cell surface. *Biochemical Engineering Journal* 140: 1–8.

Santek, M. I., Beluhan, S., and Santek, B. 2018. Production of microbial lipids from lignocellulosic biomass. *Advances in Biofuels and Bioenergy*: 137–164.

Sathishkumar, R. S., Sundaramanickam, A., Srinath, R., Ramesh, T., Saranya, K., Meena, M., and Surya, P. 2019. Green synthesis of silver nanoparticles by bloom forming marine microalgae *Trichodesmium erythraeum* and its applications in antioxidant, drug-resistant bacteria, and cytotoxicity activity. *Journal of Saudi Chemical Society* 23: 1180–1191.

Scordia, D., Cosentino, S. L., Lee, J. W., and Jeffries, T. W. 2012. Bioconversion of giant reed (Arundo donax L.) hemicellulose hydrolysate to ethanol by Scheffersomyces stipitis CBS6054. *Biomass and Bioenergy* 39: 296–305.

Seffati, K., Esmaeili, H., Honarvar, B., and Esfandiari, N. 2020. AC/CuFe2O4@ CaO as a novel nanocatalyst to produce biodiesel from chicken fat. *Renewable Energy* 147: 25–34.

Seo, J. Y., Praveenkumar, R., Kim, B., Seo, J. C., Park, J. Y., Na, J. G., Jeon, S. G., Park, S. B., Lee, K., and Oh, Y. K. 2016. Downstream integration of microalgae harvesting and cell disruption by means of cationic surfactant-decorated Fe_3O_4 nanoparticles. *Green Chemistry* 18: 3981–3989.

Shanab, S. M. M., Partila, A. M., Ali, H. E. A., and Abdullah, M. A. 2019. Characterization and Impact of Silver nanoparticles on cell growth, lipid, carbohydrate and fatty acids of Chlorella vulgaris and *Dictyochloropsis splendida. Beilstein Archives* 2019: 91.

Shanmugam, S., Krishnaswamy, S., Chandrababu, R., Veerabagu, U., Pugazhendhi, A., and Mathimani, T. 2020. Optimal immobilization of Trichoderma asperellum laccase on polymer coated Fe3O4@ SiO2 nanoparticles for enhanced biohydrogen production from delignified lignocellulosic biomass. *Fuel* 273: 117777.

Shariati, F., and Shirazi, M. A. 2019. Effect of SiO2 Nanoparticles on chlorophyll, carotenoid and growth of green micro-algae Dunaliella salina. *Nanomedicine Research Journal* 4: 164–175.

Sharma, D., Kanchi, S., and Bisetty, K. 2019. Biogenic synthesis of nanoparticles: a review. *Arabian Journal of Chemistry* 12: 3576–3600.

Sharma, R. K., Saxena, M., O'Neill, C. A., Ramos, H. A., and Griebenow, K. 2018. Synthesis of Rhizopus arrhizus lipase nanoparticles for biodiesel production. *ACS Omega* 3: 18203–18213.

Shi, K., Gao, Z., Shi, T. Q., Song, P., Ren, L. J., Huang, H., and Ji, X. J. 2017. Reactive oxygen species-mediated cellular stress response and lipid accumulation in oleaginous microorganisms: the state of the art and future perspectives. *Frontiers in Microbiology* 8: 793.

Sivagnanam, S. P., Getachew, A. T., Choi, J. H., Park, Y. B., Woo, H. C., and Chun, B. S. 2017. Green synthesis of silver nanoparticles from deoiled brown algal extract via Box-Behnken based design and their antimicrobial and sensing properties. *Green Processing and Synthesis* 6: 147–160.

Subhash, G. V., and Mohan, S. V. 2014. Deoiled algal cake as feedstock for dark fermentative biohydrogen production: an integrated biorefinery approach. *International Journal of Hydrogen Energy* 39: 9573–9579.

Suwanno, S., Rakkan, T., Yunu, T., Paichid, N., Kimtun, P., Prasertsan, P., and Sangkharak, K. 2017. The production of biodiesel using residual oil from palm oil mill effluent and crude lipase from oil palm fruit as an alternative substrate and catalyst. *Fuel*, 195:82–87.

Teo, S. H., Islam, A., Chan, E. S., Choong, S. T., Alharthi, N. H., Taufiq-Yap, Y. H., and Awual, M. R. 2019. Efficient biodiesel production from Jatropha curcus using CaSO4/Fe2O3–SiO2 core-shell magnetic nanoparticles. *Journal of Cleaner Production* 208: 816–826.

Teo, S. H., Islam, A., Taufiq-Yap and Y. H. 2016. Algae derived biodiesel using nanocatalytic transesterification process. *Chemical Engineering Research and Design* 111: 362–370.

Vashist, V., Chauhan, D., Bhattacharya, A., and Rai, M. P. 2020. Role of silica coated magnetic nanoparticle on cell flocculation, lipid extraction and linoleic acid production from *Chlorella pyrenoidosa*. *Natural Product Research* 34: 2852–2856.

Tran, D. T., Chen, C. L., and Chang, J. S. 2012. Immobilization of *Burkholderia* sp. lipase on a ferric silica nanocomposite for biodiesel production. *Journal of Biotechnology* 158: 112–119.

Valenstein, J. S., Kandel, K., Melcher, F., Slowing, I. I., Lin, V. S. Y., and Trewyn, B. G. 2012. Functional mesoporous silica nanoparticles for the selective sequestration of free fatty acids from microalgal oil. *ACS Applied Materials & Interfaces* 4: 1003–1009.

Welker, C. M., Balasubramanian, V. K., Petti, C., Rai, K. M., DeBolt, S., and Mendu, V. 2015. Engineering plant biomass lignin content and composition for biofuels and bioproducts. *Energies* 8: 7654–7676.

Xie, W., Han, Y., and Tai, S. 2017. Biodiesel production using biguanide-functionalized hydroxyapatite-encapsulated-γ-Fe2O3 nanoparticles. *Fuel* 210: 83–90.

Xie, W., and Huang, M. 2018. Immobilization of *Candida rugosa* lipase onto graphene oxide Fe3O4 nanocomposite: characterization and application for biodiesel production. *Energy Conversion and Management* 159: 42–53.

Xu, L., Guo, C., Wang, F., Zheng, S., and Liu, C. Z. 2011. A simple and rapid harvesting method for microalgae by in situ magnetic separation. *Bioresource Technology* 102: 10047–10051.

Xu, C., Ouyang, W., Muñoz-Batista, M. J., Fernández-García, M., & Luque, R. (2018). Highly Active Catalytic Ruthenium/TiO_2 Nanomaterials for Continuous Production of γ-Valerolactone. *ChemSusChem*, 11 (15), 2604–2611.

Yang, Y., Hou, J., Wang, P., Wang, C., Miao, L., Ao, Y., Xu, Y., Wang, X., Lv, B., You, G., and Yang, Z. 2018. Interpretation of the disparity in harvesting efficiency of different types of Microcystis aeruginosa using polyethylenimine (PEI)-coated magnetic nanoparticles. *Algal Research* 29: 257–265.

Yellapu, S. K., Kaur, R., Kumar, L. R., Tiwari, B., Zhang, X., and Tyagi, R. D. 2018. Recent developments of downstream processing for microbial lipids and conversion to biodiesel. *Bioresource Technology* 256: 515–528.

Yu, Q., Liu, Z., Xu, H., Zhang, B., Zhang, M., and Li, M. 2015. TiO_2 nanoparticles promote the production of unsaturated fatty acids (UFAs) fighting against oxidative stress in *Pichia pastoris*. *RSC Advances* 5: 41033–41040.

Zhang, L., He, X., Zhang, Z., Cang, D., Nwe, K. A., Zheng, L., Li, Z. and Cheng, S. 2017. Evaluating the influences of ZnO engineering nanomaterials on VFA accumulation in sludge anaerobic digestion. *Biochemical Engineering Journal* 125: 206–211.

Zhao C, Yang L, Xing S Luo, W., Wang, Z., and Lv, P., 2018. Biodiesel production by a highly effective renewable catalyst from pyrolytic rice husk. *Journal of Cleaner Production* 199: 772–780.

Zhao, F., Chu, H., Zhang, Y. Y., Jiang, S., Yu, Z., Zhou, X., and Zhao, J. 2017. Increasing the vibration frequency to mitigate reversible and irreversible membrane fouling using an axial vibration membrane in microalgae harvesting. *Journal of Membrane Science* 529: 215–223.

Zhao, Y., Liang, W., Liu, L. Li, F., Fan, Q., and Sun, X. 2015. Harvesting *Chlorella vulgaris* by magnetic flocculation using Fe3O4 coating with polyaluminium chloride and polyacrylamide. *Bioresource Technology*198: 789–796.

Zhong, L., Feng, Y., Wang, G. Wang, Z., Bilal, M., Lv, H., Jia, S., and Cui, J. 2020. Production and use of immobilized lipases in/on nanomaterials: a review from the waste to biodiesel production. *International Journal of Biological Macromolecules* 152: 207–222.

10 Gas Fermentation
A Potential Platform for Production of Low-Carbon Fuels and Chemicals

Chandan Mukherjee, Debashish Ghosh, and Thallada Bhaskar

CONTENTS

10.1 INTRODUCTION

Energy derived from fossil fuel sources like coal, oil, and natural gas results in substantial damage of the ecosystem, including air and water pollution, harm to public health, wildlife and habitat loss, water use, land use, and global warming. The world oil reserves will be exhausted within the next 50 years at the present global consumption rate [Vasudevan et al., 2007]. Thus, there is an urgent demand for producing affordable and environmentally sustainable fuels/chemicals/materials, which could reduce the deleterious environmental effects arising from the build-up of atmospheric carbon dioxide (CO_2) from the consumption of fossil fuels. Many nations have responded to this challenge

DOI: 10.1201/9781003335740-10

by adopting mandates and legislation to promote research and development (R&D) and marketing for low-cost, low-carbon fuel-emitting technologies. The global renewable energy market, which converts the energy from different natural sources such as the sun, tides, wind, and biofuels, was valued at $928.0 billion in 2017, and is expected to reach $1,512.3 billion by 2025, a 6.1% increase from 2018 to 2025 (https://www.alliedmarketresearch.com/renewable-energy-market). Among these sources, biofuels are particularly important, as they are derived from biomass and have been attracting maximum attention worldwide due to their vast potential and ability to solve the pollution problem. The term "biofuels" is used to describe solid fuels (biochar), liquid fuels, (bioethanol, biobutanol, and biodiesel), and gas fuels, primarily biomass derivatives (biogas, biosyngas, and biohydrogen). The United States, followed by Brazil and the European Union (EU), are the world's leading manufacturers of biofuels.

Low-carbon fuels are popularly made from the fermentation of the starch/sugar derived from different lignocellulosic biomasses [Naik et al., 2010]. Several pretreatment and hydrolysis steps are required to isolate cellulose and hemicellulose from the recalcitrant lignin and for the polysaccharides to be broken into fermentable monosaccharides. Measures include steam explosion, extension of ammonia filament, extrusion, extraction of ionic fluid and diluting, and enzyme hydrolysis, all of which are costly and energy intensive. Additional difficulties include the elimination of hexose (C6) and pentose (C5) sugars and isolation of a solid biocatalytic agent that can ferment all of them [Geddes et al., 2011]. Employment of cellulolytic microorganisms capable of hydrolysis of the lignocellulotic content and sugar fermentation has been successful in the laboratory, but not in the manufacturing environments, and conversion rates and efficiency problems persist.

One of the solutions is to directly collect C1 gases (carbon) like CH_4, CO, and CO_2, as feedstock through the use of microbes, before incorporation into. This process is referred to as the gas fermentation process. These gases are low-cost waste products of a broad range of manufacturing processes, with global emissions of around 9.7 gigatons in 2014 [EPA, 2017]. Syngas, the combination of mainly hydrogen (H_2), carbon monoxide (CO), and carbon dioxide (CO_2) gases, is used by certain anaerobic microorganisms and converted into fuels or chemicals like acetic acid, ethanol, 2,3-butanediol, butyric acid, and butanol. Similarly, biogas, mainly methane (CH_4), is used as a substrate by methanogenic bacteria and is converted to methanol.

This chapter has been designed to review the manufacturing process of biofuels and other value-added chemicals, using biogas, syngas, and industrial waste gas as the feedstock, through the gas fermentation technique. It addresses the benefits of gas fermentation over traditional sugar and thermochemical fermentation methods, and its freedom to use a variety of feedstocks to produce the desired chemicals sets the tone for the next sections.

10.2 FEEDSTOCK FOR GAS FERMENTATION

Gasification is the complete thermal breakdown of biomass into a combustible gas, volatiles, char, and ash in an enclosed reactor or gasifier. Practically every carbon content compound may be used as a gasifying feedstock [Liew et al., 2013]. Syngas, with a typical composition of 30–60% CO, 25–30% H_2, 0–5% CH_4, 5–15% CO_2, and some amount of water vapor, has already been established as a useful power generator and precursor of many fuel products. Syngas is usually a product of gasification and can be produced from many sources, including natural gas, coal, and biomass, or virtually any hydrocarbon feedstock by its reaction with steam or oxygen. Nonfood biomass like farming waste, residual crops, woodland residue, and municipal solid waste, and also glycerol, can be used as a gasifying material. The use of existing waste sources such as urban organic waste, which are available at commercially cheaper rates, requires minimum use of land and is widely available, and is also a good feedstock option for syngas production. Again, nonbiomass sources such as shale (its biomass), cokes, shallow oil, tar, sewage sludge (its biomass), heavy oil processing waste, and refined natural gas are also widely available and can be used for gasification for syngas production (not all are good for syngas) [Liew et al., 2013]. In addition, a large amount of CO and/or carbon-rich gas streams is produced as waste by industries such as steel processing, oil refining, and chemical manufacturing. Using gas fermentation techniques, these hazardous waste streams can be effectively turned into useful resources such as biofuels. Again, lignin, which can constitute up to 40% of plant biomass, is wasted with the traditional biomass processing needed to generate cellulosic ethanol [Liew et al., 2013]. Converting whole biomass into syngas by gasification enables the usage of this wasted lignin fraction as well, and also bypasses the biomass pretreatment expenditure. This valorization of biomass to a gaseous combination of CO, CO_2, H_2, and N_2 adds a significant perspective to the sustainable production of syngas. Different sources of syngas and their gas compositions are listed in Table 10.1.

Using syngas for fuel production has been commercially done by the Fischer-Tropsch process (FTP), wherein the syngas is passed through several processes, which are energy intensive, to maintain the required ratio ($2.05 < H_2/CO < 2.15$) for the process. Absorption techniques with monoethyl and diethyl amines for CO_2 removal requires high energy input for the steps of absorption, desorption, and compression. Also, the FTP technology has some complexity and is rather capital-intensive, employing high-temperature (150–350 °C), high-pressure (30 bar), and heterogeneous catalysts like cobalt [Liew et al., 2013].

Fixing gas and inorganic carbon into organic matter is also done by microorganisms [Liew et al., 2016]. Eukaryotic algae, archaea, and bacteria all fix carbon. Microorganisms like acetogens, carboxytrophs, and methanogens are able to utilize the $CO_2 + H_2$ and/or CO available in syngas as their sole source of carbon and energy for growth, as well as the production of biofuels and other valuable products. Microorganisms that are able to reduce CO_2 to acetate via the acetyl coenzyme A (acetyl-CoA) or Wood-Ljungdahl pathway are termed acetogens. Acetogens are strictly anaerobic bacteria that can grow by the conversion of C1 compounds such as H_2-CO_2, CO, or formate to acetate. However, only acetogens have been known to synthesize metabolic end

TABLE 10.1
Composition of Syngas from Different Sources

	Approximate Volume of Gaseous Components (%)		
Syngas Sources	**CO**	**CO_2**	**H_2**
Nonliving source			
Gasification of coal	60	10	30
Gas from coal oven	6	1.5	56
Partial oxidation of fuel oil	47	4.5	46
Reforming reactions of natural gas	15.5	8	76
Reforming reactions of naphtha compounds	7	16	66
Water gas	30	3.5	32
Gases from steel mill	44	22	2
Biomass and waste source			
Wood waste and sludge from sewage	10.5	15	8
Shell of cacao	8	16	9
Biomass generated from dairy waste	9	16	19
Wasted switchgrass	15	16.5	4.5
Wasted bluegrass straw	13	17.5	3
Wasted willow	9.5	17	7.5

Source: Liew et al. (2013)

products that have potential as liquid transportation fuels. While biological processes are generally considered slower than chemical reactions, the use of these microbes to carry out syngas fermentation offers several key advantages over the FTP technology. First, microbial processes operate at ambient temperatures and low pressures, which offer significant energy and cost savings. Second, the ambient conditions and irreversible nature of biological reactions also avoid thermodynamic equilibrium relationships and allow near-complete conversion efficiencies [Klasson et al., 1991a and 1992b]. Third, biological conversions are commonly more specific due to high enzymatic specificities, resulting in higher product yield with the formation of fewer by-products. Fourth, unlike traditional chemical catalysts, which require a set feed gas composition like a H_2:CO ratio in the ratio of ~2:1 to yield desired products, microbial processes have freedom to operate for the production of the same suite of products across a wider range of CO:H_2 ratios in the feed gas. Fifth, biocatalysts exhibit a much higher tolerance to poisoning by tars, sulfur, and chlorine than inorganic catalysts [Mohammadi et al., 2011].

Biomass can also be anaerobically digested to produce biogas (mainly methane). Table 10.2 lists the composition of gases produced by anaerobic digestion. Methane is gaseous and expensive to contain, ship, and distribute at ambient temperatures due to its boiling point of −164 °C. Although CH_4 can be compressed into compressed natural gas (CNG) or liquid natural gas (LNG), these processes require considerable energy and capital investment. Due to their high pressure (21–25 MPa) and low temperature (−160 °C) requirements, both CNG and LNG are dangerous. In order to solve these problems, CH_4 can be converted into fluid fuels such as methanol, ethanol, and butanol. The two-stage heat conversion process with syngas as the intermittent is currently the most widely used CH_4-to-liquid fuel technology [Park and Lee, 2013]. The CH_4 molecule is difficult to activate due to its high strength of C-H bond, the high propensity for ionization, and low acidity. Then, a variety of obstacles, such as high pressure and/or temperature, costly chemical catalysts, and low efficacy, are commonly faced during thermochemical conversion. Also, since biogas includes H_2S and NH_3 pollutants capable of disabling chemical catalysts, it is important that expensive treatments be conducted before thermochemical transformations [Navarro et al., 2013]. Biological conversion is very desirable in contrast to thermochemical processes because of the underlying requirements of successful conversion reactions. In addition, biological diversity of microorganisms is remarkable and makes them ideal for diverse habitats. Two bacteria classes are present in nature which can activate the stable C-H bond of CH_4 under environmental conditions. One is aerobic methanotrophic bacteria, which activates and uses CH_4 as the source of their carbon and energy by the methane monooxygenase (MMO) enzyme. The other group is ammonia oxidizing bacteria, which uses ammonia as an energy source to partly oxidize methanol to CH_4 [Ge et al., 2014]. Figure 10.1 presents an overview of different feedstocks and product spectrums of microbial gas fermentation.

TABLE. 10.2
Composition of Biogas and Natural Gas from Different Sources

	Approximate Amounts of Components								
Source	**CH_4**	**CO_2**	**N_2**	**O_2**	**H_2**	**Hydrocarbons**	**H_2S**	**NH_3**	**Siloxane**
	%						ppm		mg/gm of dry weight
Biogas from anaerobic digester	53–70	30–50	2–6	0–5	-	-	0–2000	<100	<0.08–0.5
Biogas from landfilling	30–65	25–47	<1–17	<1–3	0–3	-	30–500	0–5	<0.3–36
Natural gas	81–89	0.67–1	0.28–14	0	NA	3.5–9.4	0–2.9	NA	NA

Source: Yang and Ge (2016)

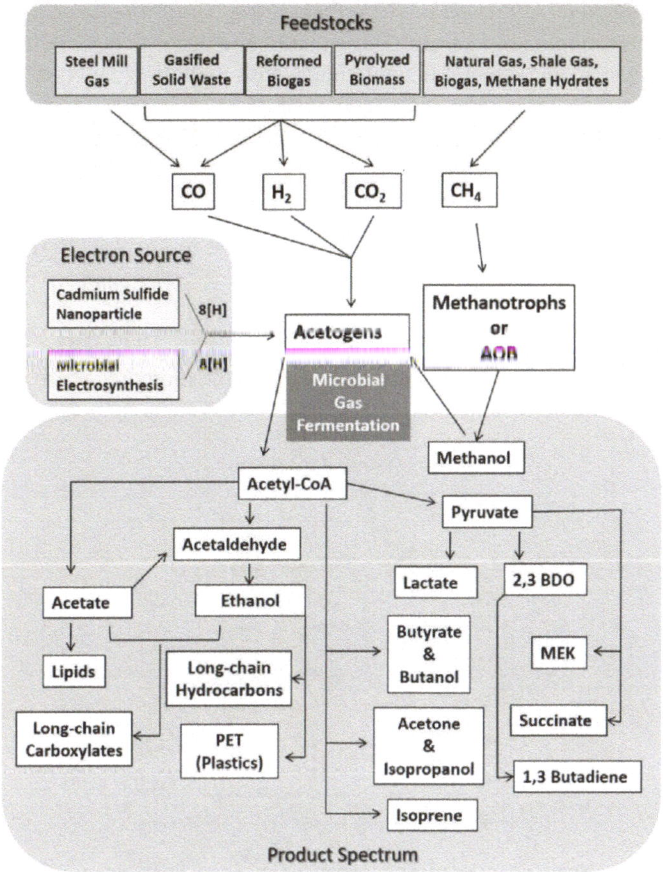

FIGURE 10.1 Overview of different feedstocks and product spectrums of microbial gas fermentation. Acronyms: AOB, Ammonia oxidizing bacteria; 2,3-BDO, 2,3-Butanediol; MEK, methylethylketone; PET, Polyethylene terephthalate.

10.3 MICROBIAL CONVERSION OF GASES INTO FUEL

Syngas and biogas can be converted into fuels by two processes. Detailed information about the metabolic pathways, microbes involved, and product ranges are discussed next.

10.3.1 Anaerobic Syngas Fermentation in Acetogens by the Wood-Ljungdahl Pathway

Acetogens, an obligate anaerobic species, adopt the reductive acetyl-CoA cycle for the reduction of CO_2 to acetyl moiety of acetyl-coenzyme A for energy storage and assimilation of CO_2 into cell carbon. In addition to the reductive acetyl-CoA pathway, autotrophic fixation of CO_2 occurs via four other biological pathways, namely, the Calvin cycle, tricarboxylic acid (TCA) cycle, 3-hydroxypropionate/malyl-CoA cycle, and the 3-hydroxypropionate/4-hydroxybutyrate cycle. In honor of the two legends, Lars G. Ljungdahl and Harland G. Wood, who elucidated the chemical and enzymological mechanism of the reductive acetyl-CoA pathway using *Moorella thermoacetica* (formerly *Clostridium thermoaceticum*), the CODH/ACS pathway (named after the main enzyme of the carbon monoxide dehydrogenase/acetyl-CoA synthase pathway) is often referred to as the "Wood-Ljungdahl" pathway. This pathway is widely spread among various bacterial genera like *Acetitomaculum, Bryantella, Butyribacterium, Caloramator, Clostridium, Moorella, Ruminococcus, Thermoanaerobacter*, and *Treponema*. Acetogens can use $CO_2 + H_2$ gases and/or CO to generate acetic acid and ethanol in compliance with the following equations:

$$2\,CO_2 + 4\,H_2 \Rightarrow CH_3COOH + 2\,H_2O \quad \Delta H = -75.3\,kJ\,/\,mol \tag{10.1}$$

$$2\,CO_2 + 6\,H_2 \Rightarrow C_2H_5OH + 3\,H_2O \quad \Delta H = -97.3\,kJ\,/\,mol \tag{10.2}$$

$$4\,CO + 2\,H_2O \Rightarrow CH_3COOH + 2\,CO_2 \quad \Delta H = -154.9\,kJ\,/\,mol \tag{10.3}$$

$$6\,CO + 3\,H_2O \Rightarrow C_2H_5OH + 4\,CO_2 \quad \Delta H = -217.9\,kJ\,/\,mol \tag{10.4}$$

Figure 10.2 shows the overview of the Wood-Ljungdahl pathway. The acetyl-CoA route is basically a terminal electron acceptance route that absorbs CO_2 into the biomass. There is an Eastern branch

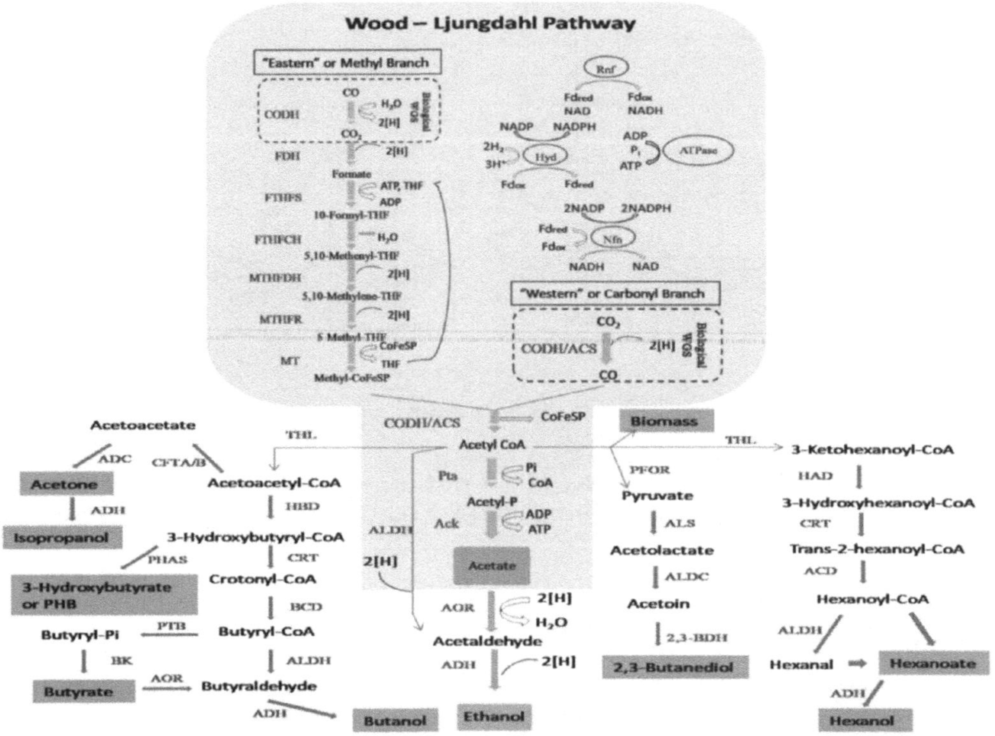

FIGURE 10.2 Overview of CO, H_2, and CO_2 metabolism in acetogenic bacteria via Wood-Ljungdahl pathway, its energy-conserving mechanisms, and different product formation pathways. Enzymes are indicated by red text, the enzymes involved in energy conservation are shown in purple circles, and products are shown by blue boxes. Acronyms: FTHFS, formyl THF synthase; FTHFCH, formyl-THF-cyclohydrolase; MTHFDH, methylene-THF-dehydrogenase; MTHFR, methylene-THF-reductase, MT, methyltransferase, Ack, acetate kinase; ACS, acetyl-CoA synthase; CODH, carbon monoxide dehydrogenase; CoFeSP, corrinoid iron sulfur protein; FDH, formate dehydrogenase; Pta, phosphotransacetylase; THF, tetrahydrofolate; Nfn, transhydrogenase; Rnf, *Rhodocbacter* nitrogen fixation; WGS, water-gas shift reaction; Hyd, hydrogenase; AOR, aldehyde ferredoxin oxidoreductase; ALDH, aldehyde dehydrogenase; ADH, alcohol dehydrogenase, THL, thiolase; HBD, 3-hydroxybutyryl-CoA dehydrogenase; BCD, butyryl-CoA dehydrogenase; PTB, phosphotransbutyrylase; BK, butyrate kinase, HAD: 3-hydroxyacyl-CoA dehydrogenase; ACD: acyl-CoA dehydrogenase, PFOR, pyruvate ferredoxin oxidoreductase; ALS, acetolactate synthase; ALDC, acetolactate decarboxylase; 23BDH, 2,3-butanediol dehydrogenase, CTFA/B, acetoacetyl-CoA:acetate/butyrate-CoA-transferase; ADC, acetone decarboxylase, PHBS, poly-3-hydroxybutyrate synthase.

(or carbonyl) and a Western branch (or methyl) in this pathway. The Eastern route uses a sequence of enzymes to reduce CO_2 by six electrons to the methyl group of acetyl-CoA. When CO, the methyl group, and CoASH are conjugated by ACS, acetyl-CoA is finally generated. The oxidation of molecular hydrogen under autotrophic growth or NADH and reduced ferredoxin under heterotrophic growth gives the reducing power for CO_2 fixation.

$$H_2 \Leftrightarrow 2\,H^+ + 2\,e^- \qquad (10.5)$$

The redox potential of the part capable of interacting with the enzyme determines the course of the hydrogenase reaction. The formation of hydrogen is caused by the electron donor, and the presence of electron acceptors contributes to oxidation of the hydrogen.

Most acetogens may also use CO as a source of electrons, which does not inherently require hydrogen. CO is around 1000 times capable of creating exceedingly low potential electrons than NADH, which can minimize cellular electron transports such as ferredoxin and flavodoxin, with its CO_2/CO potential reducted from –524 to –558 mV. The reduced equivalents of CO oxidation can be associated with the conversion of CO_2 into acetate, butyrate, and/or methane; evolution of proton molecular H_2; and reductions in nitrates, sulfides, and aldehyde reduction to alcohols. However, comparatively few microorganisms are able to use CO as the sole source of carbon and energy, possibly due to the inhibition of growth because of metalloprotein and hydrogenase sensitivity development against CO. Some other electron donors for acetogens are formate, CH_3Cl, lactate, pyruvate, alcohols, betaines, sugars, acetoin, oxalate, and citrate. CODH is able to separate water into hydrogen and electrons in a stoichiometry-based biological water-gas shift reaction:

$$CO + H_2O \Leftrightarrow CO_2 + 2\,H^+ + 2\,e^- \qquad (10.6)$$

In addition to ethanol and acetic acid, some acetogens, such as *C. carboxidivorans*, can convert CO, CO_2, and H_2 to butyric acid [Eqs. (10.7) and (10.8)], butanol [Eqs. (10.9) and (10.10)], hexanoic acid [Eqs. (10.11) and (10.12)], and hexanol [Eqs. (10.13) and (10.14)] by the following stoichiometries:

$$10\,CO + 4\,H_2O \rightarrow C_3H_7COOH + 6\,CO_2 \quad (\Delta G^o = -421\ kJ/mol) \qquad (10.7)$$

$$10\,H_2 + 4\,CO_2 \rightarrow C_3H_7COOH + 6\,H_2O \quad (\Delta G^o = -220\ kJ/mol) \qquad (10.8)$$

$$12\,CO + 5\,H_2O \rightarrow C_4H_9OH + 8\,CO_2 \quad (\Delta G^o = -486\ kJ/mol) \qquad (10.9)$$

$$12\,H_2 + 4\,CO_2 \rightarrow C_4H_9OH + 7\,H_2O \quad (\Delta G^o = -245\ kJ/mol) \qquad (10.10)$$

$$16\,CO + 6\,H_2O \rightarrow C_5H_{11}COOH + 10\,CO_2 \quad (\Delta G^o = -663\ kJ/mol) \qquad (10.11)$$

$$16\,H_2 + 6\,CO_2 \rightarrow C_5H_{11}COOH + 10\,H_2O \quad (\Delta G^o = -341\ kJ/mol) \qquad (10.13)$$

$$18\,H_2 + 6\,CO_2 \rightarrow C_6H_{13}OH + 11\,H_2O \quad (\Delta G^o = -397\ kJ/mol) \qquad (10.14)$$

In order to get energy through substrate-level phosphorylation, acetogens have to produce acetate to balance the energy incurred in formate activation in the Western branch of the acetyl-CoA pathway. Only acetate as a key product was found with CO_2 and H_2 as the substrates [Allen et al., 2010]. A variety of products such as ethanol, butanol, butyrate, 2,3-butanediol, and lactate were identified with the reduced CO substrate [Köpke et al., 2011].

10.3.2 Aerobic Microbial Metabolism for Syngas Fermentation

Table 10.3 presents the microorganisms associated with the metabolism of syngas constituents resulting in the production of different value-added chemicals. Metabolic pathway modifications,

TABLE 10.3
Substrate and Their Products in the Microorganisms Associated with Syngas Fermentation

Organism	Substrates	Product(s)
Anaerobic mesophilic microorganisms		
Acetobacterium woodii	H_2/CO_2, CO	Acetate
Acetonema longum	H_2/CO_2	Acetate, butyrate
Alkalibaculum bacchi	H_2/CO_2, CO	Acetate, ethanol
Butyribacterium methylotrophicum	H_2/CO_2, CO	Acetate, ethanol, butyrate, butanol
Clostridium aceticum	H_2/CO_2, CO	Acetate
Clostridium autoethanogenum	H_2/CO_2, CO	Acetate, ethanol, 2,3-butanediol, lactate
Clostridium carboxidivorans	H_2/CO_2, CO	Acetate, ethanol, butyrate, butanol, lactate
Clostridium coskati	H_2/CO_2, CO	Acetate, ethanol
Clostridium diffificile	H_2/CO_2, CO	Acetate, ethanol, butyrate
Clostridium drakei	H_2/CO_2, CO	Acetate, ethanol, butyrate
Clostridium formicoaceticum	CO	Acetate, formate
Clostridium glycolicum	H_2/CO_2	Acetate
Clostridium ljungdahlii	H_2/CO_2, CO	Acetate, ethanol, 2,3-butanediol, lactate
Clostridium magnum	H_2/CO_2	Acetate
Clostridium mayombei	H_2/CO_2	Acetate
Clostridium methoxybenzovorans	H_2/CO_2	Acetate, formate
Clostridium ragsdalei	H_2/CO_2, CO	Acetate, ethanol, 2,3-butanediol, lactate
Clostridium scatologenes	H_2/CO_2, CO	Acetate, ethanol, butyrate
Eubacterium limosum	H_2/CO_2, CO	Acetate, butyrate
Oxobacter pfennigii	H_2/CO_2, CO	Acetate, butyrate
Blautia productus	H_2/CO_2, CO	Acetate
Anaerobic thermophilic microorganisms		
Moorella thermoacetica	H_2/CO_2, CO	Acetate
Moorella thermoautotrophica	H_2/CO_2, CO	Acetate
Aerobic microorganisms		
Ralstonia eutropha	$CO_2 + H_2 + O_2$	Poly-D-3-hydroxybutyric acid
Ideonella sp. O-1	$CO_2 + H_2 + O_2$	Poly-D-3-hydroxybutyric acid
Ralstonia eutropha	$CO_2 + H_2 + O_2$	Pentadecane/heptadecane
Ralstonia eutropha	$CO_2 + H_2 + O_2$	Botryococcene
Rhodobacter capsulatus	$CO_2 + H_2 + O_2$	Botryococcene
Ralstonia eutropha	$CO_2 + H_2 + O_2$	Methyl ketones
Ralstonia eutropha	$CO_2 + H_2 + O_2$	Cyanophycin
Ralstonia eutropha	CO_2 + electricity+ O_2	Isobutanol, 3-methyl-1-butanol
Synechocystis sp. PCC6803	CO_2 + electricity + O_2	Ethanol, lactate
Synechococcus elongatus	CO_2 + light + O_2	2,3-butanediol, isobutyraldehyde
Methylococcus capsulatus	$CH_4 + O_2$	Single cell protein
Methylosinus trichosporium	$CH_4 + O_2$	Poly-D-3-hydroxybutyric acid
Methylocystis GB25	$CH_4 + O_2$	Poly-D-3-hydroxybutyric acid
Methylotrophus capsulatus	$CH_4 + O_2$	Methanol
Methylomonas sp. 16a	$CH_4 + O_2$	Astaxanthin

Source: Liew et al. (2016)

varying and fine-tuning of the expression of genes encoding acyl-ACP reductase and aldehyde decarboxylase, and recombinant DNA technology approaches have been successfully employed in many aerobic bacteria and cyanobacteria to produce biofules and other diverse products by syngas fermentation [Lu et al., 2012]. Aerobic syngas fermentation is an adenosine triphosphate (ATP)–intensive process and thus could be used to produce complex high-value products directly from CO and CO_2, in comparison to anaerobic approaches.

10.3.3 Microbial Conversion of Biogas for Methanol Production

Table 10.4 represents microorganisms associated with biological methanol production from methane and ammonia, and Figure 10.3 depicts those pathways employed by the microbes.

TABLE 10.4
Microorganisms Associated with Biological Methanol Production

Type	Microorganisms
Methanotrophic bacteria: oxidizing methane to methanol	*Acidimethylosilex fumarolicum, Methyloacida kamchatkensis, Methylobacter tundripaludum, Methylocapsa acidiphila, Methylocella palustris, Methylocella silvestris, Methylocella tundrae, Methylococcus thermophilus, Methylococcus capsulatus, Methylohalobius crimeensis, Methylomicrobium buryatense, Methylomonas pelagica, Methylosarcina fifibrate, Methylosarcina lacus, Methylosarcina quisquiliarum, Methylosphaera hansonii, Methylothermus* sp. *HB*
Ammonia-oxidizing bacteria: oxidizing ammonia to methanol	*Nitrosococcus oceani, Nitrosococcus halophilus, Nitrosomonas aestuarii, Nitrosomonas briensis, Nitrosomonas communis, Nitrosomonas cryotolerans, Nitrosomonas europaea, Nitrosomonas eutropha, Nitrosomonas halophia, Nitrosomonas marina, Nitrosomonas mobilis, Nitrosomonas nitrosa, Nitrosomonas oligotropha, Nitrosomonas ureae, Nitrosospira multiformis, Nitrosospira tenuis.*

Source: Ge et al. (2014)

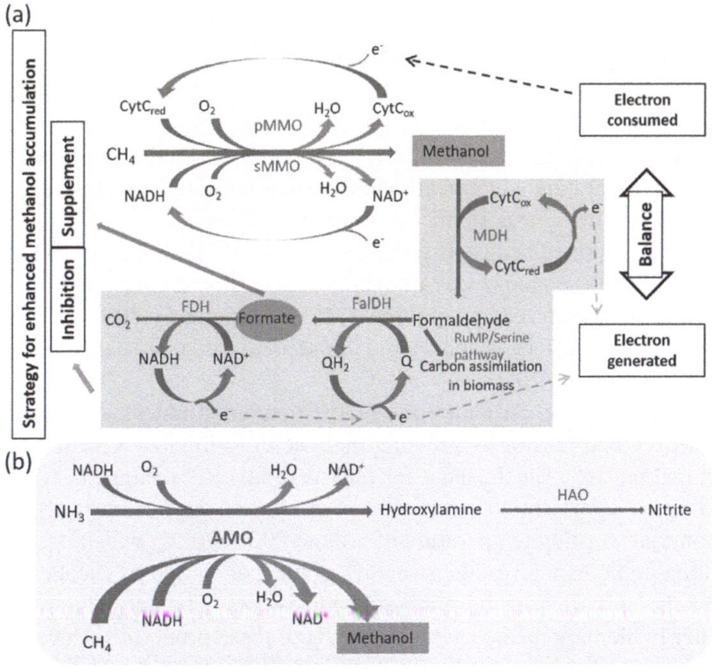

FIGURE 10.3 Pathways for the (a) oxidation of methane to methanol in methanotrophic bacteria, (b) co-oxidation of methane to methanol and ammonia to nitrite by ammonia-oxidizing bacteria. Enzymes are indicated by red text. Acronyms: $CytC_{red}$: reduced cytochrome C; $CytC_{ox}$: oxidized cytochrome C; e^-: electron; FalDH: formaldehyde dehydrogenase; FDH: formate dehydrogenase; H^+: proton; MDH: methanol dehydrogenase; NAD+: oxidized nicotinamide adenine dinucleotide; NADH: reduced nicotinamide adenine dinucleotide; pMMO: particulate methane monooxygenase; Q: ubiquinone; QH_2: ubiquinol; sMMO: soluble methane monooxygenase; AMO: ammonia monooxygenase; HAO: hydroxylamine oxidoreductase; NH3: ammonia; QH2: ubiquinol.

Methanotrophic bacteria: Methanogenic bacteria can carry out regulated partial methane oxidation, using methane as the single energy source. Methanotrophs are therefore a valuable candidate for the bioconversion of methane into methanol. However, raw biogas includes regulators of cell growth like sulfide, hydrogen, and ammonia, which can affect methanol production. Conversion of methanol from methane is achieved by the step-wise conversion of methanol to formaldehyde, formaldehyde to formate, and formate to CO_2, producing six reducing equivalents, each of them releasing two reducing equivalents, $CH_4 + O_2 + 2e^- \rightarrow CH_3OH + H_2O$. The ultimate methane-to-CO_2 conversion then generates four reducing equivalents (CH_4+O_2 / CO_2 + $4H^+$+$4e^-$) used in biosynthesis or other metabolic reactions (e.g., conservation of energy). In this pathway, the reactions are catalyzed sequentially with methane monooxygenase (MMO), methanol dehydrogenase (MDH), formaldehyde dehydrogenase (FalDH), and formate dehydrogenase (FDH). Methanotrophic cells may also be used to build up methanol by strain manipulation.

Ammonia-oxidizing bacteria (AOB): Bacteria that oxidize ammonia to nitrite using CO_2 as the source of fuel are mainly chemolithoautotrophs. Oxidation of ammonia to nitrite requires two steps: (i) oxidation of NH_3 by ammonia monooxygenase (AMO) to hydroxylamine (NH_2OH); (ii) oxidation of NH_2OH by hydroxylamine oxidoreductase (HAO) to NO_2^-. Ammonia-oxidizing organisms (AMOs) also have the potential to oxidize CH_4 to methanol, which is similar to MMO in methanotrophs [Taher and Chandran, 2013].

10.4 PROCESS OVERVIEW

10.4.1 Gasification and Anaerobic Digestion of Solid Waste

Gasification of solid waste for syngas production: Gasification is a thermal chemical reaction where in the presence of an oxidant such as air, steam, or oxygen, carbons are transformed to gaseous intermediates at a high temperature of 600–1000 °C. Prior to the process, the biomass is first typically pretreated by drying, reducing the dimensions (shaving, grinding, and cutting), and pyrolyzing, depending on the configuration of the gasifier. The resulting syngas primarily includes CO, CO_2, H_2, and N_2 with differing quantities of CH_4, water vapor, and impurities like H_2S, COS, NH_3, HCl, HCN, NOx, phenole, and light hydrocarbons as well as tar [Griffin and Schultz, 2012]. Several gasifiers used successfully for syngas synthesis are described next.

Fixed bed gasifier: The fixed bed gasifier can be further classified into an updraft or downdraft reactor in compliance with the course of carbonated fuel and oxidizing flows (air or steam). The biomass is dried (100 °C), pyrolyzed and gasified (900 °C), oxidized (1400 °C), and then extracted from the reactor. This gasifier mode is considered impractical for large-scale operations because of size limitations [Siedlecki et al., 2011].

Fluidized bed reactor: The carbonaceous fuel is combined along with the inert bed material within the fluidized bed reactor by pressing the reactor to fluidize a medium (e.g., air and/or steam). The inert bed makes it much easier for fuel materials to exchange heat and contribute to nearly isothermal operating conditions and high conversion efficiencies for feedstock. The gasifier normally has a nominal working temperature of around 800–900 °C, which is constrained by the bed material melting point. Moreover, the structure of the reactor and its excellent mixing properties also allow for the upscaling of the fluidized bed reactors. The duo-plate reactor is the most widely used gasifier in biomass feedstock today, owing to these properties. However, this mode of gasification is not ideal for feedstocks with high ash and alkaline metals, as their smelting induces the adhesivity and creation of larger bumps that eventually impact the reactor's hydrodynamics [Liew et al., 2013].

Entrained flow reactor: The entrained flow reactor shows high carbon conversion efficiency, low tar production, and is a favored route for large-scale gasification of the coal, petcoke, and refinery residues. Here, the feedstock is used to power a pneumatic transport device at a high speed to transform tar and methane at a temperature of 1200–1500 °C, resulting in improved syngas quality. Feedstock needs to be pulverized into ~50-µm fine particles before the process—not a big problem for coal but challenging and expensive for other biomass sources [Siedlecki et al., 2011].

Anaerobic digestion of solid waste for biogas production: In anaerobic digestion (AD), organic materials are processed efficiently into a beneficial biogas commodity, and its principal component is CH_4. AD is carried out in four main stages: hydrolysis, acidgenesis, acetogenesis, and methanogensis. A combination of microorganisms present at every point of the digestion process leads to the forming of digestate (decomposed feed) and a gas mixture, even in the absence of oxygen. The chemistry of the four main phases of AD is described here [Chen et al., 2008]:

Hydrolysis: Hydrolysis is a relatively small step that can restrict the rate of total digestion, in particular when solid waste substrates are used. During this step, cellulose ($C_6H_{10}O_5$) is converted to glucose ($C_6H_{12}O_6$) via addition of H_2O, giving off H_2 [Ostrem, 2004].

Acidogenesis: During this step, the monomers obtained in the previous step are converted to volatile fatty acids (VFAs) by acidogenic bacteria. The following equations describe this AD stage [Ostrem, 2004]:

$$C_6H_{12}O_6 \leftrightarrow 2\,CH_3CH_2OH + 2\,CO_2 \quad (10.15)$$

$$C_6H_{12}O_6 + 2\,H_2 \leftrightarrow 2\,CH_3CH_2COOH + 2\,H_2O \quad (10.16)$$

$$C_6H_{12}O_6 \rightarrow 3\,CH_3COOH \quad (10.17)$$

Acetogenesis: In this step, the VFAs generated previously are converted to acetic acid (CH_3COOH), CO_2, and H_2. The following equations reflect this AD stage [Ostrem, 2004]:

$$CH_3CH_2COO^- + 3\,H_2O \leftrightarrow CH_3COO^- + H^+HCO_3^- + 3\,H_2 \quad (10.18)$$

$$C_6H_{12}O_6 + 2\,H_2O \leftrightarrow 2\,CH_3COOH + 2\,CO_2 + 4\,H_2 \quad (10.19)$$

$$CH_3CH_2OH + 2\,H_2O \leftrightarrow CH_3COO^- + 3\,H_2 + H^+ \quad (10.20)$$

Methanogenesis: This is the fourth and last step of the AD process. In this step, CH_3COOH and H_2 are converted to CO_2 and CH_4 by bacteria referred to as methanogens. The reaction equation of the methanogenic stage of the AD processes is given here [Ostrem, 2004]:

$$CH_3COOH \rightarrow CH_4 + CO_2 \quad (10.21)$$

$$CO_2 + 4\,H_2 \rightarrow CH_4 + 2\,H_2O \quad (10.22)$$

$$2\,CH_3CH_2OH + CO_2 \rightarrow CH_4 + 2\,CH_3COOH \quad (10.23)$$

10.4.2 Pretreatment of Syngas and Biogas

Syngas cleaning: Impurities, particularly particulate matters, tars, sulfur, and nitrogen-containing compounds, must be eliminated from syngas before its use in biofuel production by the Fischer Tropsch synthesis (FTS) method [Woolcock and Brown, 2013].

Particulate matter removal: Alkaline metals, alkaline earth metals, silica, and other metals such as iron and magnesium may contain particulate matter as created by a gasifier. For syngas applications, complete elimination of the particulate matter is required. The elimination of syngas particulate content has been carried out by traditional air quality control technology, including cyclone, filtration, and electrostatic precipitation [Yang and Ge, 2016].

Tar removal: The tars consist of condensed organic compounds like heavy deoxidized hydrocarbons and polycyclic aromatic hydrocarbon (PAHs). High-temperature degradation, destruction by catalysts, and physical removal are techniques used to remove tar. Since techniques of physical removal such as scrubbing and electrical precipitation of tar can be done at low temperatures, they are favored [Yang and Ge, 2016].

Sulfur removal: Hydrogen sulfide (H_2S) and carbonyl sulfide (COS) are the main sulfur pollutants in syngas. Sulfa toxins are oxidized into SO_2 when the syngas is burnt. Sulfur reduction approaches include the burning of SO_2 sulfur contaminants and the use of traditional wet scrubbing technology (such as calestone scrubbing) to eliminate SO_2 in the gas stream. Another way of regulating sulfur is physically or chemically adsorbing it [Vamvuka et al., 2004].

Biogas cleaning: The use of biogas for injection in pipelines and the conversion to transport fuels has stringent composition criteria. In order to avoid dry ice formation and corrosion, biomethanol must be filtered for the processing of liquefied biomethanes in order to have CO_2, H_2S, and H_2O less than 25, 4, and 1 ppm, respectively. Pressurized water scrubbing (PWS), pressurized swing adsorption (PSA), amine or other organic material absorption, membrane permeation, and cryogenic process are widely used cleaning methods [Yang and Ge, 2016].

10.4.3 Main Microbial Fermentation Process

Bioreactor design for syngas fermentation: An optimized gas fermentation system requires efficient transfer of the mass of gaseous substrates into cultivation media and use of microbial catalysts. As a result of low-aqueous solubilities of CO and H_2, the gas-to-liquid mass transfer was defined as the rate-limiting stage at 35 °C [Kapic et al., 2006]. A bioreactor design that offers enough energy-efficient transmission of gas to liquid mass at a commercial scale for gas fermentation is thus a major engineering challenge. Table 10.5 presents a summary of syngas fermentation achieved with various types of bioreactors using acetogens, and Table 10.6 represents the different conditions for biological production of methanol by various microorganisms. A short description of the reactor set-ps for gas fermentation operations is described next.

Continuous stirred tank reactor: Gas substrates are fed continuously to the reactor in a continuous stirred tank reactor (CSTR) and mechanically sheared into smaller bubbles by deflated impulses with a wider interface for mass transfer [Mohammadi et al., 2011]. Moreover, smaller bubbles are more slow-moving and hold longer in the aqueous medium, which results in improved mass movement from gas to fluid. However, the CSTR requires the microbes to have thorough mixing and uniform delivery of gas substrates, making this method impractical for commercial gas fermentation systems due to the large volume needed to drive the burning agent [Bredwell et al., 1999].

Bubble column reactor: In the bubble column reactor, gas mixing is accomplished by gas splitting and mechanical agitation. This reactor design has fewer moving parts and therefore, in spite of exhibiting strong heat and mass transfer efficiencies, has lower capital and operating expense, rendering it an outstanding candidate for large-scale gas fermentation. However, the heterogeneous flow and back-mixing of the substrates of gas have been cited as challenges for increased mixing [Abubackar et al., 2011].

Immobilized cell column reactor: In an immobilized cell column reactor, the immobilization of microbes through linkage or adsorption of insoluble biosubstantial materials and the ensuing packaging in the column provides a number of advantages like application of high cell densities, plug-in flow operation, high mass transfer rate via direct contact between the microbe and gas, time reduction, and mechanical operation. Channeling difficulties, however, can occur when the interstitial space is overgrown and filled entirely. This reactor layout is not scalable to operate under different gas fermentation conditions due to limitations in the dimension of the column and packaging [Klasson et al., 1992].

Trickle bed reactor: A gaseous or liquid reactor, known as a trickle bed reactor, is a continuous, packed bed reactor that trickles liquid culture by the packaging of suspended or immovable pulley media. In this reactor structure, low gas and liquid flow rates are usually applied, producing a relatively low pressurized drop. The gas substratum is delivered in a copurpose or in contrast to the

TABLE 10.5
Summary of Syngas Fermentation Achieved with Various Types of Bioreactors Using Acetogens

Bioreactor[a]	Vol. L	Microorganisms	Mode of Operation	Medium	$CO:H_2:CO_2:N_2$ %	Product g/L
Stirred tank	100	*Clostridium ragsdalei*	Semi-continuous	Corn steep liquor	12–18:7–12:10–17:55–60	Ethanol: 15.0, Acetate: 2.8
Stirred tank	7.5	*Clostridium ragsdalei*	Semi-continuous	Corn steep liquor	20:5:15:60	Ethanol: 9.6, Acetate: 3.4
Stirred tank	3	Mixed culture of *Alkalibaculum bacchi* and *Clostridium propionicum*	Semi-continuous	Corn steep liquor	40:30:30:0	Ethanol: 1.1, acetate: 5.5, propanol: 1.0, propionate: 1.5, butanol: 0.1, butyrate: 0.4
Stirred tank	3	*Clostridium ragsdalei*	Semi-continuous	Activated carbon	40:30:30:0 or 20:5:15:60	Ethanol: 19.0, acetate: < 1.0
Stirred tank	7.5	*Clostridium carboxidivorans*	Semi-continuous	ATCC 1754	20:5:15:60	Ethanol: 1.8, acetate: 0.4, butanol: 0.4
Hollow fiber membrane	8	*Clostridium carboxidivorans*	Continuous	ATCC 1754 PETC	20:5:15:60	Ethanol: 23.9, acetate: 7.0
Bubble column	4	*Clostridium carboxidivorans*	Semi-continuous	Modified ATCC 1754	15:4:17:57	Ethanol: 5.0, acetate: 1.0, butanol: 0.6, butyrate: 0.4
Monolithic biofilm	8	*Clostridium carboxidivorans*	Continuous	Modified ATCC 1754	20:5:15:60	Ethanol: 5.0, acetate: 3.0
Trickle bed	0.5	*Clostridium ragsdalei*	Semi-continuous	Modified ATCC 1754	38:29:28:5	Ethanol: 5.7, acetate: 12.3
Trickle bed	0.5	*Clostridium ragsdalei*	Continuous	Modified ATCC 1754	38:29:28:5	Ethanol: 13.2, acetate: 4.3
Gas lift	3	*Clostridium carboxidivorans*	Semi-continuous	Modified ATCC 1754	13-1:20:12:27-45	Ethanol: 0.1, acetate: 0.1
Packed bed biofilm	3.3	*Clostridium carboxidivorans*	Continuous	Modified ATCC 1754	20:5:15:60	Ethanol: 7.0, acetate: 6.0
Stirred tank	2	*Clostridium carboxidivorans*	Semi-continuous	Synthetic medium	80:0:20:0	Ethanol: 1.2, butanol: 0.4

Source: Sun et al. (2019)

liquid flow, and mechanic agitation is not required [Bredwell et al., 1999]. This reactor has excellent gas conversion rates and higher productivity than other reactor types.

10.4.3.1 Syngas Fermentation Parameters

Gas composition: The composition of gas and its impurities influences the efficiency of the gas fermentation process. A greater molar ratio of H_2:CO makes it easier to turn carbon into products, since reduced equivalents are formed by the oxidation of H_2 (rather than CO). However, CO is also a recognized hydrogenase inhibitor, which can influence fermentation use of H_2. Acetogens are able to expand with CO_2 and H_2 while CO is ingested. Tar, sugars, ethane, ethylene, acetylene H_2S, NH_3, and NO are the typical impurities found during biomass gasification and have been shown to induce cell dormancy, hydrogen uptake inhibition, low cell growth, and changes between acetogens acidogenesis and solventogenesis [Abubackar et al., 2011].

TABLE 10.6
Conditions for Biological Production of Methanol by Microorganisms

Microorganisms	Medium	Incubation Temperature °C	CH_4 (Pure or Artificial Biogas): Air (Pure O_2 or Air) v/v	Methanol Concentration g/L
Methanotrophic bacteria				
Consortium	Nitrate minimal salt media (NMS)/40 mM Phosphate	30	4:6	~0.22
Consortium	NMS/100 mM NaCl	30	4:6	~0.16
Consortium	NMS/40 mM NH_4Cl	30	4:6	~0.19
Consortium	NMS/50 µM EDTA	30	4:6	~0.19
Consortium	NMS/100 mM NaCl	30	1:9	~0.03
Methylosinus trichosporium	100 mM Phosphate/5 mM $MgCl_2$/40 mM Formate/	35	1:1	0.34
Methylosinus trichosporium	12.9 mM Phosphate/67 nM Cyclopropanol/14.3 mM Formate	25	1:1.6	0.17
Methylosinus trichosporium	20 mM Phosphate/5 mM $MgCl_2$	32	1:4:1:1 (CH_4:CO_2:O_2:N_2)	0.0006
Methylosinus trichosporium	12.9 mM Phosphate/200 mM NaCl/20 mM Formate	30	1:0	0.25
Methylosinus trichosporium	12.9 mM Phosphate/100 mM NaCl/1 mM EDTA/20 mM Formate	25	1:1	0.44
Methylosinus trichosporiumd	400 mM Phosphate/10 mM $MgCl_2$/20 mM Formate	30	1:1	1.12
Ammonia-oxidizing bacteria				
Nitrosomonas europaea	3.3 g/L $(NH_4)_2SO_4$ + other inorganic salts	28	-	0.0192
Nitrosomonas europaea, Nitrosococcus oceanus	20 mM HEPES buffer (pH 7.8)+1 mg/L NH_4-N+other inorganic salts & salinity of 10–32%	25	-	0.016
Mixed nitrifying cultures	500 mg-N L^{-1} of $(NH4)_2SO_4$ and NH_2OH	22	-	0.015-0.039

Source: Ge et al. (2014)

Substrate pressure: The partial pressure of syngas components has a heavy effect on microbial growth and product profiles, as these enzymes are susceptible to the exposure of the suspended substrate. Due to low CO and H_2 solubilities in water, dense bacterial cell cultures can face mass transmission limits, which can help mitigate the issue by increased partial pressure of the gaseous substrates [Vega et al., 1989].

Medium formulation: While CO and CO_2/H_2 are available for use as carbon and energy supply by acetogens, other components such as vitamins, trace metal elements and minerals, and reducing agents are needed for sustaining the high metabolic activity of the microbes. A low redox potential is needed to produce a strict anaerobic environment, and substances like sodium thioglycolate, ascorbic acid, methyl viologen, benzyl viologen, titanium (III)-citrate, potassium ferricyanide, cysteine- HCl, and sodium sulfide can be frequently added to the fermented medium for this reduction.

Moreover, these agents direct the electron and carbon flow towards solventogenesis by increasing the usable reduction equivalents to NADH for the manufacture of alcohol. However, excessive addition can result in slower microbial growth due to the reduction in acetogenesis ATP formation, so the best concentration should be established first. Sulfur-containing gases (e.g., H_2S) in syngas are lethal to chemical catalysts but beneficial for microbial catalysts, as they reduce the medium-redox potential, which in turn is helpful for redox-sensitive enzymes such as CODH and activates the production of alcohol [Liew et al., 2013].

Medium pH: As with other species, acetogens have a small pH range for optimum growth and requires a tightly regulated pH of the fermentation medium. Extracellular pH has a clear effect on intracellular pH, membrane potential, and motive force for protons, and thus on the use of the substrate and its product profile. Also, a decrease in medium pH diverts carbon and electron flow from cell formation to the development of alcohol [Kundiyana et al., 2011].

Temperature: The optimal temperature for mesophilic and thermophilic acetogens is between 30–40 °C and 55–58 °C, respectively. In addition to the usage of substratum, growth rate, and lipid membrane structure of the acetogens, the fermentation temperature influences the supply of gas substrates because gas solubility is increased with a declining temperature [Liew et al., 2016].

10.4.3.2 Conversion of Methane to Methanol

Methanotrophic bacteria: The production of methanol by methanotrophic strains is generally carried out by giving electron donors in mineral media prepared in phosphate buffers for MDH inhibition. Phosphate (40–400 mM), formate (15–40 mM), $MgCl_2$ (5–10 mM), cyclopropanol (67 nM), NaCl (100–200 mM), EDTA (0.05–1 mM), and NH_4Cl (40 mM) act as inhibitors to form the base of MDH inhibition. When methanotrophic cultivation takes place, methane and oxygen transfer from the gas to the liquid level tends to be a limiting factor in aqueous media due to their poor solubility. In order to maximize the methane transition between a gas and aqueous phases, these chemicals have to be used to improve gaseous liquid mass transfer [Han et al., 2009].

AOB: Many microbes have been found to be inhibited by CH_4, while the use of NH_3 was not inhibited in them. It has been observed that when NH_4^+ or hydrazine was supplied as a reductant, O_2 absorption was also doubled with the addition of CH_4. Methanol was found to be produced in an atmosphere of 50% CH_4 in one hour during cultivation with NH_4^+. The addition of NH_3 has improved oxidation and cellular assimilation of CH_4. Both nitrite and nitrate have been found to increase the oxidation of CH_4 [Hyman et al., 1988].

10.4.4 Product Recovery after Bioconversion

Microbes may be maintained as biofilms attached to carrier materials to sustain high cell density in a reactor. In a fermentation rim, planktonic cells can be maintained by installing solid/liquid separators such as membrane ultra-filters, spiral woven filters, hollow fiber, cell recycling membranes, or centrifuges [Liew et al., 2013]. Gas solvent concentrations are rarely greater than 6% [w/v], requiring a cost-effective and energy-efficient method for product recovery. In the same way, a solvent recovery method can increase the efficacy of the solvents by reducing solvent concentrations (and hence the toxicity) in a fermentation vessel. In fact, acetogens often show poor tolerance to solvents such as ethanol and butanol. Distillation has been the conventional form of product recovery, but due to the related high energy costs, alternative methods such as liquid extraction, pervaporation, perstraction, and gas deletion have become more popular [Brown et al., 2011] as described next.

Liquid-liquid extraction: A water-insoluble organic extractant is combined with the fluid liquid in the fermentation container. Since the solvents are more soluble in the organic phase than in the aqueous phase, they are concentrated selectively in the extractant. Although this method would not exclude water or nutrients from the fermentation broth, some gas substrates may be extracted because of the higher solubility of CO and H_2 in organic solvents than water. Due to its nontoxic nature, oleyl alcohol is generally chosen as the extractor solvent [Ejeji et al., 2007].

Perstraction: Liquid-liquid mining is related to many issues like microbial toxicity, emulsion development, and the buildup of microbes at the interface between the extracting material and fermenting bread. Perstraction was created in an effort to remedy these issues, using a membrane to disassemble the extractant from the fermentation broth. This physical boundary not only avoids direct interaction between the microbe and the toxic extractant but also decreases its rate and is vulnerable to fouling of the solvent extraction [Ejeji et al., 2007].

Pervaporation: In a product recovery procedure called pervaporation, a membrane that is specifically in contact with the fermentation broth is used to selectively extract volatile compounds such as ethanol and butanol. The toxic compounds diffuse across the membrane as vapor and then are accumulated by condensation. A differential partial pressure across the membrane is typically preserved by vacuum or inert gases (e.g., N_2) to promote the volatilization of permeates into a vapor. Polydimethylsiloxane (PDMS), poly(1-trimethylsilyl-1-propylene), and hydrophobic zeolite can be used as the controlling content in the membranes [Ejeji et al., 2007].

Gas stripping: Gas striping is a process for product recuperation for the in situ recovery of escaped gas from the bioreactor. The wastewater and gas can be recycled back into the bioreactor after product recovery via condensation. Gas stripping has been proven to be 200% more effective than nonintegrated processes for acetone-butanol-ethanol (ABE) production, complete use of the substrate, and even total acid conversion into solvents [Ejeji et al., 2007].

10.5 CHALLENGES FACED AND EFFECTIVE STRATEGIES FOR THE WHOLE PROCESS

Mass transfer limitation: The restriction on gas-to-liquid mass transfer is still one of the main obstacles in the marketing of syngas fermentation technology. A reliable parameter can be used to compare the mass transfer capability in different reactor setups as volumetric mass transfer coefficients. Advances in impeller technologies, fluid flow patterns, power quality aeration, mixing time, blast configuration, and use of microscopic dispersers are typical methods to improve mass propagation limitations [Munasinghe and Khanal, 2010].

Quality of syngas: Syngas, which is used for the fermentation process, should be free of impurities such as tar, smoke, ethylene, ethane, acetylene, and sulfur-containing gases (SOx) and nitrogen (NOx). Several gas clean-up strategies such as cyclones, diverse types of filters (fabric, ceramic, or bag), particle rotary separators, water scrubbers, and wet-static precipitator sare currently available for fermenting syngas. Additionally, impurity generation can be minimized significantly at the source itself by using efficient gasification and preprocessing methods such as frying and liquidation [Hasler and Nussbaumer, 1999].

Microbial catalysts: Another essential feature of commercialization of the process is the isolation and maintenance of anaerobic microorganisms capable of transforming syngas into ethanol and other by-products. A significant problem is the genetic alteration of the current syngas fermentation microbes to create high-yield anaerobic organisms [Munasinghe and Khanal, 2010].

Product recovery: The conventional approach to separate ethanol from a mixture of water and other industrial fermentation by-products is threatened by high energy costs. Alternative approaches like recompression of vapor, ultrasonic ionization, substantive distillation, and selective water adsorption are being investigated to minimize the cost of the recovery of ethanol [Sato et al., 2001].

Redirection of metabolic pathway: Syngas fermentation is often related to the production of acid as a side product, which decreases the pH of the cultivation medium. Low pH results in an unfavorable atmosphere for solvent production. Redirecting the metabolic pathway to the production of solvents by blocking the production of acid can increase ethanol production [Munasinghe and Khanal, 2010].

Conversion of biogas to liquid fuels: Methanotrophs, AOBs, and acetogens act as "cell factories" for the generation of liquid fuels from CH_4. However, there are several anticipated obstacles that interfere with the production of liquid CH_4 fuels at an industrial scale: (i) cell growth inhibition by H_2S (or other compounds) when CH_4 is used in biogas; (ii) high cost of the electron donors usually needed for conversions; (iii) limitations related to gas-liquid mass transfer; and (iv) lack of microorganisms for direct conversion of methane into other fuels rather than methanol. The latest existing technology will potentially fix these problems [Ge et al., 2014].

10.6 GENETIC ENGINEERING FOR A TARGETED PRODUCT OR ENHANCED PRODUCT TITER

For industrial-scale gas fermentation, the microorganisms that are to be used as a biocatalyst must show the following characteristics: high product titer, low substrate/product inhibition, no strain degradation, not producing spores, longer cell viability, high tolerance for air, high growth rate, maximum biomass productivity, and maximum utilization of syngas as substrates. These can be sorted based on adaptive evolution, random mutation, and/or genetic modification. Conventionally, chemical mutagenesis [Hermann et al., 1985] and strategies for adaptation [Soucaille et al., 1987] have been applied to observe these traits. Table 10.7 presents the summary of genetic modifications of syngas-fermenting bacteria and their effects.

10.6.1 Genetic Modification of Syngas-Fermenting Microbes

The genome sequencing of several solventogenic *Clostridia*, including gas-fermenting species, has been done since 2001 [Köpke et al., 2010], and there is an increased availability of transcriptomic [Wang et al., 2012], proteomic [Janssen et al., 2010] array, and systems analysis [Haus et al., 2011]. However, to generate genetically modified *Clostridia*, certain difficulties are encountered while introducing foreign DNA into cells due to the lack of well-developed genetic tools for this specific bacterium. There is very little information available for metabolic engineering of acetogens. Many genetic modification approaches have been generated to increase the product range of syngas fermentation such as DNA transfer, knockdown of target gene via antisense RNA, intron-based gene inactivation, stable genome insertion or deletion via homologous recombination, precise genetic manipulation via allelic-exchange, precise genetic manipulation via triple cross, transposon mutagenesis, CRISPR/Cas9, and availability of varieties of genetic parts.

In the beginning of the 1990s, Prof. Terry Papoutsakis [Walter et al., 1994] attempted the first targeted genetic alteration of *Clostridia*, using plasmids for (over)expression of genes in *C. acetobutylicum*. For downregulation of genes, antisense RNA (asRNA) was utilized, wherein single-stranded RNA prevented the translation of messenger RNA (mRNA) by binding to a complementary target, thus inhibiting interactions of the ribosome-binding site [Desai et al., 1999]. Several procedures were applied using homologous recombination for inserting or deleting genes in many sugar-utilizing *Clostridia*. Earlier, for the generation of knockout mutants, only a single crossover event was applied, but the mutants could revert back to the wild type [O'Connor et al., 2006]. So, for more flexible screening of the rare second recombination for *C. acetobutylicum* [Soucaille et al., 2008] and *C. thermocellum* [Argyros et al., 2011], counter selectable markers were developed. For the disruption of genes, specificity of mobile group II intron Ll.*ltrB* of *Lactoccocus lactis* was utilized by ClosTron for insertion into a specified genome site via an RNA-mediated, retro-homing mechanism [Karberg et al., 2001]. This method was then successfully applied to many solventogenic and pathogenic *Clostridia*, including *C. acetobutylicum, C. difficile, C. sporogenes, C. perfringens*, and *C. botulinum* [Camiade et al., 2010]. Another method for integration of DNA into the genome, namely allele-coupled exchange (ACE), has been proposed where the employment of a counter-selective marker to detect the rare second recombination event is not needed. Rather, a selectable phenotype for activation or inactivation of gene(s) is utilized to direct the order of recombination events [Heap et al., 2012]. While some critical metabolic engineering resources such as successful inducible promoters are still lacking, the array of available tools has dramatically enhanced the ABE method and the cellulolytic fermentation process of *Clostridia*.

TABLE 10.7
Summary of Genetic Modifications of Syngas-Fermenting Bacteria and Their Effects

Microorganisms	Approaches Taken for Genetic Modifications	Effects of the Modifications
Modifications for improved product titer		
Acetobacterium woodii	Plasmid-based expression of acetone biosynthetic (*thlA-ctfAB-adc*) genes in *Clostridium acetobutylicum* using different combinations of promoter genes (ackA, pta-ack) and plasmid origins (pIP404, pBP1, pCB102, pCD6)	Continuous production of 26.4 mg/L/h acetone from synthetic syngas in CSTR
Clostridium aceticum	Plasmid-based expression of acetone biosynthetic (*thlA-ctfAB-adc*) genes in *Clostridium acetobutylicum*	Production of up to 9 mg/L acetone from synthetic syngas in bottles
Clostridium aceticum	Plasmid-based expression of synthetic acetone operon for thiolase and acetoacetate decarboxylase (*thl-adc*) genes in *Clostridium acetobutylicum*, thioesterase (*teII*) genes in *Bacillus subtilis*, and acyl-CoA thioesterase (*ybgC*) genes in *Haemophilus influenzae*	Production of up to 59 mg/L acetone
Clostridium autoethanogenum, *Clostridium ljungdahlii*	Plasmid-based expression of butanol biosynthetic (*thlA-crt-hbd-bcd-etfAB*) genes in *Clostridium acetobutylicum*	Production of up to 1.93 g/L butanol from steel mill gas and syngas in bottles
Clostridium autoethanogenum, *Clostridium ljungdahlii*	Plasmid-based expression of acetone and isopropanol biosynthetic genes in *Clostridium acetobutylicum (thlA)* and *Clostridium beijerinckii (ctfA-ctfB-adc)*	Production of up to 300 mg/L acetone and 25 mg/L isopropanol from steel mill gas and syngas in bottles; continuous production of 700 mg/L/d isopropanol from steel mill gas in CSTR
Clostridium autoethanogenum	Plasmid-based expression of malonyl-coenzyme A reductase in *Chloroflexus aurantiacus*	Production of low levels of 3-hydroxypropionate from steel mill gas and H_2/CO_2
Clostridium autoethanogenum	Plasmid-based expression of *Klebsiella pneumoniae (S)*–specific butanediol dehydrogenase and *Klebsiella oxytoca* diol dehydratase *pddABC* in *Clostridium autoethanogenum* with inactivated native 2,3-butanediol dehydrogenase	Production of up to 370 mg/L meso-2,3-butanediol, MEK, and 2-butanol from steel mill gas in bottles
Clostridium autoethanogenum	Plasmid-based expression of unspecific acyltransferase in *Acinetobacter baylyi*	Production of low levels of butanoic acid and butyl ester from steel mill gas
Clostridium autoethanogenum	Plasmid-based overexpression of DOXP synthase, expression of mevalonate pathway, *Clostridium beijerinckii* isopentenyl diphosphate isomerase, and either *Poplar* isoprene synthase or farnesene synthase	Production of low levels of mevalonate, isoprene, and franesene from steel mill gas and syngas
Clostridium ljungdahlii	Plasmid-based expression of *Clostridium acetobutylicum* butanol biosynthetic genes (*thlA-crt-hbd-bcd, adhE-bdhA*)	Production of up to 148 mg/L *n*-butanol from synthetic syngas in bottles; conversion to butyrate at the end of growth
Clostridium ljungdahlii	Plasmid expression of *Clostridium acetobutylicum* acetone biosynthetic genes (*thlA-ctfAB-adc*) under lactose-inducible promoter	Production of up to 871 mg/L acetone from CO in bottles
Clostridium ljungdahlii	Chromosomal integration (single cross-over recombination) of *Clostridium acetobutylicum* butyrate biosynthetic genes (*thl-crt-bcd-etfB-etfA-hbd-ptb-buk*) into the *pta* promoter region	Production of up to 881 mg/L butyrate using H_2/CO_2 in bottles

(*Continued*)

TABLE 10.7 (Continued)
Summary of Genetic Modifications of Syngas-Fermenting Bacteria and Their Effects

Microorganisms	Approaches Taken for Genetic Modifications	Effects of the Modifications
Clostridium ljungdahlii	Plasmid-based expression of mevalonate pathway, *Escherichia coli* isopentenyl diphosphate isomerase, and *Poplar* isoprene synthase	Production of up to 68 mg/L mevalonate and low levels of isoprene from syngas in bottles
Moorella thermoacetica	Genome insertion of *ldh* from *Thermoanaerobacter pseudethanolicus* into *pyrF* locus	Heterotrophic production of up to 613 mg/L lactate from glucose in bottles
Modifications for enhanced production		
Acetobacterium woodii	Plasmid-based expression of formyl-THF-synthetase, methenyl-THF-cyclohydrolase, methylene-THF-dehydrogenase, and methylene-THF-reductase of *Clostridium ljungdahlii*	Increase in volumetric acetate production rate by 14% under H_2/CO_2 in CSTR
Clostridium autoethanogenum	Inactivation of acetolactate decarboxylase gene *budA*	Abolishment of 2,3-butanediol production along with enhanced ethanol production by 79%; small levels of succinate and lactate produced during growth on steel mill gas
Clostridium autoethanogenum	Inactivation of lactate dehydrogenase gene *ldhA*	Abolishment of lactate production
Clostridium autoethanogenum	Plasmid expression of vitamin biosynthetic genes, *thiC* of *Clostridium ragsdalei*, and *panBCD* of *Clostridium beijerinckii*	Successful complementation of thiamine and panthothenate biosynthesis pathways; strains independent of vitamin B_1 and B_5 supplementation during growth on steel mill gas
Clostridium autoethanogenum	Plasmid expression of native *groES* and *groEL*	Increased cell viability when challenged with ethanol during growth on steel mill gas
Clostridium difficile	Generation of more than 70,000 unique mutants via transposon mutagenesis, coupled with transposon-directed insertion site sequencing	Identification of 404 essential genes for growth; identification of 798 genes that are likely to affect sporulation
Clostridium ljungdahlii	Chromosomal deletion of *adhE1 and/or adhE2*	Six-fold reduction in ethanol concentration of Δ*adhE1* mutant (but not Δ*adhE2* mutant) under heterotrophic conditions
Clostridium ljungdahlii	Plasmid expression of formate dehydrogenase from *Escherichia coli*	4.3-fold increase in intracellular NADH concentration; 2.3-fold improvement in maximum power density in a sodium formate–fed microbial fuel cell

Source: Reproduced from Liew et al. (2016)

Only a few acetogenic *Clostridia* have been successfully modified to date. A nonbutanol producer species, *C. ljungdahlii*, was genetically modified with butanol biosynthetic genes (*thlA, hbd, crt, bcd, adhE*, and *bdhA*) from *C. acetobutylicum* ATCC 824 resulting in 2 mM of butanol production using syngas [Köpke et al., 2010]. Production of 140 μM acetone was achieved by sending a plasmid with acetone biosynthesis genes *ctfA, ctfB, adc*, and *thlA* in *C. aceticum* and using gas mixes (80% H_2/20% CO_2 and 67% H_2/33% CO_2) [Lederle, 2011]. Butanol biosynthetic genes *thlA, hbd, crt, bcd, etfA*, and *etfB* were heterologously expressed in *C. ljungdahlii* and *C. autoethanogenum*, resulting in up to 26 mM butanol production using steel mill gas [Koepke and Liew, 2011]. Genetically modified *C. autoethanogenum* had shown better alcohol resistance than the wild strain when their native *groESL* operon was overexpressed [Simpson et al., 2011]. A reported 8.9 mM isobutanol were produced from cellulose in *C. cellulolyticum* by overexpressing *kivD, yqhD, alsS, ilvC,* and *ilvD* genes [Higashide et al., 2011]. However, in view of the unsolved detailed energy study of acetogens, it is uncertain if the development of such high-energy liquid fuels by gas fermentation could be feasible.

10.6.2 Genetic Modification of Methanogenic Bacteria

Harwood et al. [1972] used N-methyl-N'-nitro-N-nitrosoguanidine, N-nitroso-N-methyl urethane, ethyl methanesulfonate, and ultraviolet (UV) light to increase mutagenesis frequency and performed the first research on methanotroph mutagenesis to obtain one auxotrophic mutant requiring p-aminobenzoic acid for growth. Williams et al. [1977] further studied the effects of various repair-dependent (UV, γ, methyl methanesulphonate) and repair-independent mutagens (ethyl methanesulphonate and N-methyl-N'-nitro-N-nitrosoguanidine) on *Methylococcus capsulatus* and *Methylomonas albus* and isolated mutants resistant to trimethoprim. Dichloromethane proved to be more effective in conjunction with other mutagens [Murrell, 1992] and was used to separate *Methylosinus trichosporium* OB3b, *Methylomonas albus* BG8WM, and *Methylosinus sporium* 2C10P MMO-deficient mutants [Bowman and Sayler, 1996].

For specific genetic modifications like knockout and homologous recombination between chromosomal and plasmid DNA, conjugation has been extensively used in methanotrophs. A reported 10^{-2} to 10^{-9} transfer frequencies were noticed among the strains and the plasmids used for such studies [Ye and Kelly, 2012]. Electroporation has also been used to send plasmids into methanotrophs [Baani and Liesack, 2008]. The IncP broad-host-range promoter probe vector pMHA203 was transferred by both conjugation and electroporation into *M. silvestris* [Crombie and Murrell, 2011], and almost equal transfer efficiency was observed. Bacteriophages can also be used for genetic manipulation of methanotrophs [Murrell, 1992].

Methylomonas sp. 16a has been widely examined as the platform for the processing of aromatic components, sugars, terpenoids, and pigments [Koffas et al., 2007]. It has several mechanisms to integrate carbon into three-carbon units, which can be used by conjugation in genetic engineering. *Methylocella silvestris* BL2 is the only characterized facultative methanotrophic bacteria, and a system for genetic modification of this strain has been developed by Crombie and Murrell [2011]. A kanamycin-resistance gene cassette was substituted for the isocitrate lyase gene (Msil, 1641), which resulted in deletion of the unmarked gene upon removal of that cassette. This method has shown tremendous promise for exposing the metabolic pathways of this strain, for investigating its growth regulatory mechanisms on various substrates, and for modeling strains for desirable properties.

10.6.3 Genetic Modification of Ammonia-Oxidizing Bacteria

It is very challenging to genetically modify AOB due to many reasons, like stunting growth, problem of growth agar media, poor selection of mutant chemolithoautotrophic bacteria, high antibiotic resistance property, and poor transformation ability [Sayavedra-Soto and Stein, 2011]. *Nitrosomonas europaea* has been studied mostly for genetic modification of AOB, and its complete gene map has been reported by Chain et al. [2003], making this organism perfect for efficient genetic modification. Electroporation and conjugation are the most widely used methods for sending genes into this organism. *Nitrosospira multiformis* has also been genetically modified by conjugation, which is the only mode of its transformation due to failure of electroporation using varieties of buffers, temperature, and substrate concentration [Sayavedra-Soto and Stein, 2011]. Vectors were built and used for AOB gene inactivation. By incorporating the *lacZ* and *kan* genes into amoA, Hommes et al. [1998] created a cassette, and this gene cassette was incorporated into the chromosomal amoA loci in

N. europaea. An internal fragment of the *nirK* gene was incorporated into a suicide vector (pRVS3) by Beaumont et al. [2002], which was then introduced by homologous recombination into the chromosomal copy of the *nirK* gene, resulting in the destruction of this gene. Another approach for disrupting a gene is replacing a target gene with an antibiotic-resistance gene [Berube et al., 2007]. As this gene has three copies in the strain, Hommes et al. [2002] constructed three double-hao mutants of *N. europaea*. Using broad-host set plasmids and reporter genes, expression in gene level has been examined in *N. europaea*. Fusing vectors (such as pPROBE) with the green fluorescent protein gene (gfp) by transcription, Gvakharia et al. [2009] constructed plasmids and tested the promoters in *N. europaea*. Hirota et al. [2006] examined the expression in *Nitrosomonas* ENI-11 by building broad-host range plasmids in pkz27 with hao-lacZ fusions.

10.7 COMMERCIALIZATION OF BIOLOGICAL GAS FERMENTATION

Ineos Bio, established in 2008, started to produce ethanol at a capacity of 8 million gallons per year (MMGY) by 2012 [IBTP, 2016]. The company employed patented natural biocatalysts in the gasification process, having high ethanol selectivity and resistance to most impurities. The processed ethanol had a purity of 99.7% and could be combined with petrol. Alliance Bio-Products, Inc., purchased the Ineos Bio plant in 2017 to develop the method of cellulose to sugar conversion [ABP, 2017]. Lanzatech has two operating plants, one in China and the other in Belgium with a 16 MMGY and 21 MMGY ethanol production capability, respectively. The company wanted to commercialize a method for transforming steel mill gas to ethanol and purified and extracted the coproducts via a hybrid water recycled separation system, which was produced using a patented microbe strain by the bioconversion of syngas. Lanzatech aims to transform syngas into a wide range of products, such as jet fuel, butadiene, 2,3-butanediol, isopropanol, and isobutylene [LanzaTech, 2016]. Another company, Coskata, developed a semi-commercial method for fermenting syngas for chemical fuels and was acquired by Synata Bio in 2015 [Coskata, 2016].

10.8 CONCLUSION AND FUTURE PROSPECTS

In recent years, major improvements in the gene modification techniques for gas-fermenting microorganisms have led to increased production of new chemicals via contemporary routes of gas fermentation in only single steps. These improvements in the gas fermentation methods for the manufacture of fuels, chemicals, and thermoplastics at economically appropriate rates will slowly result in their substitution of fossil fuels, limiting climate change at a global level. Continued advancement in research and development, especially in the areas of bioprocess and genetic engineering, would make biological gas conversion to biofuels and other value-added chemicals theoretically and economically feasible in the long run.

ACKNOWLEDGMENT

The authors thankfully acknowledge Dr. Anjan Ray, director CSIR-Indian Institute of Petroleum, for his constant encouragement and support.

CONFLICT OF INTEREST

The authors declare that they have no competing interests

REFERENCES

Abubackar HN, Veiga MC, Kennes C, Coruña L. Biological conversion of carbon monoxide: rich syngas or waste gases to bioethanol. *Biofuels, Bioproducts & Biorefining*. 2011:93–114.

Allen TD, Caldwell ME, Lawson Pa, Huhnke RL, Tanner RS. Alkalibaculum bacchi gen. nov., sp. nov., a CO-oxidizing, ethanol-producing acetogen isolated from livestock- impacted soil. *International Journal of Systematic and Evolutionary Microbiology*. 2010;60(Pt 10):2483–2489.

Alliance Bio-Products gains USDA approval to buy Ineos Bio plant, 2017. http://www.biomassmagazine.com/articles/14517/alliance-bio-products-gains-usdaapproval-to-buy-ineos-bio-plant (accessed 12.05.2021).

Argyros DA, Tripathi SA, Barrett TF, Rogers SR, Feinberg LF, Olson DG, et al. High ethanol titers from cellulose by using metabolically engineered thermophilic, anaerobic microbes. *Applied and Environmental Microbiology*. 2011 Dec;77(23):8288–8294.

Baani M, LiesackW. Two isozymes of particulatemethane monooxygenase with different methane oxidation kinetics are found in Methylocystis sp. strain SC2. *Proceedings of the National Academy of Sciences of the United States of America*. 2008;105:10203–10208.

Beaumont HJ, Hommes NG, Sayavedra-Soto LA, Arp DJ, Arciero DM, Hooper AB, et al. Nitrite reductase of Nitrosomonas europaea is not essential for production of gaseous nitrogen oxides and confers tolerance to nitrite. *Journal of Bacteriology*. 2002;184:2557–2560.

Berube PM, Samudrala R, Stahl DA. Transcription of all amoC copies is associated with recovery of *Nitrosomonas europaea* from ammonia starvation. *Journal of Bacteriology*. 2007;189:3935–3944.

Bowman JP, Sayler GS. Development of copper-tolerant mutant of soluble methane monoxygenase-producing methanotrophs for application in bioremediation. *DTIC Document*. 1996. https://apps.dtic.mil/sti/citations/ADA317429

Bredwell MD, Srivastava P, Worden RM. Reactor design issues for synthesis-gas fermentations. *Biotechnology Progress*. 1999 Sep–Oct;15(5):834–844. PubMed PMID: WOS:000082936200007.

Brown SD, Guss AM, Karpinets TV, Parks JM, Smolin N, Yang S, et al. Mutant alcohol dehydrogenase leads to improved ethanol tolerance in Clostridium thermocellum. *Proceedings of the National Academy of Sciences of the United States of America*. 2011;108(33):13752–13757.

Camiade E, Peltier J, Bourgeois I, Couture-Tosi E, Courtin P, Antunes A, et al. Characterization of Acp, a Peptidoglycan hydrolase of Clostridium perfringens with Nacetylglucosaminidase activity That is implicated in cell separation and stress induced autolysis. *Journal of Bacteriology*. 2010 May;192(9):2373–2384. PubMed PMID: WOS:000276685800009.

Chain P, Lamerdin J, Larimer F, Regala W, Lao V, Land M, et al. Complete genome sequence of theammonia-oxidizing bacteriumand obligate chemolithoautotroph Nitrosomonas europaea. *Journal of Bacteriology*. 2003;185:2759–2773.

Chen Y, Cheng JJ, Creamer KS. Inhibition of anaerobic digestion process: a review. *Bioresource Technology*. 2008;99:4044–4064.

Coskata's technology re-emerges as Synata Bio, 2016. http://www.biofuelsdigest.com/bdigest/2016/01/24/coskatas-technology-re-emerges-as synata-bio/ (accessed 12.05.2021).

Crombie A, Murrell JC. Development of a system for genetic manipulation of the facultative methanotroph Methylocella silvestris BL2. *Methods in Enzymology*. 2011;495:119–133.

Desai RP, Papoutsakis ET. Antisense RNA strategies for metabolic engineering of Clostridium acetobutylicum. *Applied and Environmental Microbiology*. 1999 Mar; 65(3):936–945. PubMed PMID: WOS:000078882000010.

EPA. 2017. https://www.epa.gov/ghgemissions/global-greenhouse-gas-emissions-data.

Ezeji TC, Qureshi N, Blaschek HP. Bioproduction of butanol from biomass: from genes to bioreactors. *Current Opinion in Biotechnology*. 2007;18(3):220–227.

Ge X, Yang L, Sheets J P., Yu Z, Li Y. Biological conversion of methane to liquid fuels: status and opportunities. *Biotechnology Advances*. 2014;32(8):1460–1475. ISSN 0734-9750, https://doi.org/10.1016/j.biotechadv.2014.09.004.

Geddes CC, Nieves IU, Ingram LO. Advances in ethanol production. *Current Opinion in Biotechnology*. 2011;22:312–319. https://doi.org/10.1016/j.copbio.2011.04.012.

Griffin DW, Schultz MA. Fuel and chemical products from biomass syngas: a comparison of gas fermentation to thermochemical conversion routes. *Environmental Progress & Sustainable Energy*. 2012 Jul;31(2):219–224. PubMed PMID: WOS: 000302794000008.

Gvakharia BO, Bottomley PJ, Arp DJ, Sayavedra-Soto LA. Construction of recombinant Nitrosomonas europaea expressing green fluorescent protein in response to cooxidation of chloroform. *Applied Microbiology and Biotechnology*. 2009;82:1179–1185.

Han B, Su T, Wu H, Gou Z, Murrell JC. Paraffin oil as a methane vector for rapid and high cell density cultivation of Methylosinus trichosporium OB3b. *Applied Microbiology and Biotechnology*. 2009;83:669e677.

Harwood JH, Williams E, Bainbridge BW. Mutation of the methane oxidizing bacterium Methylococcus capsulatus. *Journal of Applied Microbiology*. 1972;35:99–108.

Hasler P, Nussbaumer T. Gas cleaning for IC engine applications from fixed bed biomass gasification. *Biomass and Bioenergy*. 1999;16:385–395.

Haus S, Jabbari S, Millat T, Janssen H, Fischer R-J, Bahl H, et al. A systems biology approach to investigate the effect of pH-induced gene regulation on solvent production by Clostridium acetobutylicum in continuous culture. *BMC Systems Biology*. 2011;5(1):10.

Heap JT, Ehsaan M, Cooksley CM, Ng YK, Cartman ST, Winzer K, et al. Integration of DNA into bacterial chromosomes from plasmids without a counter-selection marker. *Nucleic Acids Research*. 2012 Apr 1;40(8):e59. PubMed PMID: 22259038. Pubmed Central PMCID: 3333862. Epub 2012/01/20.eng.

Hermann M, Fayolle F, Marchal R, Podvin L, Sebald M, Vandecasteele JP. Isolation and characterization of butanol-resistant mutants of Clostridium acetobutylicum. *Applied Environmental Microbiology*. 1985;50:1238–1243.

Higashide W, Li YC, Yang YF, Liao JC. Metabolic engineering of clostridium cellulolyticum for production of isobutanol from cellulose. *Applied and Environmental Microbiology*. 2011 Apr;77(8):2727–2733. PubMed PMID: WOS:000289459300020.

Hirota R, Kuroda A, Ikeda T, Takiguchi N, Ohtake H, Kato J. Transcriptional analysis of the multicopy hao gene coding for hydroxylamine oxidoreductase in Nitrosomonas sp.strain ENI-11. *Biosci Biotechnol Biochem*. 2006;70:1875–1881.

Hommes NG, Sayavedra-Soto LA, Arp DJ. Mutagenesis and expression of amo, which codes for ammonia monooxygenase in Nitrosomonas europaea. *Journal of Bacteriology*. 1998;180:3353–3359.

Hommes NG, Sayavedra-Soto LA, Arp DJ. The roles of the three gene copies encoding hydroxylamine oxidoreductase in Nitrosomonas europaea. *Archives of Microbiology*. 2002;178:471–476.

Hyman MR, Murton IANB, Arp DJ. Interaction of ammonia monooxygenase from Nitrosomonas europaea with alkanes, alkenes, and alkynes. *Applied and Environmental Microbiology*. 1988;54:3187e3190.

Ineos Bio Technology Platform. 2016. http://www.ineos.com/businesses/ineosbio/technology (accessed 20.05.2021).

Janssen H, Döring C, Ehrenreich A, Voigt B, Hecker M, Bahl H, et al. A proteomic and transcriptional view of acidogenic and solventogenic steady-state cells of Clostridium acetobutylicum in a chemostat culture. *Applied Microbiology and Biotechnology*. 2010;87(6):2209–2226.

Kapic A, Jones ST, Heindel TJ. Carbon monoxide mass transfer in a syngas mixture. *Industrial & Engineering Chemistry Research*. 2006 Dec;45(26):9150–9155. PubMed PMID: WOS:000242786200049.

Karberg M, Guo HT, Zhong J, Coon R, Perutka J, Lambowitz AM. Group II introns as controllable gene targeting vectors for genetic manipulation of bacteria. *Nature Biotechnology*. 2001 Dec;19(12):1162–1167. PubMed PMID: WOS:000172524400030.

Klasson KT, Ackerson MD, Clausen EC, Gaddy JL. Bioreactor design for synthesis gas fermentations. *Fuel*. 1991a May;70(5):605–614. PubMed PMID: WOS:A1991FL49400009.

Klasson KT, Ackerson MD, Clausen EC, Gaddy JL. Bioconversion of synthesis gas into liquid or gaseous fuels. *Enzyme and Microbial Technology*. 1992b Aug;14(8):602–608. PubMed PMID: WOS: A1992JE44900001.

Koepke M, Liew FM, Inventors. *Recombinant microorganism and methods of production thereof.* U.S. 2011. https://patents.google.com/patent/US20110236941A1/en.

Koffas M, Odom JM, Schenzle A. High growth methanotrophic bacterial strain methylomonas 16a. *European Patent Application Pub. No.: EP1320579 B1*; 2007. https://patents.google.com/patent/EP1320579A2/en.

Köpke M, Dürre P. Biochemical production of biobutanol. In: Luque R, Campelo J, Clark JH, editors. *Handbook of biofuel production—Processes and technologies*. Cambridge, UK: Woodhead Publishing Ltd; 2011. p. 221–257.

Köpke M, Held C, Hujer S, Liesegang H, Wiezer A, Wollherr A, et al. Clostridium ljungdahlii represents a microbial production platform based on syngas. *Proceedings of the National Academy of Sciences of the United States of America*. 2010;107(29):13087–13092.

Kundiyana DK, Wilkins MR, Maddipati P, Huhnke RL. Effect of temperature, pH and buffer presence on ethanol production from synthesis gas by "Clostridium ragsdalei". *Bioresource Technology*. 2011;102(10):5794–5799.

LanzaTech. 2016. http://www.lanzatech.com/ (accessed 30.04.2021).

Lederle S. *Heterofermentative acetonproduktion*. Ulm, BW: University of Ulm; 2011.

Liew FM, Köpke M, Simpson SD. *Gas fermentation for commercial biofuels production, liquid, gaseous and solid biofuels – conversion techniques*, Zhen Fang, IntechOpen; 2013. doi:10.5772/52164. Available from: https://www.intechopen.com/books/liquid-gaseous-and-solid-biofuels-conversion-techniques/gas-fermentation-for-commercial-biofuels-production

Liew FM, Martin ME, Tappel RC, Heijstra BD, Mihalcea C, Köpke M. Gas fermentation—A flexible platform for commercial scale production of low-carbon-fuels and chemicals from waste and renewable feedstocks. *Frontiers in Microbiology*. 2016;7:694. doi:10.3389/fmicb.2016.00694

Lu J, Brigham CJ, Gai CS, Sinskey AJ. Studies on the production of branched-chain alcohols in engineered Ralstonia eutropha. *Applied Microbiology and Biotechnology*. 2012;96:283–297.

Mohammadi M, Najafpour GD, Younesi F, Lahijani P, Uzir MH, Mohamed AR. Bioconversion of synthesis gas to second generation biofuels: a review. *Renewable & Sustainable Energy Reviews*. 2011 Dec;15(9):4255–4273. PubMed PMID: WOS: 000298764400005.

Munasinghe PC, Khanal SK. Biomass-derived syngas fermentation into biofuels: opportunities and challenges. *Bioresource Technology*. 2010;101(13):5013–5022.

Murrell JC. Genetics and molecular biology of methanotrophs. *FEMS Microbiology Letters*. 1992;88:233–248.

Naik SN, Goud VV, Rout PK, Dalai AK. Production of first and second generation biofuels: a comprehensive review. *Renewable and Sustainable Energy Reviews*. 2010;14:578–597, doi:10.1016/j.rser.2009.10.003

Navarro R, Pawelec B, Alvarez-Galván MC, Guil-Lopez R, Al-Sayari S, Fierro JLG. Renewable syngas production via dry reforming of methane. In: Falco MD, Iaquaniello G, Centi G, editors. *CO2: a valuable source of carbon*. London: Springer; 2013. p. 45–66.

O'Connor JR, Lyras D, Farrow KA, Adams V, Powell DR, Hinds J, et al. Construction and analysis of chromosomal Clostridium difficile mutants. *Molecular Microbiology*. 2006 Sep;61(5):1335–1351. PubMed PMID: WOS:000239701400020.

Ostrem, K. *Greening waste: Anaerobic digestion for treating the organic reaction of municipal solid wastes*. Master's Thesis, Columbia University, New York, NY, 2004.

Park D, Lee J. Biological conversion of methane to methanol. *Korean Journal of Chemical Engineering*. 2013;30:977–987.

Sato M, Matsuura K, Fujii T. Ethanol separation from ethanol-water solution by ultrasonic atomization and its proposed mechanism based on parametric decay instability of capillary wave. *Journal of Chemical Physics*. 2001;114(5):2382–2386.

Sayavedra-Soto LA, Stein LY. Chapter seventeen–genetic transformation of ammoniaoxidizing bacteria. In: Martin GK, editor. *Methods in enzymology research on nitrification and related processes, part A*. USA: Academic Press; 2011. p. 389–402.

Siedlecki M, de Jong W, Verkooijen AHM. Fluidized bed gasification as a mature and reliable technology for the production of bio-syngas and applied in the production of liquid transportation fuels-A review. *Energies*. 2011 Mar;4(3):389–434. PubMed PMID: WOS:000288788000002.

Simpson SD, Koepke M, Liew FM, Inventors. *Recombinant microorganisms and methods of use thereof.* U.S. 2011. https://patents.google.com/patent/US20110236941A1/en.

Soucaille P, Joliff G, Izard A, Goma G. Butanol tolerance and autobacteriocin production by Clostridium acetobutylicum. *Current Microbiology*. 1987;14(5):295–299. PubMed PMID: WOS:A1987G253100011.

Soucaille P, Rainer F, Christian C, Inventors. *Process for chromosomal integration and DNA sequence replacement in clostridia*. EU. 2008. https://patents.google.com/patent/EP1908840A1/en.

Sun X, Atiyeh HK, Huhnke RL, Tanner RS. Syngas fermentation process development for production of biofuels and chemicals: a review. *Bioresource Technology Reports*. 2019;7:100279, ISSN 2589-014X, https://doi.org/10.1016/j.biteb.2019.100279.

Taher E, Chandran K. High-rate, high-yield production of methanol by ammonia oxidizing bacteria. *Environmental Science & Technology*. 2013;47:3167e3173.

Vamvuka D, Arvanitidis C, Zachariadis D. Flue gas desulfurization at high temperatures: a review. *Environmental Engineering Science*. 2004;21(4):525e547.

Vasudevan R, Nigam SK, Velkennedy R, Ramalinga Chandra Sekar A, Sundarakannan B. Utilization of waste polymers for flexible pavement and easy disposal of waste polymers. *Proceedings of the International Conference on Sustainable Solid Waste Management*. 5–7 September 2007, Chennai, India. p. 105–111.

Vega JL, Holmberg VL, Clausen EC, Gaddy JL. Fermentation parameters of *Peptostreptococcus* productus on gaseous substrates (CO, H2/CO2). *Archives of Microbiology*. 1989;151(1):65–70. PubMed PMID: WOS:A1989R169500012.

Walter KA, Mermelstein LD, Papoutsakis ET. Studies of recombinant Clostridium acetobutylicum with increased dosages of butyrate formation genes. *Recombinant DNA Technology Ii.* 1994;721:69–72. PubMed PMID: WOS:A1994BA43Z00006.

Wang Y, Li X, Mao Y, Blaschek HP. Genome-wide dynamic transcriptional profiling in Clostridium beijerinckii NCIMB 8052 using single-nucleotide resolution RNA-Seq. *BMC Genomics.* 2012;13(1):102.

Williams E, Shimmin MS, Bainbridge BW. Mutation in the obligate methylotrophs Methylococcus capsulatus and Methylomonas albus. *FEMS Microbiology Letters.* 1977;2:293–296.

Woolcock PJ, Brown RC. A review of cleaning technologies for biomass-derived syngas. *Biomass and Bioenergy.* 2013;52:54e84.

Yang L, Ge X, Chapter Three—Biogas and syngas upgrading. In: Li Y, Ge X, editors. *Advances in bioenergy,* volume 1. Elsevier; 2016. p. 125–188, ISSN 2468-0125, ISBN 9780128095225, https://doi.org/10.1016/bs.aibe.2016.09.003.

Ye R, Kelly K. Construction of carotenoid biosynthetic pathways through chromosomal integration in methane-utilizing bacterium Methylomonas sp. Strain 16a. In: Barredo JL, editor. *Microbial carotenoids from bacteria and microalgae.* 892 ed. USA: Humana Press, 2012; p. 185–195.

Index